HYBRID NANOFLUIDS

HYBRID NANOFLUIDS

Preparation, Characterization and Applications

Edited by

ZAFAR SAID

Department of Sustainable and Renewable Energy Engineering, University of Sharjah, Sharjah, United Arab Emirates

Research Institute for Sciences and Engineering, University of Sharjah, Sharjah, United Arab Emirates

U.S.-Pakistan Center for Advanced Studies in Energy (USPCAS-E), National University of Sciences and Technology (NUST), Islamabad, Pakistan

Elsevier
Radarweg 29, PO Box 211, 1000 AE Amsterdam, Netherlands
The Boulevard, Langford Lane, Kidlington, Oxford OX5 1GB, United Kingdom
50 Hampshire Street, 5th Floor, Cambridge, MA 02139, United States

Library of Congress Cataloging-in-Publication Data
A catalog record for this book is available from the Library of Congress

British Library Cataloguing-in-Publication Data
A catalogue record for this book is available from the British Library

ISBN: 978-0-323-85836-6

For information on all Elsevier publications
visit our website at https://www.elsevier.com/books-and-journals

Publisher: Matthew Deans
Acquisitions Editor: Edward Payne
Editorial Project Manager: Leticia M. Lima
Production Project Manager: Prem Kumar Kaliamoorthi
Cover Designer: Greg Harris

Typeset by STRAIVE, India

Contents

Contributors ix
Preface xi
Acknowledgments xv

Chapter 1 Introduction to hybrid nanofluids 1
Zafar Said and Maham Aslam Sohail
1.1 Introduction 2
1.2 Preparation of hybrid nanofluids 10
1.3 Properties of hybrid nanofluids 13
1.4 Applications of hybrid nanofluids 19
1.5 Challenges and outlook 22
1.6 Conclusion 23
References 23

Chapter 2 Preparation and stability of hybrid nanofluids 33
Neeti Arora, Munish Gupta, and Zafar Said
2.1 Introduction 33
2.2 Stability of nanofluids 37
2.3 Challenges and outlook 57
2.4 Summary 58
References 59

Chapter 3 Thermophysical, electrical, magnetic, and dielectric properties of hybrid nanofluids 65
E. Venkata Ramana, L. Syam Sundar, Zafar Said, and Antonio C.M. Sousa
3.1 Thermophysical properties 65
3.2 Conclusion 87

Acknowledgments 88
References 88

Chapter 4 Hydrothermal properties of hybrid nanofluids 93
L. Syam Sundar, E. Venkata Ramana, Zafar Said, and Antonio C.M. Sousa
4.1 Introduction 93
4.2 Surface tension 94
4.3 Friction factor 96
4.4 Pressure drop 99
4.5 Pumping power 101
4.6 Fouling factor of nanofluid 101
4.7 Conclusions and challenges 105
Acknowledgments 106
References 106

Chapter 5 Rheological behavior of hybrid nanofluids 111
Abdulla Ahmad Alshehhi, Zafar Said, and Maham Aslam Sohail
5.1 Introduction 111
5.2 Experimental and numerical studies on rheology 113
5.3 Effects of various parameters on the rheology of hybrid nanofluids 117
5.4 Conclusion and future outlook 122
References 123

Chapter 6 Radiative transport of hybrid nanofluid 131
Arun Kumar Tiwari, Amit Kumar, and Zafar Said
6.1 Introduction 132
6.2 Optical properties 132
6.3 Radiative transfer 140
6.4 Effect of different parameters on optical properties 143
6.5 Challenges and outlook 144
6.6 Summary 145
References 145

Chapter 7 Theoretical analysis and correlations for predicting properties of hybrid nanofluids 149
Arun Kumar Tiwari, Amit Kumar, and Zafar Said
7.1 Introduction 149
7.2 Different theoretical models 150
7.3 Different correlations to predict the properties of hybrid nanofluid 152
7.4 Challenges and summary 164
References 166

Chapter 8 Brief overview of the applications of hybrid nanofluids . . 171
M. Sheikholeslami, Elham Abohamzeh, Z. Ebrahimpour, and Zafar Said
8.1 Introduction 172
8.2 Electronics cooling 172
8.3 Solar collectors 177
8.4 Heat exchangers 181
8.5 Engine cooling 185
8.6 Refrigeration 188
8.7 Machining 190
8.8 Desalination 192
8.9 Challenges and outlook 194
8.10 Summary 195
References 196

Chapter 9 Recent advances in the prediction of thermophysical properties of nanofluids using artificial intelligence 203
Mehdi Jamei and Zafar Said
9.1 Introduction 203
9.2 Modeling structure using AI methods 208
9.3 Sensitivity analysis 225
9.4 Summary 226
References 226

Chapter 10 Challenges and difficulties in developing hybrid nanofluids and way forward 233

Zafar Said and Maham Aslam Sohail

10.1 Introduction 233
10.2 Foam formation 234
10.3 Stability 237
10.4 Safety and environmental concerns 239
10.5 High cost 242
10.6 Degradation of original properties 244
10.7 Increased friction factor, pumping power, and pressure drop 245
10.8 Selecting suitable hybrid nanofluids 248
10.9 Predicting models for thermophysical properties 248
10.10 Challenges and outlook 252
10.11 Conclusion 253
References 254

Index 261

Contributors

Elham Abohamzeh Department of Energy, Materials, and Energy Research Center (MERC), Karaj, Iran

Abdulla Ahmad Alshehhi Space Missions Department, UAE Space Agency, Abu Dhabi, United Arab Emirates

Neeti Arora Department of Mechanical Engineering, Guru Jambheshwar University of Science and Technology, Hisar, Haryana, India

Z. Ebrahimpour Department of Mechanical Engineering; Renewable Energy Systems and Nanofluid Applications in Heat Transfer Laboratory, Babol Noshirvani University of Technology, Babol, Iran

Munish Gupta Department of Mechanical Engineering, Guru Jambheshwar University of Science and Technology, Hisar, Haryana, India

Mehdi Jamei Faculty of Engineering, Shohadaye Hoveizeh Campus of Technology, Shahid Chamran University of Ahvaz, Dashte Azadegan, Iran

Amit Kumar Mechanical Engineering Department, Institute of Engineering & Technology, Dr. A.P.J. Abdul Kalam Technical University, Uttar Pradesh, Lucknow, India

E. Venkata Ramana I3N, Department of Physics, University of Aveiro, Aveiro, Portugal

Zafar Said Department of Sustainable and Renewable Energy Engineering; Research Institute for Sciences and Engineering, University of Sharjah, Sharjah, United Arab Emirates; U.S.-Pakistan Center for Advanced Studies in Energy (USPCAS-E), National University of Sciences and Technology (NUST), Islamabad, Pakistan

M. Sheikholeslami Department of Mechanical Engineering; Renewable Energy Systems and Nanofluid Applications in Heat Transfer Laboratory, Babol Noshirvani University of Technology, Babol, Iran

Maham Aslam Sohail Department of Sustainable and Renewable Energy Engineering, University of Sharjah, Sharjah, United Arab Emirates

Antonio C.M. Sousa Centre for Mechanical Technology and Automation (TEMA-UA), Department of Mechanical Engineering, University of Aveiro, Aveiro, Portugal

L. Syam Sundar Centre for Mechanical Technology and Automation (TEMA-UA), Department of Mechanical Engineering, University of Aveiro, Aveiro, Portugal

Arun Kumar Tiwari Mechanical Engineering Department, Institute of Engineering & Technology, Dr. A.P.J. Abdul Kalam Technical University, Uttar Pradesh, Lucknow, India

Preface

"Seek knowledge from the Cradle to the Grave."

Prophet Muhammad (peace be upon him)

Hybrid nanofluid is introduced as a new class for engineering applications comprising solid particles with a size typically ranging from 1 to 100 nm dispersed in base fluids. Nanoparticles suspended in traditional heat transfer fluid enhance thermal conductivity. The addition of these nanoparticles to the conventional heat transfer fluids enhances the heat transfer rate. In this work, the history of hybrid nanofluids, preparation techniques, thermoelectrical properties, rheological behavior, optical properties, theoretical modeling and correlations, and the effect of all these factors on the potential applications such as solar energy, electronics cooling, heat exchangers, machining, and refrigeration are discussed in detail. In addition, future challenges and future work scope have been included. The information from this book will enable the readers develop novel techniques, resolve existing research limitations, and come up with novel hybrid nanofluids, which can be implemented for heat transfer applications. The subject of the book is the current comprehensive research and development of hybrid nanofluids, their implementation in various applications, and directions for various research gaps that are still required in preparations, stability, characterization, and applications to overcome the challenges being faced by both researchers and the industry for large-scale applications. Thus, this book is the most recent source of guidelines for future trends.

Chapter 1 focuses on hybrid or composite nanofluids, which are developed as a novel class of nanofluids synthesized by combining two or more nanoparticles containing metal or metal oxide or combining both particles in a base fluid. The preparation process is a significant step in the nanocomposites to further improve the thermal conductivity of heat transfer fluids. Thermophysical properties of hybrid nanofluids show good enhancement as compared with mono nanofluids. Higher volume fractions result in enhanced values of thermophysical properties. Several investigations have been reported on hybrid nanofluids' thermal conductivity and viscosity, but research on other properties like density, specific heat, thermal diffusivity, and magnetic is limited. Hybrid nanofluids are promising and can be utilized in several applications such as heat transfer, electrical and engine cooling, refrigeration, machining, desalination, nuclear PWR, heat exchangers, and solar collectors. However, some challenges still need to be identified, such as stability, increased pumping power, and production cost to be employed in industrial applications.

Chapter 2 provides insight into the preparation, stability, and characterization of hybrid nanofluids. The chapter presents the synthesis, stability evaluation, and stability enhancement methods of hybrid nanofluids in brief. Proper preparation of hybrid nanofluids is necessary for enhancing thermophysical properties and their stability. The stability of nanofluids plays a vital role in their proper working in thermal systems. Various stability evaluation methods like sedimentation, zeta potential, spectral absorbance, and electron microscopy are deliberated to gain important indicative information about the stability of nanofluids. For improving the stability of nanofluids, various stability enhancement techniques such as ultrasonication, surfactant addition, surface modifications of nanoparticles, and pH change are also described in the chapter. Proper selection of nanomaterials is obligatory for preparing hybrid nanofluids according to their synergy level. At the end of the chapter, some challenges and outlooks about hybrid nanofluids have been discussed for their worldwide applications.

Chapter 3 outlines the promising thermophysical, electrical, magnetic, and dielectric properties of hybrid nanofluids that display considerable potential in heat transfer applications. Hybrid nanofluids exhibit a remarkable improvement in heat transfer performance as compared with mono nanofluids. The chapter presents the recent advancements in the augmentation of hybrid nanofluids' thermophysical properties. At augmented particle loadings and temperatures of hybrid nanofluids, the thermophysical properties of thermal conductivity and specific heat are enhanced. Moreover, the dynamic viscosity and density of hybrid nanofluids decrease with temperature and increase with volume concentration. Thus, temperature and volume concentration are the significant parameters that affect the thermophysical properties of hybrid nanofluids. The chapter also presents studies on the properties such as magnetic and dielectric properties.

Chapter 4 outlines hybrid nanofluids' hydrothermal properties, demonstrating that they possess higher heat transfer rates over the single nanoparticle-based nanofluids. Properties like surface tension, pumping power, pressure drop, friction factor, and fouling factor are discussed. Surface tension, friction factor, pressure drop, and pumping power are augmented with higher particle loadings. Increased fouling factor indicates a reduction in heat transfer coefficient with the use of hybrid nanofluids. Recent challenges are also presented, and it is reported that further investigation is required to study various types of hybrid nanoparticles, the stability of the nanofluid, thermophysical properties, heat transfer, less friction factor, pumping power, and pressure drop characteristics.

Chapter 5 focuses on the rheological behavior of hybrid nanofluids. Recently, the field of nanofluids has gained enormous interest due to their great advantages over conventional fluids. The presence of nanosized particles within conventional fluids led to an increase in the thermal conductivity coefficients compared with base fluids. Researchers across the globe continuously work toward further advancement and improvements of the nanofluids' behavior. The rheological behavior of a nanofluid is described by the relationship between the shared stress and its rate. The chapter outlines the rheological behavior of hybrid nanofluids with a review of past and recent studies and findings of a mixture of dual nanometer-sized (<100 nm) particles with variant base fluids and variant volume fraction. It is observed that the size of the nanoparticle, shear rate, and volume fraction of the nanoparticles affect the rheological behavior of the nanofluids significantly.

Chapter 6 focuses on the radiative transport of hybrid nanofluids. These hybrid nanofluids can be used for direct absorption solar thermal systems as a working fluid. So, it becomes essential to study the optical properties of the hybrid nanofluids. Different nanoparticles have different properties. For direct solar absorption applications, it is essential to estimate the optical properties of the hybrid nanofluids. Different theories predict the extinction coefficient, out of which Mie scattering theory is the most suitable theory. Different theories have been discussed in the present study to calculate the extinction coefficient theoretically, and these theoretical values of the extinction coefficient are compared with experimentally obtained values.

Chapter 7 discusses the theoretical analysis and correlations for predicting the properties of hybrid nanofluids. The main objective of the chapter is to provide a comprehensive review of the thermophysical properties of hybrid nanofluids up to date and the correlation used for predicting those properties. The main contributing factors that affect the thermophysical properties, such as stability, nanoparticle type, size, volume concentration, type of base fluid, temperature, surfactant, pH value, and sonication time, are also addressed. In addition, various empirical correlations developed by researchers for the thermophysical properties of the hybrid nanofluids are compiled and reported.

Finally, challenges with the stability and thermophysical properties of their correlations are summarized.

Chapter 8 provides a brief overview of the applications of hybrid nanofluids. The chemical and physical features of different materials are combined in the hybrid material simultaneously, providing the characteristics in a homogeneous phase. The effective viscosity and density of hybrid nanofluids may be of the same order as that of mono nanofluids, while their thermal conductivity might be substantially higher than that of mono nanofluids considering synergistic effects. The outstanding enhancement in thermal transfer properties of nanofluids led researchers to use them in various engineering applications, including nuclear cooling, desalination, machining, refrigeration, engine cooling, heat exchangers (HEX), solar collectors, and electronics cooling. This chapter gives a brief overview of the applications of hybrid nanofluids, the challenges associated with them, and the way forward for research gaps that still need attention.

Chapter 9 sheds light on the recent advances in predicting nanofluids' thermophysical properties using artificial intelligence (AI). Recently, methods have been widely welcomed due to the weakness of traditional regression-based methods and their low accuracy in nonlinear problems related to the study of thermophysical properties of nanofluids. In recent years, various investigations have been devoted to applying AI in estimating the thermophysical properties of nanofluids and energy applications, most of which have focused on single nanofluids. AI-based investigations on thermophysical properties of nanofluids demonstrated that most of the applications of machine learning and data-driven models are related to thermal conductivity and viscosity of mono fluids, and limited research has been conducted to model hybrid nanofluids. However, given the increasing capabilities of AI methods and their integration with robust optimization algorithms, it can be hoped to solve nonlinear problems of hybrid nanofluids with many input variables to achieve promising results. In this direction, the chapter provides an overview of the recent advances in AI for predicting the thermophysical properties of nanofluids.

Chapter 10 focuses on the challenges being faced by the researchers and scholars in the commercialization of nanofluids and the way forward. Several key challenges such as foam formation, stability, high cost, increased friction factor, pressure drop, pumping power, degradation of original properties, predicting models for thermophysical properties, safety, environmental concerns, and suitable hybrid materials are selected and discussed. Future directions and possible research gaps are provided as well.

Acknowledgments

This book would not have been possible without the dedicated and insightful work of the chapter authors. I am grateful to the authors for their precious time and effort to this adventure. I appreciate their kindness, dedication, and excellence in providing high-quality chapters that summarize the main characteristics, challenges, and applications of hybrid nanofluids. It was a pleasure and an honor working with you all on this crucial milestone in this emerging area. The book would not have been possible without the continuous dedication and support from the University of Sharjah and my family, especially my parents, wife and son.

Next, I genuinely thank the excellent support from the Elsevier team. Their kindness, patience, continuous support, technical expertise, and insights were essential to making this book a reality. It has been an absolute pleasure to be associated with you!

Finally, being the first book in this highly dynamic area, I hope that this work will become a milestone to further foster increasing scientific and engineering efforts and that hybrid nanofluids will increasingly become implemented as a new class of high-performance heat transfer fluids supporting a more sustainable future.

Zafar Said

1

Introduction to hybrid nanofluids

Zafar Said[a,b,c,*] and Maham Aslam Sohail[a]

[a]Department of Sustainable and Renewable Energy Engineering, University of Sharjah, Sharjah, United Arab Emirates. [b]Research Institute for Sciences and Engineering, University of Sharjah, Sharjah, United Arab Emirates. [c]U.S.-Pakistan Center for Advanced Studies in Energy (USPCAS-E), National University of Sciences and Technology (NUST), Islamabad, Pakistan

**Corresponding author: zsaid@sharjah.ac.ae, zaffar.ks@gmail.com*

Chapter outline

1.1 Introduction 2
1.1.1 Development of nanomaterials and nanofluids 4
1.1.2 Drawbacks of mono nanofluids 7
1.1.3 Development of hybrid nanofluids 7
1.2 Preparation of hybrid nanofluids 10
1.3 Properties of hybrid nanofluids 13
1.3.1 Thermal conductivity 14
1.3.2 Viscosity 16
1.3.3 Density 17
1.3.4 Specific heat capacity 17
1.3.5 Thermal diffusivity 17
1.3.6 Electrical, magnetic, dielectric 18
1.4 Applications of hybrid nanofluids 19
1.4.1 Electronic cooling 19
1.4.2 Solar collectors 20
1.4.3 Heat exchangers 20
1.4.4 Nuclear PWR 21
1.4.5 Engine cooling 21
1.4.6 Refrigeration 21
1.4.7 Machining 21
1.4.8 Desalination 22
1.5 Challenges and outlook 22
1.6 Conclusion 23
References 23

Hybrid Nanofluids: Preparation, Characterization and Applications. https://doi.org/10.1016/B978-0-323-85836-6.00001-6

1.1 Introduction

Energy resources are strongly dominant with the largest share on fossil fuels for energy production, which significantly impact the environment. Carbon footprint modification is a critical issue due to the increasing concerns about global climate change. In-depth studies are being investigated to obtain sustainable solutions as it is regarded to be one of the major drivers [1–3]. A continuous decrease in fossil fuel resources and increased carbon emissions made the developed nations progress toward renewable energy sources [4]. Energy specialissts and policymakers suggest that if suitable investments are made to develop renewable energy for power generation, the economies presently supported on fossil fuels will become independent from nonrenewable sources sooner or later [3, 5]. Renewable energy sources are forecasted to deliver 70%–85% of power by 2050, significantly reducing carbon emissions (Fig. 1.1) [6]. Renewable energy consists of a sequence of infinite and environment-friendly energy sources without human involvement, such as solar energy, wind energy, biomass, hydropower, geothermal, and energy storage.

With the reduction of fossil fuel reserves, solar energy is considered a plentiful renewable source [7]. Solar energy is a main element in clean energy technologies because it delivers infinite, clean, and environmentally friendly energy [8]. The earth accommodates solar radiation of about 170 PW, in which 30% of this reflects back to space, and the remainder is absorbed by the earth and sea [9]. It can be stored and employed in various technologies such as solar thermal and photovoltaic (PV) systems. Solar photovoltaics (PV) is considered a reliable technology that converts solar radiation directly to electricity [10]. Solar thermal technology consists of the solar radiation harnessed for useful thermal energy and includes application areas in solar desalination [11], solar-thermal power plants [12], residential or commercial heating [13], absorption cooling [14], and so on. Therefore, how to efficiently transfer or store solar energy has a novel point to consider for the scientific and research community in the present day and future. Photovoltaic/thermal (PV/T) systems consist of photovoltaic (PV) and solar thermal component systems that can generate heat and electricity. They have a promising potential for energy savings and remarkable efficiency. PV/T systems are emerging as a strong candidate for power generation shortly.

Another serious challenge that researchers and engineers face in today's practical applications is heat exchange between several devices. It is certain to apply heat exchangers and heat sinks for heat transfer in several applications [15]. Electronic components

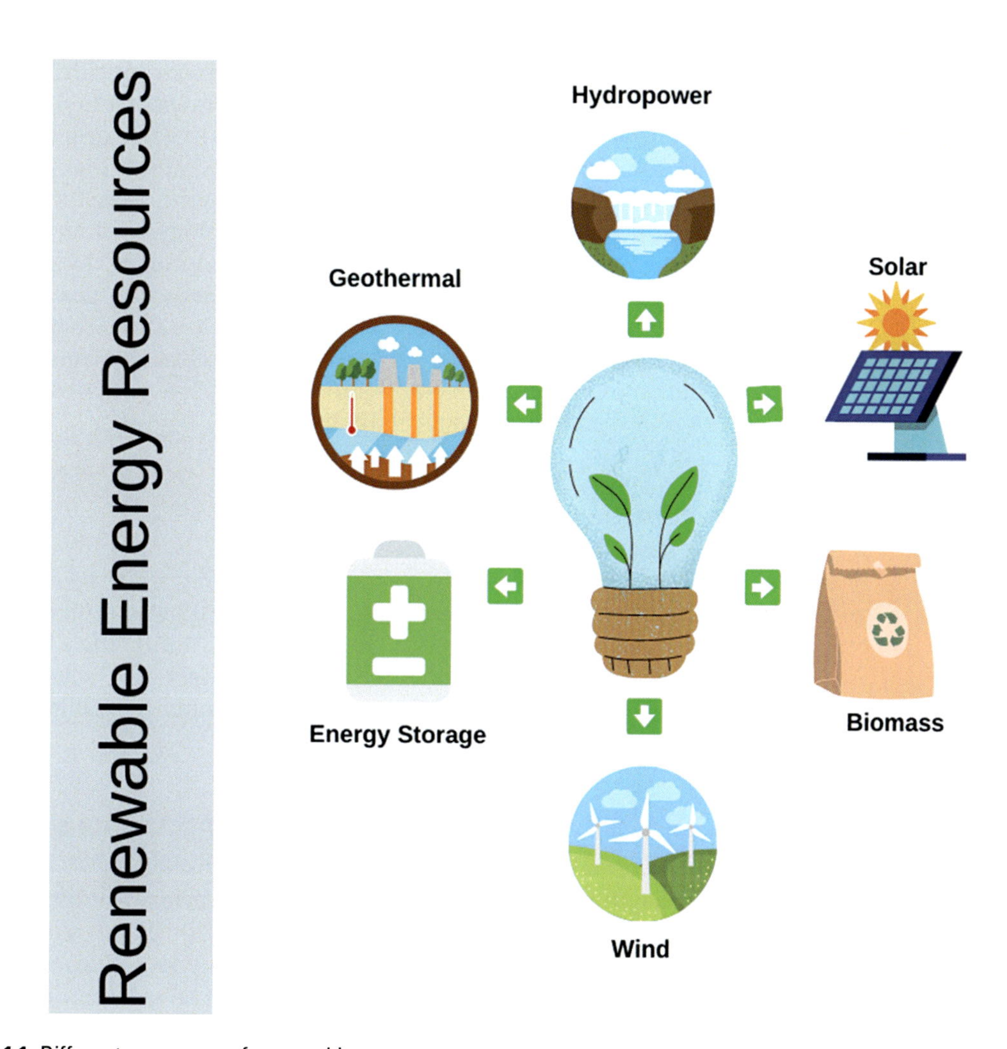

Fig. 1.1 Different resources of renewable energy.

in such devices produce undesirable heat by decreasing the efficiency or cause device failure. Advancement in heat transfer plays a significant part in industrial applications for cost savings and energy savings as the demand for high-performance devices is increasing nowadays with the progress in science and technology [16]. In the past decade, several investigations were developed to enhance heat transfer devices' efficiency by various techniques such as different shapes of fins and their optimization. The performance and efficiency of heat transfer can potentially impact the boundary conditions and thermophysical properties such as thermal conductivity, viscosity, density, and specific heat of the working fluid. This working fluid performs a substantial role in the heat transfer rate. Conventional fluids such as water, ethylene glycol, and oil are widely used in several heat transfer applications and possess low thermal conductivity [17]. Due to low thermal conductivity, these fluids could not meet the ever-increasing demand for enhanced heat transfer technology as their performance was not effective and caused high pumping losses [18]. Nobel scientist Maxwell [19] proposed an idea by introducing millimeter or micrometer-sized particles in the base fluid to enhance the thermal conductivity and performance, but this caused the following challenges in the liquids due to large particles, which are unfavorable for practical applications:

- Sedimentation
- Clogging
- Increased pumping power
- Corrosion

Masuda et al. [20] observed similar challenges of sedimentation, greater pumping power losses by investigating micro-sized solid particles dispersed in the base fluid. From then, several investigations were conducted by researchers and scholars to boost the poor thermal conductivity of fluids by adding solid particles.

1.1.1 Development of nanomaterials and nanofluids

Nanomaterials have gained extensive significance in the present day, having the promising potential to perform an innovative role in practical applications. Based on the International Organization for Standardization (ISO), the prefix nano defines as a size ranging from 1 to 100 nm. For instance, the carbon atom is about 0.25 nm in diameter, and the distance between carbon atoms is about 0.15 nm. Therefore, nanomaterials are bigger than individual atoms or small groups of atoms. They have distinctive properties such as thermal, magnetic, electrical, optical, and mechanical properties due to their large specific area, lattice structure, altered electronic states, etc. [21]. Nanomaterials are extensively investigated in various applications such as solar cells [22], water treatment [23], improved heat transfer [24], batteries [25], biosensors [26], etc. Nanomaterials are classified into four categories [27–31]:

- Zero-dimensional (0D): This includes quantum dots, fullerenes (hollow spheres), gold nanoparticles, etc.
- One-dimensional (1D): This includes nanofibers, nanotubes, nanowires, and nanorods.
- Two-dimensional (2D): This includes thin films, nanocoatings, and nanoplates.

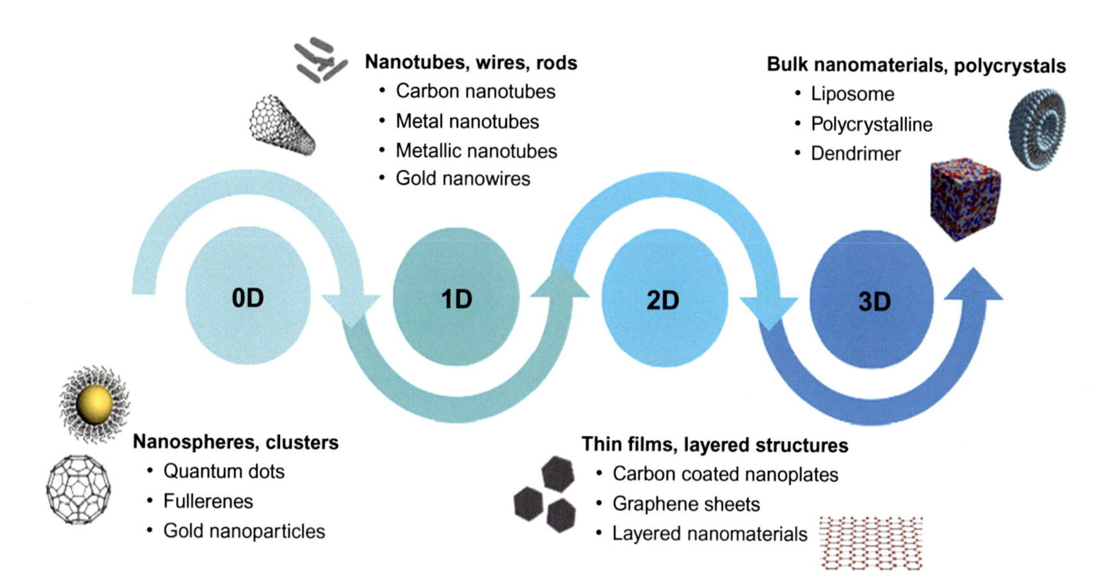

Fig. 1.2 Classification of nanomaterials.

- Three-dimensional (3D): This includes nanocomposites, nanostructured materials, and polycrystals.

Nanomaterials have a wide range of applications in different sectors such as aerospace, chemicals, construction, cosmetics, energy, electronics, automobile, engineering, environment, medicine, military, and sports (Fig. 1.2).

With the significant development of modern nanotechnology and nanomaterials, the discovery of nanofluids that contained nanosized particles dispersed in base fluids completely changed the picture. Choi and his team [32] proposed the term "nanofluid," a novel kind of nanotechnology-based heat transfer fluids defining as an engineered colloidal suspension of nanometer-sized particles typically ranging from 1 to 100 nm dispersed in the base fluid (Fig. 1.3). The progress in nanofluid-based technologies has gained significant attention from several researchers due to their excellent thermal conductivity and stability in several scientific fields such as energy-based applications [33–35], air conditioning [36], electronics, medicine, and energy and fuel management [37–39]. Apart from the high thermal conductivity, nanoparticles are desirable for heat transfer applications as they reduce clogging pumping power, which helps in energy savings [40]. Nanoparticles occupy a higher surface area

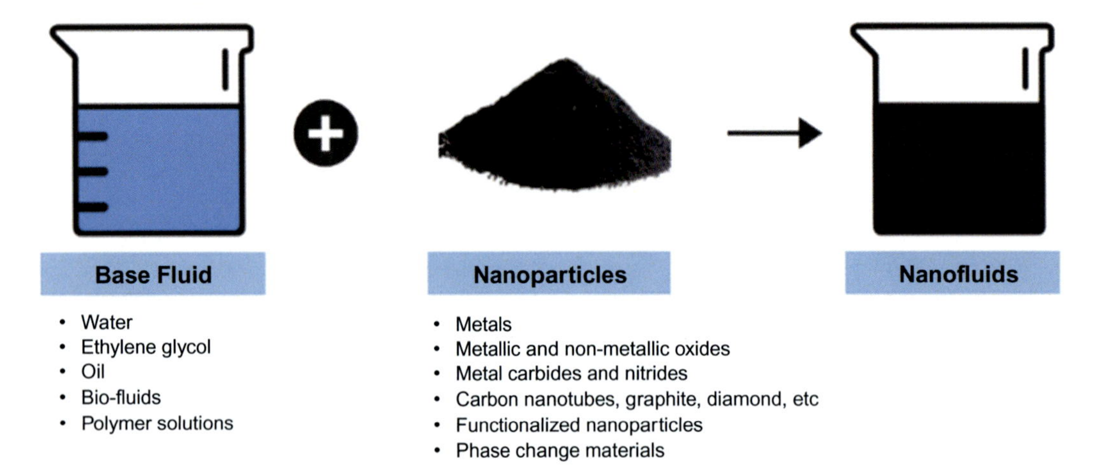

Fig. 1.3 Graphical representation of nanofluids.

than microparticles, increasing heat conduction of nanofluids, and possess remarkable stability. The nanoparticles consist of metals such as copper (Cu) [41], nickel (Ni) [42], gold (Au) [43], and silver (Ag) [44], metal oxides such as aluminum oxide (Al_2O_3), zinc oxide (ZnO), titanium dioxide (TiO_2), copper oxide (CuO), silicon dioxide (SiO_2), iron (III) oxide (Fe_2O_3), and many more, oxide ceramics, metal carbides and nitrides like aluminum nitride (AlN), silicon carbide (SiC), boron nitride (BN), carbon nanotubes, graphite, diamond, and functionalized nanoparticles [45]. Research from universities and research centers around the globe has studied or investigated nanofluids in various contexts and has analyzed the effects of various factors on heat transfer, thermal conductivity, viscosity, and boiling heat transfer [46, 47]. Numerous researchers and scientists have made potential breakthroughs in the development of thermal properties of nanofluids, proposing exceptional nanofluid analysis models [48, 49]. Michael and Iniyan [50] made a significant advancement in nanofluids' thermal performance compared with water. Choi et al. [51] noticed an increment in nanofluids' thermal conductivity up to two times compared with conventional fluids at a low volume fraction of less than 1 vol.%. Yu et al. [52] examined the thermal conductivity and viscosity of EG-based nanofluids experimentally. Nanofluids can flow smoothly without clogging at such low particle concentration, and high heat transfer efficiency can be attained. The promising properties of nanofluids depend on the following parameters:

- Temperature [53, 54]
- Volume concentration of nanoparticles [55, 56]

- Size and shape of nanoparticles [57, 58]
- Ultrasonication duration [59, 60]
- pH value of nanofluids [61–63]
- Surfactant-based or nonsurfactant-based nanofluids [64, 65]

1.1.2 Drawbacks of mono nanofluids

Aluminum oxide (Al_2O_3), also known as alumina, is one of the most widely investigated and promising nanofluids. It has been observed that Al_2O_3-based nanofluids exhibit excellent improvement in thermal conductivity ranging from 0.3% to 38% [66, 67]. Sundar et al. [68] investigated Al_2O_3-water/EG-based nanofluids without using surfactant and observed 32.3% enhancement for 1.5 vol.% concentration for 20:80% EG/H_2O and at a temperature of 60 °C. With the prolongation in nanofluids research, researchers and scholars studied further to enhance the properties of nanofluids. Mono nanofluids do not retain all favorable properties that are requisite for specialized applications. For instance, aluminum oxide (Al_2O_3) has a property of better chemical inertness and stability, but it has a drawback of low thermal conductivity. Metallic nanoparticles such as copper, silver, and aluminum exhibit remarkable thermal conductivity, but they are chemically reactive and unstable; therefore, they have either a better thermal property or a better rheological property. To trade-off between properties, hybrid nanofluids have been introduced to acquire enhanced properties, suitable for applications that involve remarkable thermal, optical, and rheological properties of the working fluid [69].

1.1.3 Development of hybrid nanofluids

Hybrid nanofluids as an advanced group of nanofluids are engineered by dispersing composite nanoparticles or two different types of nanoparticles in the base fluid, representing noteworthy physicochemical properties that are absent in individual nanoparticles [70]. In recent investigations, hybrid nanofluids are considered to increase thermal conductivity and the performance of solar energy systems, i.e., PVT systems, solar collectors, and storage systems, and this makes them remarkable due to the synergistic impact of individual nanoparticles [71, 72]. The principal purpose of hybrid nanofluids is to obtain promising properties such as excellent thermal conductivity and stability, physical strength, mechanical resistance, reduced pumping power losses, heat transfer performance rate, better aspect ratio, and reduced production cost of nanofluids.

Researchers from the past 2 decades have investigated hybrid nanofluids, mostly focused on the synthesis, characterization, properties, and applications for several industrial and commercial applications [73]. Several papers are published from the last decade with numerical and experimental investigations, and the growth can be clearly observed from Fig. 1.4. The current progress of hybrid nanofluids in this developing area is most apparent from the increasing number of keywords as compared with the nanofluids. The data for Fig. 1.4 is taken from Elsevier for the years 2015–2020. The nanoparticles or composites investigated so far are graphene-Ag, MWCNT-MgO, Fe_3O_4-graphene, Al_2O_3-CNT, Al_2O_3-SiO_2, Al_2O_3-Cu, Al_2O_3-MEPCM, diamond-Ni, SiO_2-CNT, Ag-MnO, Ag-TiO_2, Ag-CNT, Cu-TiO_2, Cu-Zn, Fe_2O_3-CNT, SWCNT-MgO, and many more in various heat transfer applications (Fig. 1.5). The metallic or metal oxide-based hybrid nanofluids show exceptional thermophysical and rheological properties [74, 75].

Based on the promising properties, the following are the work done by different scholars and researchers in the field of hybrid nanofluids.

- Turcu et al. [76] was the first one who investigated the fabrication of hybrid nanocomposites containing polypyrrole-carbon nanotube (PPY-CNT) and MWCNT on magnetic Fe_2O_3 nanoparticles.
- Niihara [77] and Oh et al. [78] investigated Al_2O_3-Cu hybrid nanocomposites and observed improvement in mechanical and thermal properties.

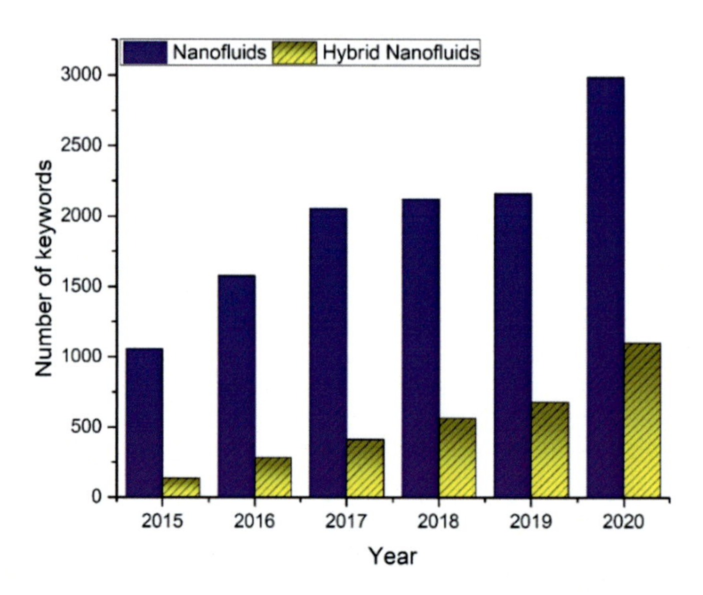

Fig. 1.4 Number of keywords used yearly for mono and hybrid nanofluids. The data is taken from Elsevier.

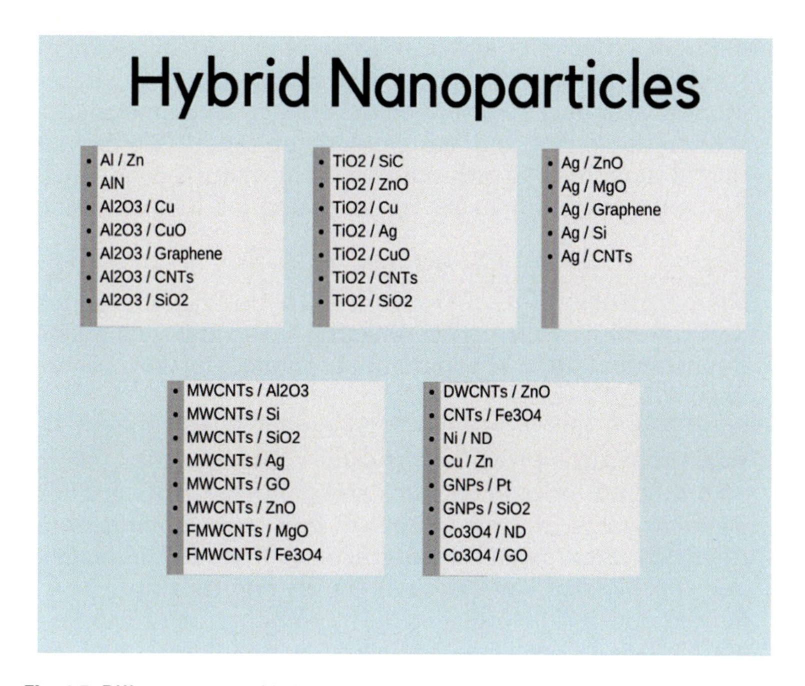

Fig. 1.5 Different types of hybrid nanoparticles.

- Jha and Ramprabhu [79] investigated experimentally MWCNT-Au/H_2O-based hybrid nanofluids and noticed enhancement in thermal conductivity of around 28%, as compared with CNT/H_2O-based nanofluid where the enhancement was around 15%.
- Madhesh et al. [80] investigated Cu-TiO_2 aqueous hybrid nanofluids and observed an improvement of 52% in heat transfer coefficient at a particle loading of 1 vol.% and pressure drop of 14.7%. They also observed that by increasing the volume concentration, the heat transfer coefficient is reduced due to nanoparticles' agglomeration into the base fluid.
- Baghbanzadeh et al. [81] examined thermophysical properties of SiO_2-MWCNTs/H_2O hybrid nanofluids and observed improvement in thermal conductivity up to 22% at 1 vol.%.
- Akilu et al. [82] analyzed the thermophysical characteristics of glycerol and EG mixture-based SiO_2-CuO/C hybrid nanofluid, and it was observed that the thermal conductivity enhanced up to 26.9%, indicating that hybrid is a potential heat transfer fluid for solar energy transportation.
- Wei et al. [83] investigated the thermophysical property of SiC-TiO_2/diathermic oil hybrid nanofluids and noticed excellent thermal conductivity as compared with mono nanofluids.
- Tiwari et al. [48] studied experimentally the thermal conductivity of CeO_2 + MWCNT/H_2O-based hybrid nanofluid with

different surfactants, and improvement in thermal conductivity of 27.38% was observed for 1.50 vol.%.

- Munkhbayar et al. [84] studied the impact of volume concentration up to 3% and temperature on Ag-MWCNTs/water hybrid nanofluids. It was observed that thermal conductivity is augmented up to 14.5% by increasing the temperature by 40 °C and at 3%.
- Esfe et al. [85] studied experimentally the thermophysical characteristics of Cu + TiO_2/water-EG hybrid nanofluids from 30 °C to 60 °C with particle loading up to 2%. Remarkable improvement up to 44% in thermal conductivity was observed at 60 °C and 2 vol.%.
- Sundar et al. [86] investigated experimentally the thermophysical properties of rGO-Co_3O_4/water-based hybrid nanofluids with different volume fractions and temperature ranges from 30 °C to 60 °C. It was observed that as the temperature increases, there is an improvement in thermal conductivity and specific heat and a drop in density and viscosity.
- Sundar et al. [87] investigated the thermophysical properties of MWCNT-Fe_3O_4/water-based hybrid nanofluids at 0.1% and 0.3% concentration and temperature ranges from 20 °C to 60 °C and observed the highest thermal conductivity of 0.8389 W/m K at 0.3 vol.% and at a temperature of 60 °C.
- Sundar et al. [88] investigated thermophysical properties, heat transfer, and exergy efficiency of nanodiamond + Fe_3O_4/H_2O-EG-based hybrid nanofluids. The highest thermal conductivity was improved to 12.79% at 0.2 vol.%, and the Nusselt number was increased to 15.65% at $Re = 7505$.
- Harandi et al. [89] observed an increment of 30% in thermal conductivity of f-MWCNTs-Fe_3O_4/water-based hybrid nanofluids.
- Askari et al. [90] experimentally investigated Fe_3O_4-graphene/Kerosene's thermal conductivity and observed a remarkable enhancement of 31% at 1 wt.% and 50 °C. Fig. 1.6 shows the graphical presentation of hybrid nanofluids. Table 1.1 shows the differences between the mono and hybrid nanofluids.

1.2 Preparation of hybrid nanofluids

Preparation of hybrid nanofluids is a significant phase for the enhancement of thermal conductivity. Chapter 2 presents the preparation and characterization techniques, including stability analysis of hybrid nanofluids. The preparation methods are of utmost importance for excellent performance and to produce stable hybrid nanofluids. The two methods followed for preparing

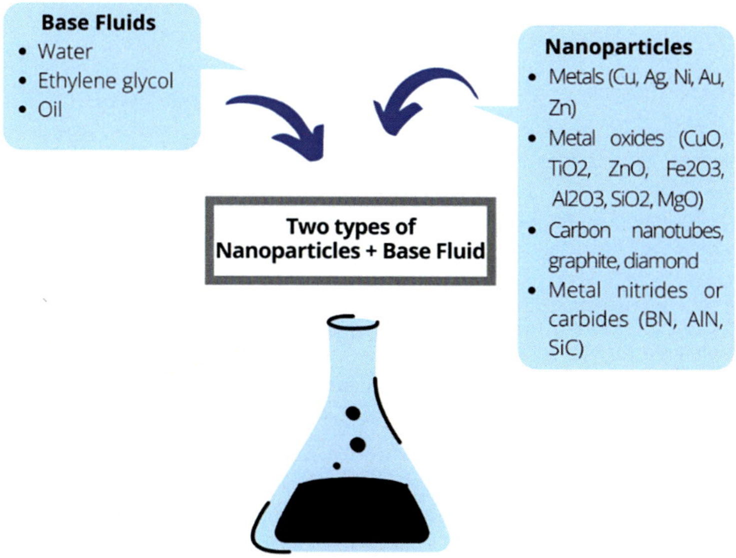

Fig. 1.6 Graphical representation of hybrid nanofluids.

Table 1.1 Difference between mono and hybrid nanofluids.

Parameters	Mono nanofluid	Hybrid nanofluid
Preparation method	• Two techniques such as one-step and two-step techniques have been developed; two-step is favorable for oxide nanoparticles and one-step for metal nanoparticles • Nanoparticles' size and shape can be modified • They are readily accessible commercial	• Researchers prepared hybrid nanofluids via a two-step technique • Size and shape of hybrid or nanocomposite cannot be controlled. Two nanoparticles may overlap, or one particle is coated over the other • Composite nanoparticles are not readily available, so they are synthesized
Characterization techniques	• Following characterization techniques are used: ◦ XRD, SEM, FT-IR, TEM, TGA, DLS, and zeta potential analysis	• Following characterization techniques are applied: ◦ XRD, SEM, FT-IR, TEM, TGA, DLS, and zeta potential ◦ SEM with EDAX and FESEM
Thermophysical properties	• Several hypothetical analysis models are offered for thermophysical properties of mono nanofluids	• Inadequate theoretical analysis models are offered for hybrid nanofluids

Continued

Table 1.1 Difference between mono and hybrid nanofluids—cont'd

Parameters	Mono nanofluid	Hybrid nanofluid
	• Mono nanofluids have exhibited reliable improvement in properties as compared with base fluids • Improvement of viscosity in internal flow results in large pressure drop, which boosts the pumping power	• They have shown remarkable improvement in thermophysical properties compared with mono nanofluids • Improvement of viscosity in internal flow has elevated pressure drop, which slightly raises the pumping power as compared with that of mono nanofluids
Stability	It can be controlled for an extended period in mono nanofluids	It is complex because of the dispersion of two different nanoparticles. Nanocomposites have shown great stability in various aqueous solvents
Cost	Carbon nanotubes and metallic nanoparticles are relatively expensive than metal oxide nanoparticles	Small amount of metallic nanoparticles and oxide nanoparticles exhibits better performance as great as mono nanofluids. Therefore, the cost of hybrid nanofluids is slightly less than metallic nanoparticle-based mono nanofluids

high-quality and stable nanofluids and to apply them in thermal applications are known as single-step and two-step methods [91, 92]. The two-step method is superior to one-step and widely implemented by scholars and researchers in terms of cost and the synthesis process, but in the aspect of stability, one-step is one of the promising methods. Fig. 1.7 represents the schematic of single-step and two-step preparation methods. As the stability of hybrid nanofluids is a significant parameter, various techniques are discovered to investigate the stability such as centrifugation and sedimentation approach, zeta potential analysis, spectral analysis technique, electron microscopy, and light scattering technique [91]. Hybrid nanofluids can miss their ability to transfer heat because of their proneness to settling. Weak stability of hybrid nanofluids can change the thermophysical properties, which results in unsatisfactory performance in heat transfer applications. Another important key challenge is the size of the hybrid nanoparticles, which should be below 100 nm to obtain stable hybrid nanofluids. Water, ethylene glycol, engine oil, and ethylene/water mixtures are frequently utilized base fluids to prepare hybrid nanofluids, as described in Table 1.2.

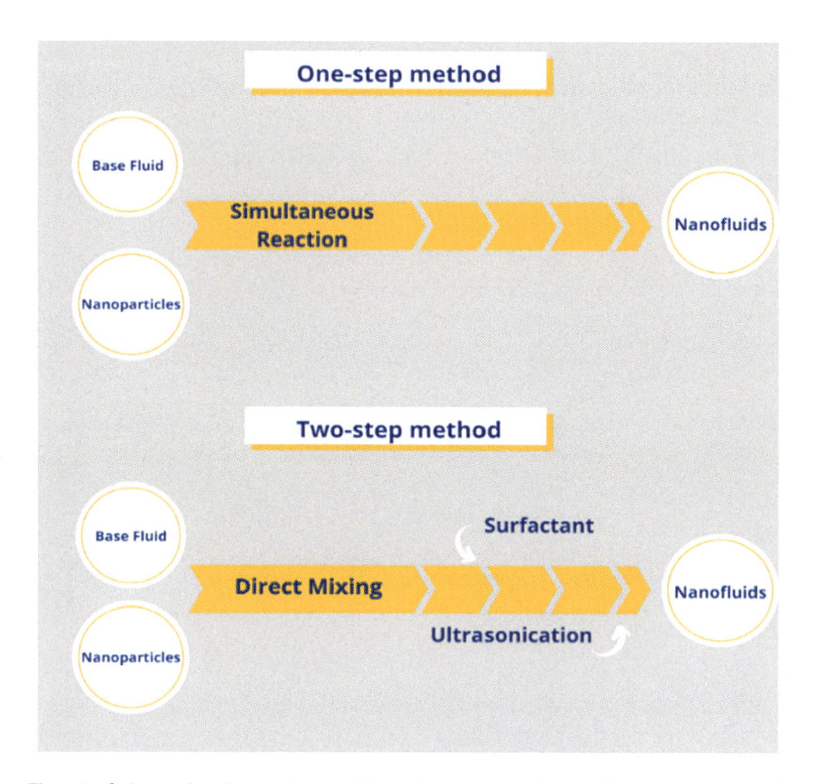

Fig. 1.7 Schematic of one-step and two-step preparation methods.

1.3 Properties of hybrid nanofluids

For the nanofluids application in energy-related systems, the basic understanding of different phenomena is essential. The properties of hybrid nanofluids depend on the preparation and application. Thermophysical, rheological, and photothermal characteristics of hybrid nanofluids are characterized for heat transfer applications. Thermal conductivity, viscosity, density, specific heat, and thermal diffusivity are the most widely investigated thermophysical properties. They are computed to determine the performance factor, for example, pressure drop, heat transfer coefficient, and energy efficiency of a thermal system [111, 112]. Since carbon nanotube-based nanofluids possess excellent thermophysical properties, adding up them to base fluid can have a remarkable effect on the fluids' thermal conductivity [113, 114]. Adding different types of oxide or metal nanoparticles has better and long-lasting stability and enhancement in the

Table 1.2 Hybrid nanocomposites investigations for several base fluids to form hybrid nanofluids.

Water-based hybrid nanofluids	Nanocomposites	References
	Ag-MWCNT	[93]
	Cu-TiO_2	[80]
	Graphene-MWCNT	[94]
	Al_2O_3-MWCNT	[95]
	Al_2O_3-Cu	[96]
	Fe_2O_3-MWCNT	[97]
	Fe_2O_3-nanodiamond	[98]
	Zn-Fe_2O_4	[73]
	rGO-Co_3O_4	[99]
	CuO-MWCNT, MgO-MWCNT, SnO_2-MWCNT	[100]
	CeO_2-MWCNT	[101]
Ethylene glycol-based hybrid nanofluids	Al_2O_3-SWCNT	[102]
	rGO-Fe_3O_4-TiO_2	[56]
	MgO-MWCNT	[103]
	SiO_2-MWCNT	[104]
	Graphene-CNT	[105]
	ZnO-DWCNT	[106]
	Al-Zn	[107]
Oil-based hybrid nanofluids	SiC/TiO_2	[83]
	Cu-Zn	[108]
	CeO_2-CuO	[109]
	f-MWCNTs-MgO	[110]

metal nanoparticles or carbon nanotubes suspensions [115]. Therefore, enhancement in thermophysical properties can be achieved by adding nanoparticles to the base fluid [116].

1.3.1 Thermal conductivity

Thermal conductivity is regarded as the most important property of nanofluids for heat transfer applications. It depends on the size, shape, and the type of nanomaterial. For instance, metallic nanoparticles have superior thermal conductivity as compared with nonmetallic ones. Improvement in the overall heat transfer rate is due to the high thermal conductivity, which increases the thermal system's efficiency. Nanofluids are regarded as suitable heat transfer fluids compared with conventional fluids because of their superior thermal conductivity and single-phase heat transfer coefficient. Nanofluids' thermal conductivity is a function of several parameters such as temperature, type of nanoparticles, composition of base fluid, the size and shape of nanoparticle, and volume concentration of nanoparticles [117]. The increment in thermal conductivity of a nanofluid is due to the improved Brownian motion of nanoparticles. The main goal of using hybrid nanofluids is to further improve the thermal conductivity, as its thermal conductivity is predicted to be higher than mono nanofluids. Recently, the thermal conductivity can be characterized by several measurement techniques, which is summarized in Chapter 3, along with the numerical simulations. Table 1.3 displays the thermal conductivity of base fluids, metals, and metal oxide nanoparticles, carbon materials, and hybrid nanofluids.

Table 1.3 Thermal conductivity values of base fluids, metals and metal oxide nanoparticles, carbon materials, and hybrid nanofluids.

Base fluid	Thermal conductivity (W/m K)
Water	0.613
Ethylene glycol	0.252
Engine oil	0.145
Nanoparticles	
Diamond	2300
CNT	2000–6000
Ti	21.9
Al	237
TiO_2	8.4
Al_2O_3	36
SiO_2	1.38
Hybrid nanofluids	
TiO_2/SiO_2 hybrid	0.517
ZnO-Ag/water [119]	0.788
TiO_2-CuO/C-EG [120]	0.301
Al_2O_3-TiO_2/water [121]	0.753
ZnO-TiO_2/EG [122]	0.35
FMWCNTs-MgO/EG [123]	1.211
Al_2O_3-MWCNT/thermal oil [124]	0.182
MWCNTS-Al_2O_3/EG [125]	1.17
CuO-MgO-TiO_2/water [126]	0.95
Al_2O_3-SiO_2/water [127]	0.77

Several researchers and scholars investigate theoretical and experimental research work related to thermal conductivity of hybrid nanofluids. Theoretically, to estimate the effective thermal conductivity, the Maxwell analysis model is presented for hybrid nanofluids as [118]

$$k_{hnf} = k_{bf} \frac{\frac{\left(\phi_{p1}k_{p1} + \phi_{p2}k_{p2}\right)}{\phi_{tot}} + 2k_{bf} + 2\left(\phi_{p1}k_{p1} + \phi_{p2}k_{p2}\right) - 2\phi_{tot}k_{bf}}{\frac{\left(\phi_{p1}k_{p1} + \phi_{p2}k_{p2}\right)}{\phi_{tot}} + 2k_{bf} - \left(\phi_{p1}k_{p1} + \phi_{p2}k_{p2}\right) + \phi_{tot}k_{bf}} \tag{1}$$

1.3.2 Viscosity

Viscosity is a significant parameter of nanofluids to evaluate the pumping power requirements due to frictional effects. It is defined as the liquid resistant to distortion produced by shear or longitudinal stresses. It is a function of temperature, volume concentration of nanoparticles, and shape and size of the nanoparticles. Nanofluids normally have much larger viscosity than base fluids. The viscosity is reduced with increase in temperature and the reduced viscosity of hybrid nanofluids is favorable for heat transfer applications [128]. By increasing the nanoparticles' concentration, nanofluids' viscosity increases due to Brownian motion and a large surface-to-volume ratio. The fluid behavior is characterized by Newtonian or non-Newtonian in which Newtonian fluid reveals a linear relation between shear stress and strain rate at a given temperature and non-Newtonian fluid represents nonlinear relation. Several investigations were performed by various scholars and researchers on the hybrid nanofluids' viscosity. Some observed that the hybrid nanofluid's viscosity reduces with the decrease in concentration [87, 103, 129–131]. While the nanofluids' viscosity is increased, hybrid nanofluids show a remarkable heat transfer rate than nanofluids. Limited proposed correlations are available to evaluate the viscosity of hybrid nanofluids. Esfe et al. [132] estimated the correlation to predict the viscosity of MgO-Ag/water hybrid nanofluids as

$$\frac{\mu_{nf}}{\mu_{bf}} = 1.123 + 0.3251\phi - 0.08994T + 0.002552T^2 - 0.00002386T^3 + 0.9695\left(\frac{T}{\phi}\right)^{0.01719} \tag{2}$$

1.3.3 Density

Density is another essential thermophysical property of mono and hybrid nanofluids. It is temperature-dependent and a function of nanofluid concentration. It declines with the increased temperature, and it rises with increased concentration [133]. The density of nanofluids is generally assumed to be a mixed property of the density of base fluid ρ_{bf} and nanoparticles ρ_p. Density of the mono and hybrid nanofluids strongly influences the Reynolds number, frictional factor, pumping power, stability, and other heat transfer properties. Sundar et al. [98, 134] presented the expression to analyze hybrid nanofluids' density (ρ_{hnf}) for MWCNT/Fe_3O_4, ND-Fe_3O_4 nanoparticles:

$$\rho_{hnf} = (1-\phi)\rho_{bf} + \phi \times \rho_p \tag{3}$$

where ρ_{bf} represents the density of base fluid, ρ_p represents density for nanoparticles, and ϕ is the volume concentration percentage.

1.3.4 Specific heat capacity

Heat capacity is another thermophysical property that describes the thermal storage capability of a material and is significant for hybrid nanofluids' heat transfer performance. It is described as the heat required to raise the temperature of 1 g of a substance by 1°C. It reduces with increasing volume concentration and temperature. Some scholars found an improvement with particle concentration while some of them observed contrary. A limited number of studies are available in the literature on the heat capacity of hybrid nanofluids. It is measured by calorimeters, and differential scanning calorimetry (DSC) is one of the effective instruments to define the specific heat of hybrid nanofluids. Sundar et al. [98, 134] expressed the following equation to estimate the specific heat of hybrid nanofluid ($C_{p,hnf}$):

$$C_{p,hnf} = (1-\phi)C_{p,bf} + \phi \times C_{p,p} \tag{4}$$

where $C_{p,bf}$ represents the specific heat of base fluid, $C_{p,p}$ represents the nanoparticles' specific heat, and ϕ is the volume concentration percentage.

1.3.5 Thermal diffusivity

Thermal diffusivity is an important property that is directly related to the nanofluid heat transfer performance, which measures how fast the nanofluid dissipates the heat energy. The thermal diffusivity increases as the particle loading increases. The thermal diffusivity of nanofluids is generally greater than the thermal diffusivity of base fluids [135]. It depends on the nature of thermal transport between the nanoparticles and the surrounding medium. The thermal properties of nanofluids are affected by several factors such as Brownian motion of nanoparticles, formation of liquid nanolayer at liquid particle interface, and the collision of nanoparticles with the base fluid. Thermal diffusivity is important in convective heat transfer applications, and only a few investigations are reported on the thermal diffusivity of nanofluids. Researchers have utilized the thermal lens approach to measure the thermal diffusivity of metallic nanofluids [136, 137].

1.3.6 Electrical, magnetic, dielectric

Unlike other properties, electrical conductivity is a limited investigated property, even it is significant for various applications. Conventional fluids have weak electrical conductivity whereas liquids such as sea water, molten metals, and electrolytes are better conductors. Studies were performed on the good conducting fluids from the past years due to their potential engineering applications. The investigation of electrical and magnetic performance of nanofluids is of utmost significance. Limited investigations have been published on the electrical conductivity of nanofluids. Dielectric properties of nanofluids depend on the nanoparticles' concentration. Subramaniyan et al. [138] investigated the dielectric properties of TiO_2 nanofluids for direct absorption solar collector (DASC) using the Maxwell model. Baby and Ramaprabhu [139] investigated the electrical conductivity of graphene/EG nanofluid and observed an increment of about 220%. Several researchers noticed that electrical conductivity increases with increasing nanoparticle loading due to the electrical double layer formation and the synergy between base fluid and nanoparticles [140–143]. Electrical conductivity depends on the base fluid type and polarity, and it can be affected by surfactants. It presents an effective tool to calculate the stability of nanofluids together with zeta potential. Remarkable electrical conductivity shows better stability, and low conductivity shows poor stability [144]. Still, several gaps exist to achieve correlations to describe electrical conductivity improvement and the various factors affecting the conductivity. Maxwell theoretical correlation was the first method used for estimating the effective electrical conductivity (σ_{eff}), expressing the concentration of the nanomaterials as [19]

$$\sigma_{eff} = \left[1 + \frac{3(\alpha - 1)\varphi}{(\alpha + 2) - (\alpha - 1)\varphi}\right]\sigma_{bf} \tag{5}$$

where $\alpha = \frac{\alpha_p}{\alpha_{bf}}$ represents the effective electrical conductivity of nanoparticles and base fluid.

Magnetic nanofluids (ferrofluids or magnetic fluids) are a special category of smart nanomaterials, in particular magnetically controllable nanofluids [145]. Magnetic particles not only improve the thermal conductivity of nanofluids but also exhibit phenomenal behavior due to their magnetic nature. Such nanofluids contain colloids of magnetic nanoparticles, such as Fe_3O_4, γ-Fe_2O_3, Co, Fe, or Fe-C, dispersed in a carrier liquid [146].

1.4 Applications of hybrid nanofluids

Hybrid nanofluids are smart fluids, containing suspensions of two or more types of nanoparticles in the base fluid, representing a remarkable improvement in their properties even at low concentrations. They are promising and utilized in applications such as heat transfer, electrical and engine cooling, refrigeration, machining, desalination, nuclear PWR, heat exchangers, solar collectors, etc. Several publications are developed on the preparation and properties of hybrid nanofluids so that they can be utilized in several industrial applications.

1.4.1 Electronic cooling

The dawn of the miniaturization of electronic devices has established innovative research areas. Efficiency and miniaturization are characteristics of novel electrical devices, demanding high heat flux removal by reducing the hydraulic diameter and/or by increasing surface area [147]. Large surface area enhances the heat transfer coefficient of the microstructure [148]; therefore, heat removal is favorable in smaller heat sinks, channels, and tubes [149]. Rapid heat dissipation is the major drawback to build small-sized microchips. Methods like using PCM materials, modeling a heat sink, broadening the heat transfer surface through fins, etc., are applied to reduce electronic devices' temperature [150]. Cooling and thermal management of such electronic devices is essential because of their lifetime and efficiency. The promising design of thermal management systems in such devices is crucial for effective and consistent operation. Hybrid nanofluids can be employed for the liquid cooling of electronic processors because of their excellent thermal conductivity. The oscillating heat pipe (OHP) cooling system will emerge as the next generation cooling device to control the heat dissipation [151]. The microchannel heat exchangers are one of the substantial possibilities accessible for electronic cooling compared with traditional heat exchangers and attracted attention by many researchers to investigate various applications such as semiconductor power devices and large-scale integration circuits, water coolants in microchannels, and many others [152–154]. A limited number of studies are available on the hybrid nanofluids employed in cooling electronic chips. Thermal properties of fluids provide the possibility to be used to control the operating modes of devices to obtain maximum output. As a result, hybrid nanofluids are recommended to use in different heat sinks in the near future research.

1.4.2 Solar collectors

The advancement of solar collectors for water heating and heat engines has gained promising attention from many scholars and researchers. Several investigations have been introduced to design collectors and analyze the characteristics to achieve remarkable system performance. Renewable energy is becoming more widespread and environmentally friendly, and solar energy is the most used in electricity generation and water heating [155, 156]. The nanofluids confronted some difficulties in large-scale applications of direct absorption solar collectors such as instability and complexity [157]. Researchers have explored hybrid nanofluids to use as a working fluid to enhance the performance of solar collectors. Li et al. [158] investigated the stability, optical characteristics, and solar-thermal behavior of SiC-MWCNTs hybrid nanofluids for direct absorption solar collectors. It was observed that SiC-MWCNTs hybrid nanofluids presented remarkable stability, and the solar-thermal conversion efficiency is enhanced to 97.3% with 1 wt.%.

1.4.3 Heat exchangers

Heat exchangers are a medium to transfer heat between two fluids with temperature differences without physical contact between the fluids. It should be designed with the maximum possible heating load. Investigators reported the impact of using nanofluids on heat exchangers' heat transfer rate and observed remarkable performance [128, 159]. Many studies have been developed to analyze hybrid nanofluids' heat transfer performance in the heat exchanger medium. Furthermore, numerical studies have been investigated by introducing a novel design for the heat exchanger [160, 161].

1.4.4 Nuclear PWR

The utilization of hybrid nanofluids in nuclear power plants seems like a prospective upcoming application. Kim et al. [162, 163] conducted an experimental study to evaluate the feasibility of using nanofluids in nuclear applications carried out at the Massachusetts Institute of Technology (MIT). Adding nanofluids enhanced the water-cooled-based nuclear performance heat removal. The applications include a pressurized water reactor (PWR) primary coolant, accelerator, and many others [164]. The hybrid nanofluids in the pressurized water reactor (PWR) enhance the critical heat flux (CHF) and become substantially productive by preventing the formation of the vapor layer as the fuel rods are covered with nanoparticles.

1.4.5 Engine cooling

There is a need to enhance the aerodynamic vehicle design and the fuel economy; engineers and manufacturers must decrease the energy demands to surmount the wind resistance. Using hybrid nanofluids as coolants would permit the smaller size and positioning of the engine's radiators.

1.4.6 Refrigeration

Investigators have studied hybrid nanocomposites to be used in refrigeration systems to employ valuable characteristics of nanoparticles to improve their efficiency. Wang et al. [165] observed that TiO_2 nanoparticles can act as additives to improve the solubility of the mineral oil in hydrofluorocarbon (HFC) refrigerant.

1.4.7 Machining

Nanofluids with their cooling and lubricating characteristics have appeared as a potential coolant in machining [166]. Heat transfer coefficients were assessed using the correlation for flow over flat plates in all lubricating conditions [167]. Researchers have stated the characterizing adjustments in the heat transfer capacities with nanoparticles in the cutting fluids. Cutting fluids with nanoparticles have increased heat transfer capacity of around 6% [168].

1.4.8 Desalination

Nanotechnology plays a significant part to obtain specific energy consumption decline as nanofluids application boosts the overall heat transfer coefficient enabling the production of more water for the same size desalination plant. Desalination technology can be classified into either single-phase or phase transition methods. Nanofluids support the advancement of sustainable desalination approaches as these fluids are recognized to overcome the thermodynamic boundaries, which are encountered by present desalination systems [169].

1.5 Challenges and outlook

Considerable research has been reported on mono nanofluids as a working medium. The hybrid nanofluids are a novel class of nanofluids and they are still in the development phase as far as their applications are related. Limited investigations have been studied in the field of hybrid nanofluids. Several challenges to employing hybrid nanofluids include long-term stability, increased pumping power and pressure drop, increased viscosity, thermal performance in turbulent flow and fully developed region, and high cost. All these challenges are mentioned in Chapter 7.

Long-term stability is the key concern for hybrid nanofluids and is considered a technical challenge because of the formation of particle aggregation due to strong van der Waals interactions. Various physical and chemical treatments such as surface modification, addition of surfactants, or applying strong force on the suspended particles' clusters have been employed [170]. The stability of hybrid nanofluids has a direct connection with thermal conductivity improvement. The superior the dispersion behavior, the greater the thermal conductivity of hybrid nanofluids [171].

Synthesis of hybrid nanofluids will be more complicated resulting in high cost, which may restrain the application and can be prepared by one-step or two-step techniques. Both techniques require complex and sophisticated equipment.

Extensive investigational work should be carried out using hybrid nanofluids to study the influence of various suspended nanoparticles, nanoparticles' concentration, surface morphology, and their preparation technique on the thermal behavior of heat transfer applications. Selection of appropriate hybrid nanocomposites with enhanced stability should be discovered. The advancement of the nanocomposite production technique will be valuable for the research community.

1.6 Conclusion

Renewable sources provide eco-friendly sustainable alternatives to replace carbon-intensive energy sources. Nanotechnology has been developed to optimize and improve energy systems and efficiency of the energy conversion processes. Nanotechnology-based thermal fluids are introduced to harness solar and geothermal energy. Nanomaterials have gained extensive interest in recent years, encouraging to perform an innovative role in sophisticated applications. With the significant development of modern nanotechnology and nanomaterials, the discovery of nanofluids that contained nanosized particles dispersed in base fluids completely changed the picture. Nanofluids do not retain all favorable properties that are requisite for specialized applications. To trade-off between properties, hybrid nanofluids have been introduced to acquire enhanced properties, suitable for applications that involve remarkable thermal, optical, and rheological properties of the working fluid.

Hybrid nanofluids are the novel high-class working fluids considered as a potential candidate for several applications. Hybrid or composite nanofluids are developed as a novel class of nanofluids synthesized by combining two or more nanoparticles containing metal or metal oxide or a combination of both particles in a base fluid. The preparation process is a significant step in the nanocomposites to further expand the thermal conductivity of heat transfer fluids. Thermophysical properties of hybrid nanofluids show promising enhancement as compared with mono nanofluids, especially thermal conductivity. Increment in nanoparticle loading results in enhanced values of thermophysical properties like thermal conductivity, viscosity, heat capacity, and density. Several investigations have been reported on the thermal conductivity and viscosity of hybrid nanofluids, but research on other properties like density, specific heat, thermal diffusivity, and magnetic properties are limited.

Hybrid nanofluids are promising and can be utilized in several applications such as heat transfer, electrical and engine cooling, refrigeration, machining, desalination, nuclear PWR, heat exchangers, solar collectors, etc. Some challenges still need to be identified such as stability, increased pumping power, and production cost to be employed in industrial applications.

References

[1] U. Soytas, R. Sari, Energy consumption, economic growth, and carbon emissions: challenges faced by an EU candidate member, Ecol. Econ. 68 (6) (2009) 1667–1675.

[2] Z. Said, et al., Central versus off-grid photovoltaic system, the optimum option for the domestic sector based on techno-economic-environmental assessment for United Arab Emirates, Sustain. Energy Technol. Assess. 43 (2021) 100944.
[3] Z. Said, A.A. Alshehhi, A. Mehmood, Predictions of UAE's renewable energy mix in 2030, Renew. Energy 118 (2018) 779–789.
[4] O. Ellabban, H. Abu-Rub, F. Blaabjerg, Renewable energy resources: current status, future prospects and their enabling technology, Renew. Sustain. Energy Rev. 39 (2014) 748–764.
[5] A. Armin Razmjoo, A. Sumper, A. Davarpanah, Energy sustainability analysis based on SDGs for developing countries, Energy Sources Pt A Recov. Util. Environ. Effects 42 (9) (2020) 1041–1056.
[6] A. Razmjoo, et al., A technical analysis investigating energy sustainability utilizing reliable renewable energy sources to reduce CO2 emissions in a high potential area, Renew. Energy 164 (2021) 46–57.
[7] Z. Said, et al., Recent advances on nanofluids for low to medium temperature solar collectors: energy, exergy, economic analysis and environmental impact, Prog. Energy Combust. Sci. 84 (2021) 100898.
[8] M. Ghodbane, et al., Energy, financial, and environmental investigation of a direct steam production power plant driven by linear fresnel solar reflectors, J. Solar Energy Eng. (2021) **143**(2).
[9] P.H. Diamandis, S. Kotler, Abundance: The Future Is Better Than You Think, Simon and Schuster, 2012.
[10] Z. Said, S. Arora, E. Bellos, A review on performance and environmental effects of conventional and nanofluid-based thermal photovoltaics, Renew. Sustain. Energy Rev. 94 (2018) 302–316.
[11] Y. Zhao, et al., A nanopump for low-temperature and efficient solar water evaporation, J. Mater. Chem. A 7 (42) (2019) 24311–24319.
[12] Y. Qiu, et al., An experimental study on the heat transfer performance of a prototype molten-salt rod baffle heat exchanger for concentrated solar power, Energy 156 (2018) 63–72.
[13] O.B. Mousa, R.A. Taylor, Global solar technology optimization for factory rooftop emissions mitigation, Environ. Res. Lett. 15 (4) (2020), 044013.
[14] Q. Li, et al., Design and analysis of a medium-temperature, concentrated solar thermal collector for air-conditioning applications, Appl. Energy 190 (2017) 1159–1173.
[15] T. Gao, et al., Mechanics analysis and predictive force models for the single-diamond grain grinding of carbon fiber reinforced polymers using CNT nano-lubricant, J. Mater. Process. Technol. 290 (2021) 116976.
[16] Z. Said, et al., Heat transfer, entropy generation, economic and environmental analyses of linear fresnel reflector using novel rGO-Co3O4 hybrid nanofluids, Renew. Energy 165 (2021) 420–437.
[17] M. Sheikholeslami, et al., Recent progress on flat plate solar collectors and photovoltaic systems in the presence of nanofluid: a review, J. Clean. Prod. (2021) 126119.
[18] A.K. Tiwar, et al., A review on the application of hybrid nanofluids for parabolic trough collector: Recent progress and outlook, J. Clean. Prod. (2021) 126031.
[19] J.C. Maxwell, A Treatise on Electricity and Magnetism, vol. 1, Clarendon Press, Oxford, 1873.
[20] H. Masuda, et al., Alteration of thermal conductivity and viscosity of liquid by dispersing ultra-fine particles dispersion of Al_2O_3, SiO_2 and TiO_2 ultra-fine particles, Netsu Bussei 7 (4) (1993) 227–233.
[21] M.M. Khin, et al., A review on nanomaterials for environmental remediation, Energ. Environ. Sci. 5 (8) (2012) 8075–8109.

[22] J. Ouyang, Applications of carbon nanotubes and graphene for third-generation solar cells and fuel cells, Nano Mater. Sci. 1 (2) (2019) 77–90.
[23] M.A. Kiser, et al., Titanium nanomaterial removal and release from wastewater treatment plants, Environ. Sci. Technol. 43 (17) (2009) 6757–6763.
[24] Y. Xuan, Q. Li, Investigation on convective heat transfer and flow features of nanofluids, J. Heat Transfer 125 (1) (2003) 151–155.
[25] Y. Wang, et al., Development of solid-state electrolytes for sodium-ion battery—a short review, Nano Mater. Sci. 1 (2) (2019) 91–100.
[26] J. Wang, Nanomaterial-based electrochemical biosensors, Analyst 130 (4) (2005) 421–426.
[27] A.M. Smith, S. Nie, Semiconductor nanocrystals: structure, properties, and band gap engineering, Acc. Chem. Res. 43 (2) (2010) 190–200.
[28] P. Brown, K. Stevens, Nanofibers and Nanotechnology in Textiles, Elsevier, 2007.
[29] M. Ohring, Materials Science of Thin Films, Elsevier, 2001.
[30] F. Uddin, Clays, nanoclays, and montmorillonite minerals, Metallurg. Mater. Trans. A 39 (12) (2008) 2804–2814.
[31] P.H.C. Camargo, K.G. Satyanarayana, F. Wypych, Nanocomposites: synthesis, structure, properties and new application opportunities, Mater. Res. 12 (1) (2009) 1–39.
[32] S.U.S. Choi, Enhancing thermal conductivity of fluids with nanoparticles, in: American Society of Mechanical Engineers, Fluids Engineering Division (Publication) FED, 1995.
[33] L.S. Sundar, et al., Experimental investigation of thermo-physical properties, heat transfer, pumping power, entropy generation, and exergy efficiency of nanodiamond+ Fe3O4/60: 40% water-ethylene glycol hybrid nanofluid flow in a tube, Therm. Sci. Eng. Prog. 21 (2021) 100799.
[34] L. Syam Sundar, et al., Heat transfer of rGO/CO3O4 hybrid nanomaterial-based nanofluids and twisted tape configurations in a tube, J. Therm. Sci. Eng. Appl. (2021) **13**(3).
[35] N.S. Pandya, et al., Influence of the geometrical parameters and particle concentration levels of hybrid nanofluid on the thermal performance of axial grooved heat pipe, Therm. Sci. Eng. Prog. 21 (2021) 100762.
[36] S. Rahman, et al., Performance enhancement of a solar powered air conditioning system using passive techniques and SWCNT/R-407c nano refrigerant, Case Stud. Therm. Eng. 16 (2019) 100565.
[37] E. Khodabandeh, et al., Application of nanofluid to improve the thermal performance of horizontal spiral coil utilized in solar ponds: geometric study, Renew. Energy 122 (2018) 1–16.
[38] R.A. Dehkordi, M.H. Esfe, M. Afrand, Effects of functionalized single walled carbon nanotubes on thermal performance of antifreeze: an experimental study on thermal conductivity, Appl. Therm. Eng. 120 (2017) 358–366.
[39] R. Ranjbarzadeh, et al., An experimental study on heat transfer and pressure drop of water/graphene oxide nanofluid in a copper tube under air cross-flow: Applicable as a heat exchanger, Appl. Therm. Eng. 125 (2017) 69–79.
[40] P. Selvakumar, S. Suresh, Use of Al_2O_3–Cu/water hybrid nanofluid in an electronic heat sink, IEEE Trans. Compon. Packag. Manuf. Technol. 2 (10) (2012) 1600–1607.
[41] N.S. Akbar, M. Raza, R. Ellahi, Peristaltic flow with thermal conductivity of H2O+Cu nanofluid and entropy generation, Res. Phys. 5 (2015) 115–124.
[42] B. Li, et al., Heat transfer performance of MQL grinding with different nanofluids for Ni-based alloys using vegetable oil, J. Clean. Prod. 154 (2017) 1–11.
[43] P. Ternik, Conduction and convection heat transfer characteristics of water–Au nanofluid in a cubic enclosure with differentially heated side walls, Int. J. Heat Mass Transfer 80 (2015) 368–375.

[44] R. Mashayekhi, et al., Application of a novel conical strip insert to improve the efficacy of water–Ag nanofluid for utilization in thermal systems: A two-phase simulation, Energ. Conver. Manage. 151 (2017) 573–586.
[45] J. Sarkar, A critical review on convective heat transfer correlations of nanofluids, Renew. Sustain. Energy Rev. 15 (6) (2011) 3271–3277.
[46] Z. Said, et al., Thermophysical properties using ND/water nanofluids: an experimental study, ANFIS-based model and optimization, J. Mol. Liq. (2021) 115659.
[47] L. Sundar, et al., Heat transfer and second law analysis of ethylene glycol based ternary hybrid nanofluid under laminar flow, J. Therm. Sci. Eng. Appl. (2021) 1–45.
[48] A.K. Tiwari, et al., 4S consideration (synthesis, sonication, surfactant, stability) for the thermal conductivity of CeO2 with MWCNT and water based hybrid nanofluid: an experimental assessment, Colloids Surf. A Physicochem. Eng. Asp. 610 (2021) 125918.
[49] A.K. Tiwari, et al., 3S (sonication, surfactant, stability) impact on the viscosity of hybrid nanofluid with different base fluids: an experimental study, J. Mol. Liq. (2021) 115455.
[50] J.J. Michael, S. Iniyan, Performance analysis of a copper sheet laminated photovoltaic thermal collector using copper oxide–water nanofluid, Solar Energy 119 (2015) 439–451.
[51] S.U.S. Choi, et al., Anomalous thermal conductivity enhancement in nanotube suspensions, Appl. Phys. Lett. 79 (14) (2001) 2252–2254.
[52] W. Yu, et al., Investigation on the thermal transport properties of ethylene glycol-based nanofluids containing copper nanoparticles, Powder Technol. 197 (3) (2010) 218–221.
[53] P. Kanti, et al., Numerical study on the thermo-hydraulic performance analysis of fly ash nanofluid, J. Therm. Anal. Calorim. (2021) 1–13.
[54] L.S. Sundar, et al., Effect of core-rod diameter on wire coil inserts for heat transfer and friction factor of high-prandtl number magnetic Fe_3O_4 nanofluids ina fully developed laminar flow, Heat Transfer Res. 52 (3) (2021).
[55] Z. Said, et al., Optimizing density, dynamic viscosity, thermal conductivity and specific heat of a hybrid nanofluid obtained experimentally via ANFIS-based model and modern optimization, J. Mol. Liq. 321 (2021) 114287.
[56] N.K. Cakmak, et al., Preparation, characterization, stability, and thermal conductivity of rGO-Fe3O4-TiO2 hybrid nanofluid: an experimental study, Powder Technol. 372 (2020) 235–245.
[57] A.K. Tiwari, et al., Experimental comparison of specific heat capacity of three different metal oxides with MWCNT/water-based hybrid nanofluids: proposing a new correlation, Appl. Nanosci. (2020) 1–11.
[58] Z. Said, R. Saidur, N. Rahim, Energy and exergy analysis of a flat plate solar collector using different sizes of aluminium oxide based nanofluid, J. Clean. Prod. 133 (2016) 518–530.
[59] M. Hemmat Esfe, S. Saedodin, Turbulent forced convection heat transfer and thermophysical properties of Mgo–water nanofluid with consideration of different nanoparticles diameter, an empirical study, J. Therm. Anal. Calorim. 119 (2) (2015) 1205–1213.
[60] M. Gupta, et al., Up to date review on the synthesis and thermophysical properties of hybrid nanofluids, J. Clean. Prod. 190 (2018) 169–192.
[61] M. Hemmat Esfe, et al., Thermal conductivity of Al2O3/water nanofluids, J. Therm. Anal. Calorim. 117 (2) (2014) 675–681.
[62] Z. Said, et al., Acid-functionalized carbon nanofibers for high stability, thermoelectrical and electrochemical properties of nanofluids, J. Colloid Interface Sci. 520 (2018) 50–57.

[63] Z. Said, et al., Energy and exergy efficiency of a flat plate solar collector using pH treated Al2O3 nanofluid, J. Clean. Prod. 112 (2016) 3915–3926.
[64] A.A. Hachicha, et al., On the thermal and thermodynamic analysis of parabolic trough collector technology using industrial-grade MWCNT based nanofluid, Renew. Energy 161 (2020) 1303–1317.
[65] Z. Said, et al., Stability, thermophysical and electrical properties of synthesized carbon nanofiber and reduced-graphene oxide-based nanofluids and their hybrid along with fuzzy modeling approach, Powder Technol. 364 (2020) 795–809.
[66] A.W. Ezzat, I.M. Hasan, Investigation of alumina nano fluid thermal conductivity, Int. J. Comput. Appl. 102 (11) (2014).
[67] R.J. Issa, Effect of nanoparticles size and concentration on thermal and rheological properties of AL2O3-water nanofluids, in: Proceedings of the World Congress on Momentum, Heat and Mass Transfer, 2016.
[68] L. Syam Sundar, et al., Thermal conductivity and viscosity of stabilized ethylene glycol and water mixture Al2O3 nanofluids for heat transfer applications: an experimental study, Int. Commun. Heat Mass Transfer 56 (2014) 86–95.
[69] T.R. Shah, H.M. Ali, Applications of hybrid nanofluids in solar energy, practical limitations and challenges: a critical review, Solar Energy 183 (2019) 173–203.
[70] S. Suresh, et al., Effect of Al2O3–Cu/water hybrid nanofluid in heat transfer, Exp. Therm. Fluid Sci. 38 (2012) 54–60.
[71] A.A. Minea, W.M. El-Maghlany, Influence of hybrid nanofluids on the performance of parabolic trough collectors in solar thermal systems: Recent findings and numerical comparison, Renew. Energy 120 (2018) 350–364.
[72] S.K. Verma, et al., Performance analysis of hybrid nanofluids in flat plate solar collector as an advanced working fluid, Solar Energy 167 (2018) 231–241.
[73] M. Gupta, V. Singh, Z. Said, Heat transfer analysis using zinc Ferrite/water (Hybrid) nanofluids in a circular tube: An experimental investigation and development of new correlations for thermophysical and heat transfer properties, Sustain. Energy Technol. Assess. 39 (2020) 100720.
[74] L.S. Sundar, et al., Energy, efficiency, economic impact, and heat transfer aspects of solar flat plate collector with Al2O3 nanofluids and wire coil with core rod inserts, Sustain. Energy Technol. Assess. 40 (2020) 100772.
[75] L.S. Sundar, et al., Combination of Co3O4 deposited rGO hybrid nanofluids and longitudinal strip inserts: Thermal properties, heat transfer, friction factor, and thermal performance evaluations, Therm. Sci. Eng. Prog. 20 (2020) 100695.
[76] R. Turcu, et al., New polypyrrole-multiwall carbon nanotubes hybrid materials, J. Optoelectron. Adv. Mater. 8 (2) (2006) 643–647.
[77] K. Niihara, New design concept of structural ceramics ceramic nanocomposites, J. Cerma. Soc. Jpn. 99 (1154) (1991) 974–982.
[78] S.-T. Oh, T. Sekino, K. Niihara, Effect of particle size distribution and mixing homogeneity on microstructure and strength of alumina/copper composites, Nanostruct. Mater. 10 (2) (1998) 327–332.
[79] N. Jha, S. Ramaprabhu, Thermal conductivity studies of metal dispersed multiwalled carbon nanotubes in water and ethylene glycol based nanofluids, J. Appl. Phys. 106 (8) (2009), 084317.
[80] D. Madhesh, R. Parameshwaran, S. Kalaiselvam, Experimental investigation on convective heat transfer and rheological characteristics of Cu–TiO2 hybrid nanofluids, Exp. Therm. Fluid Sci. 52 (2014) 104–115.
[81] M. Baghbanzadeh, et al., Synthesis of spherical silica/multiwall carbon nanotubes hybrid nanostructures and investigation of thermal conductivity of related nanofluids, Thermochim. Acta 549 (2012) 87–94.

[82] S. Akilu, et al., Properties of glycerol and ethylene glycol mixture based SiO2-CuO/C hybrid nanofluid for enhanced solar energy transport, Solar Energy Mater. Solar Cells 179 (2018) 118–128.
[83] B. Wei, et al., Thermo-physical property evaluation of diathermic oil based hybrid nanofluids for heat transfer applications, Int. J. Heat Mass Transfer 107 (2017) 281–287.
[84] B. Munkhbayar, et al., Surfactant-free dispersion of silver nanoparticles into MWCNT-aqueous nanofluids prepared by one-step technique and their thermal characteristics, Ceram. Int. 39 (6) (2013) 6415–6425.
[85] M.H. Esfe, et al., Thermal conductivity of Cu/TiO2–water/EG hybrid nanofluid: experimental data and modeling using artificial neural network and correlation, Int. Commun. Heat Mass Transfer 66 (2015) 100–104.
[86] L. Syam Sundar, et al., Experimental investigation of the thermal transport properties of graphene oxide/Co3O4 hybrid nanofluids, Int. Commun. Heat Mass Transfer 84 (2017) 1–10.
[87] L.S. Sundar, M.K. Singh, A.C.M. Sousa, Enhanced heat transfer and friction factor of MWCNT–Fe3O4/water hybrid nanofluids, Int. Commun. Heat Mass Transfer 52 (2014) 73–83.
[88] L. Syam Sundar, et al., Experimental investigation of thermo-physical properties, heat transfer, pumping power, entropy generation, and exergy efficiency of nanodiamond + Fe3O4/60:40% water-ethylene glycol hybrid nanofluid flow in a tube, Therm. Sci. Eng. Prog. 21 (2021) 100799.
[89] S.S. Harandi, et al., An experimental study on thermal conductivity of F-MWCNTs–Fe3O4/EG hybrid nanofluid: effects of temperature and concentration, Int. Commun. Heat Mass Transfer 76 (2016) 171–177.
[90] S. Askari, et al., Rheological and thermophysical properties of ultra-stable kerosene-based Fe3O4/graphene nanofluids for energy conservation, Energ. Conver. Manage. 128 (2016) 134–144.
[91] M.H. Hamzah, et al., Factors affecting the performance of hybrid nanofluids: a comprehensive review, Int. J. Heat Mass Transfer 115 (2017) 630–646.
[92] H. Babar, H.M. Ali, Towards hybrid nanofluids: preparation, thermophysical properties, applications, and challenges, J. Mol. Liq. 281 (2019) 598–633.
[93] L. Chen, W. Yu, H. Xie, Enhanced thermal conductivity of nanofluids containing Ag/MWNT composites, Powder Technol. 231 (2012) 18–20.
[94] S.J. Aravind, S. Ramaprabhu, Graphene-multiwalled carbon nanotube-based nanofluids for improved heat dissipation, RSC Adv. 3 (13) (2013) 4199–4206.
[95] M.J. Nine, et al., Investigation of Al2O3-MWCNTs hybrid dispersion in water and their thermal characterization, J. Nanosci. Nanotechnol. 12 (6) (2012) 4553–4559.
[96] S. Suresh, et al., Effect of Al2O3–Cu/water hybrid nanofluid in heat transfer, Exp. Therm. Fluid Sci. 38 (2012) 54–60.
[97] L.F. Chen, et al., Enhanced thermal conductivity of nanofluid by synergistic effect of multi-walled carbon nanotubes and Fe2O3 nanoparticles, in: Applied Mechanics and Materials, Trans Tech Publishers, 2014.
[98] L.S. Sundar, et al., Nanodiamond-Fe3O4 nanofluids: preparation and measurement of viscosity, electrical and thermal conductivities, Int. Commun. Heat Mass Transfer 73 (2016) 62–74.
[99] L. Syam Sundar, et al., Combination of Co3O4 deposited rGO hybrid nanofluids and longitudinal strip inserts: thermal properties, heat transfer, friction factor, and thermal performance evaluations, Therm. Sci. Eng. Prog. 20 (2020) 100695.
[100] A.K. Tiwari, et al., Experimental comparison of specific heat capacity of three different metal oxides with MWCNT/ water-based hybrid nanofluids: proposing a new correlation, Appl. Nanosci. (2020).

[101] N.S. Pandya, et al., Influence of the geometrical parameters and particle concentration levels of hybrid nanofluid on the thermal performance of axial grooved heat pipe, Therm. Sci. Eng. Prog. (2020) 100762.
[102] M.H. Esfe, et al., Estimation of thermal conductivity of ethylene glycol-based nanofluid with hybrid suspensions of SWCNT–Al_2O_3 nanoparticles by correlation and ANN methods using experimental data, J. Therm. Anal. Calorim. 128 (3) (2017) 1359–1371.
[103] O. Soltani, M. Akbari, Effects of temperature and particles concentration on the dynamic viscosity of MgO-MWCNT/ethylene glycol hybrid nanofluid: Experimental study, Physica E 84 (2016) 564–570.
[104] M.H. Esfe, et al., Thermal conductivity enhancement of SiO2–MWCNT (85: 15%)–EG hybrid nanofluids, J. Therm. Anal. Calorim. 128 (1) (2017) 249–258.
[105] P. Van Trinh, et al., Experimental study on the thermal conductivity of ethylene glycol-based nanofluid containing Gr-CNT hybrid material, J. Mol. Liq. 269 (2018) 344–353.
[106] M.H. Esfe, et al., Experimental evaluation, new correlation proposing and ANN modeling of thermal properties of EG based hybrid nanofluid containing ZnO-DWCNT nanoparticles for internal combustion engines applications, Appl. Therm. Eng. 133 (2018) 452–463.
[107] G. Paul, et al., Synthesis, characterization, and thermal property measurement of nano-Al95Zn05 dispersed nanofluid prepared by a two-step process, Int. J. Heat Mass Transfer 54 (15–16) (2011) 3783–3788.
[108] S.K. Mechiri, V. Vasu, A. Venu Gopal, Investigation of thermal conductivity and rheological properties of vegetable oil based hybrid nanofluids containing Cu–Zn hybrid nanoparticles, Exp. Heat Transfer 30 (3) (2017) 205–217.
[109] A. Sajeeb, P.K. Rajendrakumar, Investigation on the rheological behavior of coconut oil based hybrid CeO2/CuO nanolubricants, Proc. Inst. Mech. Eng. Pt J: J. Eng. Tribol. 233 (1) (2019) 170–177.
[110] A. Alirezaie, et al., Investigation of rheological behavior of MWCNT (COOH-functionalized)/MgO-engine oil hybrid nanofluids and modelling the results with artificial neural networks, J. Mol. Liq. 241 (2017) 173–181.
[111] Z. Said, et al., Performance enhancement of a flat plate solar collector using titanium dioxide nanofluid and polyethylene glycol dispersant, J. Clean. Prod. 92 (2015) 343–353.
[112] A.K. Tiwari, et al., Experimental and numerical investigation on the thermal performance of triple tube heat exchanger equipped with different inserts with WO3/water nanofluid under turbulent condition, Int. J. Therm. Sci. (2021) 106861.
[113] Z. Said, Thermophysical and optical properties of SWCNTs nanofluids, Int. Commun. Heat Mass Transfer 78 (2016) 207–213.
[114] M. Ghodbane, et al., Performance assessment of linear Fresnel solar reflector using MWCNTs/DW nanofluids, Renew. Energy 151 (2020) 43–56.
[115] Z. Said, et al., Thermophysical properties of single wall carbon nanotubes and its effect on exergy efficiency of a flat plate solar collector, Solar Energy 115 (2015) 757–769.
[116] L. Yang, et al., An updated review on the properties, fabrication and application of hybrid-nanofluids along with their environmental effects, J. Clean. Prod. 257 (2020) 120408.
[117] L.S. Sundar, et al., Heat transfer, energy, and exergy efficiency enhancement of nanodiamond/water nanofluids circulate in a flat plate solar collector, J. Enhanced Heat Transfer 28 (2) (2021).
[118] M.A. Mansour, et al., Effects of heat source and sink on entropy generation and MHD natural convection of Al2O3-Cu/water hybrid nanofluid filled with square porous cavity, Therm. Sci. Eng. Prog. 6 (2018) 57–71.

[119] N.N. Esfahani, D. Toghraie, M. Afrand, A new correlation for predicting the thermal conductivity of ZnO–Ag (50%–50%)/water hybrid nanofluid: an experimental study, Powder Technol. 323 (2018) 367–373.
[120] S. Akilu, A.T. Baheta, K. Sharma, Experimental measurements of thermal conductivity and viscosity of ethylene glycol-based hybrid nanofluid with TiO2-CuO/C inclusions, J. Mol. Liq. 246 (2017) 396–405.
[121] G.M. Moldoveanu, et al., Al2O3/TiO2 hybrid nanofluids thermal conductivity: an experimental approach, J. Therm. Anal. Calorim. 137 (2) (2019).
[122] D. Toghraie, V.A. Chaharsoghi, M. Afrand, Measurement of thermal conductivity of ZnO–TiO2/EG hybrid nanofluid, J. Therm. Anal. Calorim. 125 (1) (2016) 527–535.
[123] M. Afrand, Experimental study on thermal conductivity of ethylene glycol containing hybrid nano-additives and development of a new correlation, Appl. Therm. Eng. 110 (2017) 1111–1119.
[124] A. Asadi, et al., Heat transfer efficiency of Al2O3-MWCNT/thermal oil hybrid nanofluid as a cooling fluid in thermal and energy management applications: an experimental and theoretical investigation, Int. J. Heat Mass Transfer 117 (2018) 474–486.
[125] A.A.A. Arani, F. Pourmoghadam, Experimental investigation of thermal conductivity behavior of MWCNTS-Al2O3/ethylene glycol hybrid Nanofluid: Providing new thermal conductivity correlation, Heat Mass Transfer 55 (8) (2019) 2329–2339.
[126] S. Mousavi, F. Esmaeilzadeh, X. Wang, Effects of temperature and particles volume concentration on the thermophysical properties and the rheological behavior of CuO/MgO/TiO2 aqueous ternary hybrid nanofluid, J. Therm. Anal. Calorim. 137 (3) (2019) 879–901.
[127] G.M. Moldoveanu, et al., Experimental study on thermal conductivity of stabilized Al2O3 and SiO2 nanofluids and their hybrid, Int. J. Heat Mass Transfer 127 (2018) 450–457.
[128] Z. Said, et al., Fuzzy modeling and optimization for experimental thermophysical properties of water and ethylene glycol mixture for Al2O3 and TiO2 based nanofluids, Powder Technol. 353 (2019) 345–358.
[129] S. Sundar, D. Agarwal, Visceral leishmaniasis—optimum treatment options in children, Pediatr. Infect. Dis. J. 37 (5) (2018) 492–494.
[130] M.F. Nabil, et al., An experimental study on the thermal conductivity and dynamic viscosity of TiO2-SiO2 nanofluids in water: ethylene glycol mixture, Int. Commun. Heat Mass Transfer 86 (2017) 181–189.
[131] A.A. Hussien, et al., Experiment on forced convective heat transfer enhancement using MWCNTs/GNPs hybrid nanofluid and mini-tube, Int. J. Heat Mass Transfer 115 (2017) 1121–1131.
[132] M.H. Esfe, et al., Experimental determination of thermal conductivity and dynamic viscosity of Ag–MgO/water hybrid nanofluid, Int. Commun. Heat Mass Transfer 66 (2015) 189–195.
[133] H. Yarmand, et al., Graphene nanoplatelets–silver hybrid nanofluids for enhanced heat transfer, Energ. Conver. Manage. 100 (2015) 419–428.
[134] L.S. Sundar, M.K. Singh, A.C. Sousa, Enhanced heat transfer and friction factor of MWCNT–Fe3O4/water hybrid nanofluids, Int. Commun. Heat Mass Transfer 52 (2014) 73–83.
[135] F.M. Ali, W.M.M. Yunus, Study of the effect of volume fraction concentration and particle materials on thermal conductivity and thermal diffusivity of nanofluids, Jpn. J. Appl. Phys. 50 (8) (2011), 085201.
[136] M. Maillard, et al., Picosecond self-induced thermal lensing from colloidal silver nanodisks, J. Phys. Chem. B. 108 (17) (2004) 5230–5234.

[137] M. Hari, et al., Thermal diffusivity of nanofluids composed of rod-shaped silver nanoparticles, Int. J. Therm. Sci. 64 (2013) 188–194.
[138] A. Subramaniyan, L.P. Sukumaran, R. Ilangovan, Investigation of the dielectric properties of TiO2 nanofluids, J. Taibah Univ. Sci. 10 (3) (2016) 403–406.
[139] T.T. Baby, S. Ramaprabhu, Investigation of thermal and electrical conductivity of graphene based nanofluids, J. Appl. Phys. 108 (12) (2010) 124308.
[140] S. Bagheli, et al., Synthesis and experimental investigation of the electrical conductivity of water based magnetite nanofluids, Powder Technol. 274 (2015) 426–430.
[141] M.F. Coelho, et al., Permittivity and electrical conductivity of copper oxide nanofluid (12 nm) in water at different temperatures, J. Chem. Thermodynam. 132 (2019) 164–173.
[142] I. Zakaria, et al., Thermo-electrical performance of PEM fuel cell using Al2O3 nanofluids, Int. J. Heat Mass Transfer 119 (2018) 460–471.
[143] F. Mashali, et al., Nanodiamond nanofluid microstructural and thermo-electrical characterization, Int. Commun. Heat Mass Transfer 101 (2019) 82–88.
[144] A.A. Minea, A review on electrical conductivity of nanoparticle-enhanced fluids, Nanomaterials 9 (11) (2019) 1592.
[145] S. Odenbach, Ferrofluids: Magnetically Controllable Fluids and Their Applications, vol. 594, Springer, 2008.
[146] S.W. Charles, The preparation of magnetic fluids, in: Ferrofluids, Springer, 2002, pp. 3–18.
[147] M.Y. Abdollahzadeh Jamalabadi, J.H. Park, C.Y. Lee, Optimal design of magnetohydrodynamic mixed convection flow in a vertical channel with slip boundary conditions and thermal radiation effects by using an entropy generation minimization method, Entropy 17 (2) (2015) 866–881.
[148] V. Nikkhah, M. Sarafraz, F. Hormozi, Application of spherical copper oxide (II) water nano-fluid as a potential coolant in a boiling annular heat exchanger, Chem. Biochem. Eng. Q. 29 (2015) 405–415.
[149] M. Sarafraz, et al., Thermal performance of a heat sink microchannel working with biologically produced silver-water nanofluid: experimental assessment, Exp. Therm. Fluid Sci. 91 (2018) 509–519.
[150] T.-U. Rehman, H.M. Ali, Experimental investigation on paraffin wax integrated with copper foam based heat sinks for electronic components thermal cooling, Int. Commun. Heat Mass Transfer 98 (2018) 155–162.
[151] H. Ma, et al., Effect of nanofluid on the heat transport capability in an oscillating heat pipe, Appl. Phys. Lett. 88 (14) (2006) 143116.
[152] A.A.A.A. Alrashed, et al., The numerical modeling of water/FMWCNT nanofluid flow and heat transfer in a backward-facing contracting channel, Phys. B Condens. Matter 537 (2018) 176–183.
[153] A. Karimipour, A. D'Orazio, M. Goodarzi, Develop the lattice Boltzmann method to simulate the slip velocity and temperature domain of buoyancy forces of FMWCNT nanoparticles in water through a micro flow imposed to the specified heat flux, Physica A Stat. Mech. Appl. 509 (2018) 729–745.
[154] Y. Won, S. Kim, S.E. Kim, Experimental assessment of on-chip liquid cooling through microchannels with de-ionized water and diluted ethylene glycol, Jpn. J. Appl. Phys. 55 (6S3) (2016), 06JB02.
[155] M. Ehyaei, et al., Energy, exergy and economic analyses for the selection of working fluid and metal oxide nanofluids in a parabolic trough collector, Solar Energy 187 (2019) 175–184.
[156] A. Kumar, Z. Said, E. Bellos, An up-to-date review on evacuated tube solar collectors, J. Therm. Anal. Calorim. (2020) 1–17.

[157] E. Bellos, Z. Said, C. Tzivanidis, The use of nanofluids in solar concentrating technologies: a comprehensive review, J. Clean. Prod. 196 (2018) 84–99.
[158] X. Li, G. Zeng, X. Lei, The stability, optical properties and solar-thermal conversion performance of SiC-MWCNTs hybrid nanofluids for the direct absorption solar collector (DASC) application, Solar Energy Mater. Solar Cells 206 (2020) 110323.
[159] H. Maddah, M. Ghazvini, M.H. Ahmadi, Predicting the efficiency of CuO/water nanofluid in heat pipe heat exchanger using neural network, Int. Commun. Heat Mass Transfer 104 (2019) 33–40.
[160] M.H. Jahangir, et al., A numerical study into effects of intermittent pump operation on thermal storage in unsaturated porous media, Appl. Therm. Eng. 138 (2018) 110–121.
[161] Z. Said, et al., Enhancing the performance of automotive radiators using nanofluids, Renew. Sustain. Energy Rev. 112 (2019) 183–194.
[162] S.J. Kim, et al., Study of pool boiling and critical heat flux enhancement in nanofluids, Bull. Pol. Acad. Sci. Tech. Sci. (2007) 211–216.
[163] S.J. Kim, et al., Surface wettability change during pool boiling of nanofluids and its effect on critical heat flux, Int. J. Heat Mass Transfer 50 (19–20) (2007) 4105–4116.
[164] B.A. Khuwaileh, et al., On the performance of nanofluids in APR 1400 PLUS7 assembly: neutronics, Ann. Nuclear Energy 144 (2020) 107508.
[165] R. Wang, et al., A refrigerating system using HFC134a and mineral lubricant appended with n-TiO2 (R) as working fluids, in: Proceedings of the 4th international symposium on HAVC, Tsinghua University Press, Beijing, China, 2003.
[166] Z. Said, et al., A comprehensive review on minimum quantity lubrication (MQL) in machining processes using nano-cutting fluids, Int. J. Adv. Manufact. Technol. 105 (5) (2019) 2057–2086.
[167] X. Wang, et al., Vegetable oil-based nanofluid minimum quantity lubrication turning: academic review and perspectives, J. Manufact. Process. 59 (2020) 76–97.
[168] R.R. Srikant, et al., Applicability of cutting fluids with nanoparticle inclusion as coolants in machining, Proc. Inst. Mech. Eng. Pt J: J. Eng. Tribol. 223 (2009) 221–225.
[169] T. Singh, et al., The role of nanofluids and renewable energy in the development of sustainable desalination systems: a review, Watermark 12 (7) (2020) 2002.
[170] J. Sarkar, P. Ghosh, A. Adil, A review on hybrid nanofluids: recent research, development and applications, Renew. Sustain. Energy Rev. 43 (2015) 164–177.
[171] D. Wen, et al., Review of nanofluids for heat transfer applications, Particuology 7 (2) (2009) 141–150.

2

Preparation and stability of hybrid nanofluids

Neeti Arora[a], Munish Gupta[a], and Zafar Said[b,c,d,*]

[a]Department of Mechanical Engineering, Guru Jambheshwar University of Science and Technology, Hisar, Haryana, India. [b]Department of Sustainable and Renewable Energy Engineering, University of Sharjah, Sharjah, United Arab Emirates. [c]Research Institute for Sciences and Engineering, University of Sharjah, Sharjah, United Arab Emirates. [d]U.S.-Pakistan Center for Advanced Studies in Energy (USPCAS-E), National University of Sciences and Technology (NUST), Islamabad, Pakistan

**Corresponding author: zsaid@sharjah.ac.ae, zaffar.ks@gmail.com*

Chapter outline

2.1 Introduction 33
- 2.1.1 One-step method 34
- 2.1.2 Two-step method 34
- 2.1.3 Comparison of one-step and two-step methods 36

2.2 Stability of nanofluids 37
- 2.2.1 Stability evaluation methods 38
- 2.2.2 Stability enhancement methods 51

2.3 Challenges and outlook 57

2.4 Summary 58

References 59

2.1 Introduction

Nanofluids having a single kind of nanoparticles do not solely meet the demand for good properties in every aspect. They have either good thermal properties or better rheological properties. For example, metal oxide nanoparticles such as Al_2O_3 possess good chemical inertness and stability but exhibit lower thermal conductivity while metallic nanoparticles such as Ag, Al, and Cu possess high thermal conductivities but are chemically reactive and unstable. In order to make nanofluids with favorable properties for some specific applications, hybridization is obligatory.

Hybrid Nanofluids: Preparation, Characterization and Applications. https://doi.org/10.1016/B978-0-323-85836-6.00002-8

By adding two or more kinds of nanoparticles, the formation of hybrid nanofluids has superior thermophysical and rheological properties. A homogeneous mixture of two or more types of nanoparticles with new physical and chemical properties is called a hybrid nanofluid. Due to the synergistic effect, heat transfer properties of hybrid nanofluids are found much better than mono nanofluids [1–3]. Proper synthesis processes of hybrid nanofluids play a vital role in enhancing its thermophysical properties, stability, and durability of nanofluids. Preparation methods of hybrid nanofluids are somewhat similar to preparation methods of mono nanofluids. These can also be categorized into one-step and two-step methods.

2.1.1 One-step method

In this method, the production and dispersion of nanoparticles are in one go, and nanofluid is prepared in a single step. This is the most appropriate method for production of highly thermal conductivity metal nanoparticles to avoid oxidation. Direct evaporation, liquid-chemical evaporation, and physical vapor deposition (PVD) are some of the methods involved in single-step technique. By employing this technique, highly stable and uniformly dispersed nanofluids can be prepared. As nanofluid is synthesized by this method completely in single step only, various events are avoided, such as drying up of nanoparticles, storing process, and transporting them from one place to another. The main disadvantage of this method is that it is a very slow process, mass production is not possible, and it also requires more cost. Low-pressure fluids are mostly produced by this method. Moreover, only a low volume fraction of nanofluids can be prepared. Thus, nanofluids made by this method cannot be used at the commercial level.

2.1.2 Two-step method

Production and dispersion of nanoparticles occur in two different steps (Fig. 2.1). That is why, this method is called two-step method. Hybrid nanofluids are prepared either by suspending two or more individual nanoparticles or by suspending synthesized nanocomposite particles in base fluids in predefined proportions. Various researchers adopt the synthesis of nanocomposite in their own ways (Table 2.1).

First, nanoparticles or nanocomposites are produced in dry form with mass production, and then these nanoparticles or nanocomposites are dispersed in host fluids with different mixing processes such as ultrasonication, magnetic stirring, and homogenization. As nanoparticles are added to base fluid, there arise the

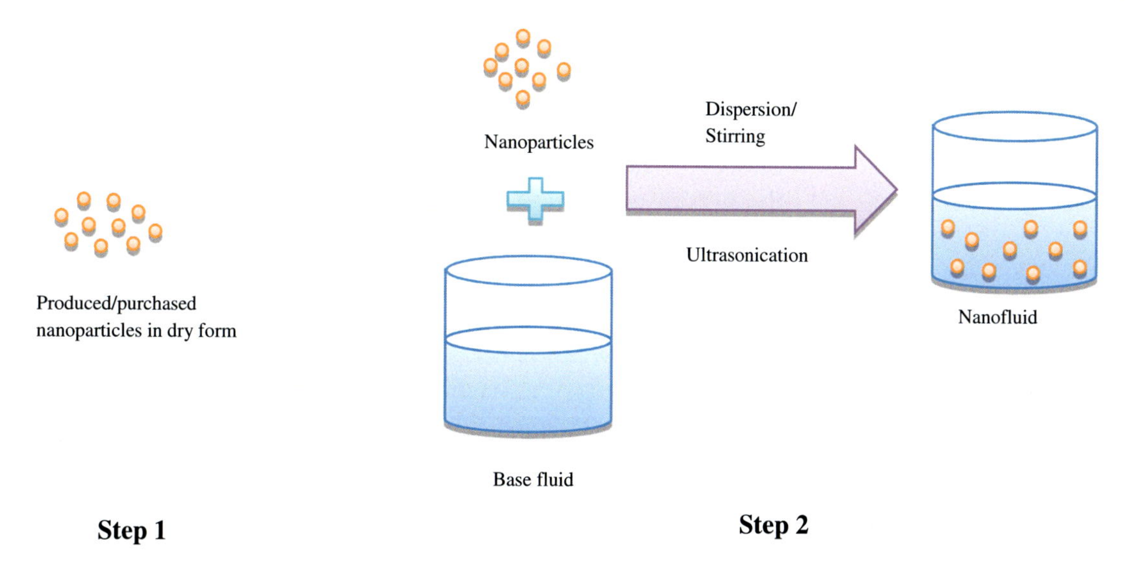

Fig. 2.1 Two-step nanofluid preparation method.

Table 2.1 Summary of various methods of hybrid nanocomposite synthesis.

Researchers	Hybrid nanocomposite	Preparation method	Precursor salts
Suresh et al. [4]	Al_2O_3-Cu	Thermochemical	Copper nitrate $Cu(NO_3)_2 \cdot 3H_2O$ and aluminum nitrate $Al(NO_3)_3 \cdot 9H_2O$
Sundar et al. [5]	MWCNT-Fe_3O_4	In situ and chemical coprecipitation	MWCNT + (HCl + HNO_3) Ferric chloride ($FeCl^{3+}$) and ferrous chloride ($FeCl^{2+}$)
Madhesh et al. [6]	Cu-TiO_2	Mechanical milling	TiO_2 and copper acetate
Baby et al. [7]	MWCNT-GO	Catalytic chemical vapor deposition (CCVD)	MWNT + hydride catalyst and graphite + H_2SO_4
Megetif et al. [8]	TiO_2-CNT	Sonication	CNTs + HNO_3-H_2SO_4 + $Ti(OBu)_4$ + N_2
Batmunkh et al. [9]	Ag-TiO_2	Mechanical stirring	Ag-TiO_2
Abbasi et al. [10]	γ-Al_2O_3-MWCNTs	Solvothermal	MWCNT + (HCl + HNO_3) and aluminum acetate + ethanol
Nine et al. [11]	Cu/Cu_2O	Wet ball milling	Cu/Cu_2O + distilled water
Chen et al. [12]	Ag-MWNT	Ball milling	MWNTs + ammonium bicarbonate + $AgNO_3$

problem of agglomeration and sedimentation. This means that less stable nanofluids are produced by this method, but it could be enhanced by adding surfactants, surface-modified nanoparticles, and other stability enhancement methods. Although the cost of drying, storing, and transporting of nanoparticles is added to the overall production cost of nanofluids, the cost of nanofluid manufacturing is less compared to the manufacturing cost of nanofluids produced by the one-step method. A variety of nanofluids can be prepared from this method at a reasonable cost because a variety of nanoparticles are directly purchased from different available companies. High concentration of nanofluids can also be prepared with this method. With taking all the mentioned benefits of this method, produced hybrid nanofluids could also be easily used commercially.

2.1.3 Comparison of one-step and two-step methods

Individual one-step and two-step hybrid nanofluid preparation methods have been discussed in the previous section. This section summarizes a concise and concrete comparative analysis of both methods in Table 2.2.

Hybrid nanofluids are not just the mixture of solid nanoparticles (or nanocomposites) in base fluids but consist of special treatment processes. Magnetic stirring, ultrasonication, and homogenization are basic techniques involved in preparing uniformly dispersed nanofluid mixture. Less agglomeration of particles leads to the stability of nanofluids. Long-time nanofluid stability is one of the basic needs for the efficient working of any thermal system. Synthesis of stable nanofluid is a topic of challenge to the researchers working in this field. Thus, the further sections of this chapter highlight the stability issues related to hybrid nanofluids.

Nomenclature

Abbreviations	
CTAB	Cetyl trimethyl ammonium bromide
EG	Ethylene glycol
GnP	Graphene nanoplatelets
IEP	Isoelectric point
MWCNT	Multiwalled carbon nanotubes
SDBS	Sodium dodecyl benzene sulfonate
SDC	Sodium deoxycholate
SDS	Sodium dodecyl sulfate
PVP	Polyvinylpyrrolidone

Table 2.2 Comparison between one-step and two-step methods.

S. no.	Comparison basis	One-step	Two-step
1.	Basic definition	Production and dispersion of nanoparticles done in a single step	Production and dispersion of nanoparticles done in two separate steps
2.	Processes involved	Direct evaporation, physical vapor deposition (PVD), liquid-chemical evaporation, etc.	First, dried nanoparticles produced or purchased in one step and then mixed up in base fluid in the second step with ultrasonication, magnetic stirring, homogenization, etc.
3.	Stability	High	Less
4.	Dispersion behavior	Uniform	Nonuniform
5.	Production level	Small scale	Mass (commercial scale)
6.	Variety of nanofluid	Cannot obtained	Can be obtained
7.	Prepared nanofluid concentration	Low	High
8.	Prepared nanofluid nature	Mostly low-pressure fluids can be prepared	Both types of nanofluids (either low or high pressure) can be prepared
9.	Preparation speed	Slow	High
10.	Additional cost (drying, storage, transportation)	Avoided	Involved
11.	Overall cost	High	Low

2.2 Stability of nanofluids

Owing to high surface charge of nanoparticles, they coagulate/agglomerate and form nanofluid unstable [1]. Particles adhere to each other and form an aggregate of large sizes, which settle down at the bottom of the container with gravity. Stability means that particles do not coagulate at a considerable rate. Generally, the rate of coagulation can be found out by the frequency of particle collisions produced by Brownian motion and cohesion probability in collisions [13]. Nanoparticle aggregations are responsible for clogging the heat transfer system channels and reducing the benefits of nanofluids by changing their thermophysical properties. Derjaguin, Landau, Verway, and Overbeek (DLVO) developed a theory that deals with colloidal suspensions' stability issues. This theory proposed that the stability of a particle in dispersion is found by the sum of Van der Walls attractive forces and electrostatic repulsive forces present between the approaching particles [2]. It gives the conditions of stability:

(i) If attractive forces are larger than repulsive forces, then particles will collide with each other and form solution unstable.

(ii) If repulsive forces dominate between particles, then the solution exists in a stable form.

For the enhancement of stability in nanofluids, firstly, critical evaluation of the stability of nanofluids is required. After this particular investigation, appropriate stability enhancement methods are applied in nanofluids.

2.2.1 Stability evaluation methods

For extending the shelf-life and for conserving thermophysical properties of nanofluids, stability evaluation methods are indispensable. Different research groups study different evaluation methods. Some of them are discussed below.

Sedimentation method

This is the most simple and reliable method used for measuring stability in nanofluids. In this method, firstly, a sample of prepared nanofluid is placed in a transparent test tube. Photographs of this prepared nanofluid sample were captured after regular intervals of time (e.g., after 10, 20, 30 days, and so on) (Fig. 2.2). The rate of sedimentation of nanoparticles can be directly visible through examining these captured photographs. The tendency of nanoparticle sediments with the effect of gravity was clearly shown in pictures taken by a camera. Nanofluid is assumed to be stable if the concentration of dispersed nanoparticles was constant throughout a regular interval of time.

Sedimentation velocity of spherical particles at steady state can also be calculated by Stokes law as given below [14]:

$$V = \frac{2r^2}{9\mu}\left(\rho_p - \rho_l\right)g \tag{2.1}$$

where V represents sedimentation velocity in the case of spherical particles, r represents the radius of nanoparticles, μ represents the viscosity of nanofluid, ρ_p and ρ_l represent the density of nanoparticles and nanofluid, respectively, and g represents the acceleration due to gravity.

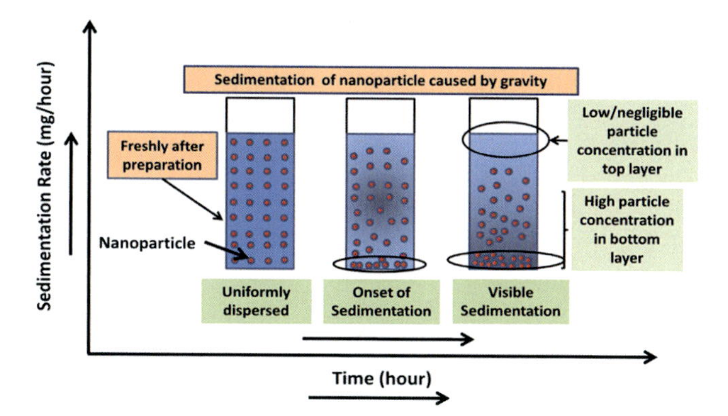

Fig. 2.2 Sedimentation process. Copyright Elsevier 2021 (Lic# 5000290015924). From S. Chakraborty, P.K. Panigrahi, Stability of nanofluid: a review, Appl. Thermal Eng. (2020) 115259.

As per Eq. (2.1), the sedimentation velocity of nanoparticles in nanofluid depends directly on the radius (or diameter) of nanoparticles. The larger the size of particles, the larger will be the sedimentation velocity. If this velocity reaches a constant value after some time, then it means that the nanofluid attains stability.

Kumar et al. [15] used a photograph capturing technique to find the stability of three different concentrations of Al_2O_3--SiO_2/water hybrid nanofluid. They observed good stability and no sedimentation just after the preparation of nanofluid samples, but after 4 weeks, visual sedimentation was observed in all three samples (Fig. 2.3). Low stability and large amount of sedimentation of nanoparticles were obtained in 0.2 wt.% nanofluid. This is due to the reason that low-concentration nanoparticles agglomerate easily. In low-concentration nanofluid, there was a lack of nanoparticles for generating adequate repulsive force to make the solution stable. High stability and less sedimentation were obtained for 0.6 wt.% of nanofluid because of a sufficient amount of nanoparticles in nanofluid.

Xian et al. [16] used the sedimentation method for finding the stability of TiO_2-(COOH-GnP)/water-EG hybrid nanofluid. They added six different kinds of surfactants (PVP, Triton X-100, CTAB, SDS, SDC, and SDBS) to enhance the sedimentation period. Their results revealed that the addition of CTAB surfactant showed less sedimentation after 40 days than other surfactants used.

Ma et al. [17] used two hybrid nanofluids (Al_2O_3-TiO_2/water and Al_2O_3-CuO/water) for analyzing the stability with this method. They showed the stability of both nanofluids with or without adding surfactants. Al_2O_3-TiO_2/water nanofluid showed clear stability even after 25 days of preparation with the help of photographs taken after 3, 11, and 25 days. Similarly,

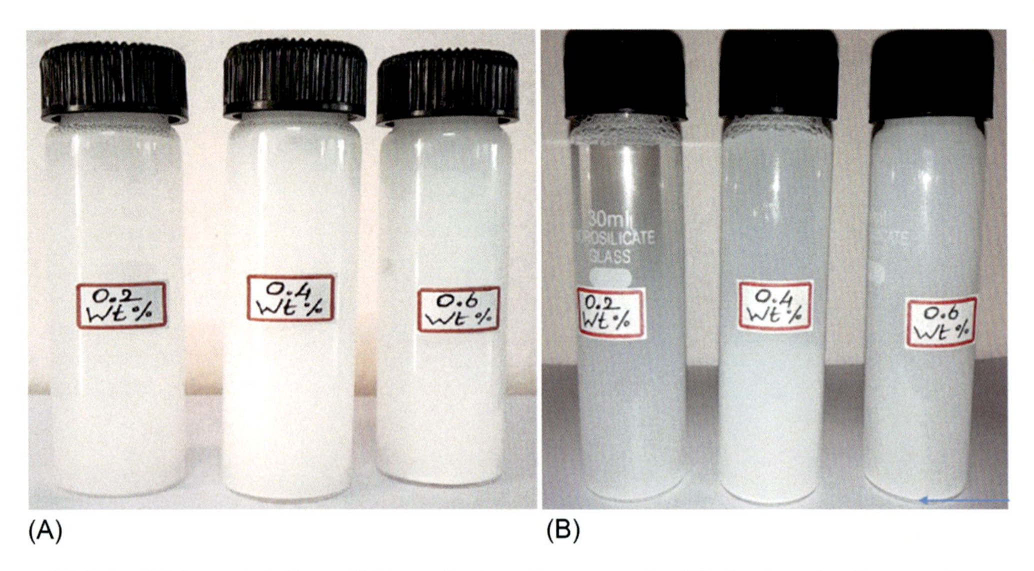

Fig. 2.3 (A) Al_2O_3-SiO_2/water hybrid nanofluid samples just after preparation (B) after 4 weeks of preparation. Copyright Elsevier 2021 (Lic# 5000291061530). From P.C.M. Kumar, K. Palanisamy, V. Vijayan, Stability analysis of heat transfer hybrid/water nanofluids, Mater. Today: Proc. 21 (1) (2020) 708–712.

Al_2O_3-CuO/water nanofluid showed good dispersion behavior even after 11 days of preparation clearly shown in Fig. 2.4.

The main disadvantages of this method are time-consuming process, not being suitable for dark-colored nanofluids such as CNTs or high concentration nanofluids, and not being able to find the partial stability of nanofluids. One of the advantage of this present method is its reliability and accuracy compared to other stability measurement methods.

The submerged tray method is another method that falls in this category, in which a tray is submerged and hanged in somewhere in between nanofluid and attached with a weighing balance. The number of nanoparticles that settle down on the tray will directly record in the reading of weighing balance. If this reading is not changing after a certain period of time interval, it means that stability occurs and no more particles are sediment. Zhu et al. [18] used a submerged tray method to measure graphite nanopowder stability dispersed in distilled water. Haghighi et al. [19] employed this technique for determining the dispersion behavior of Al_2O_3/water nanofluid.

Centrifugation method

This method is faster than the sedimentation photograph capturing method. A test tube containing the sample rotates with centrifuge effect using a dispersion centrifuge analyzer, and visual

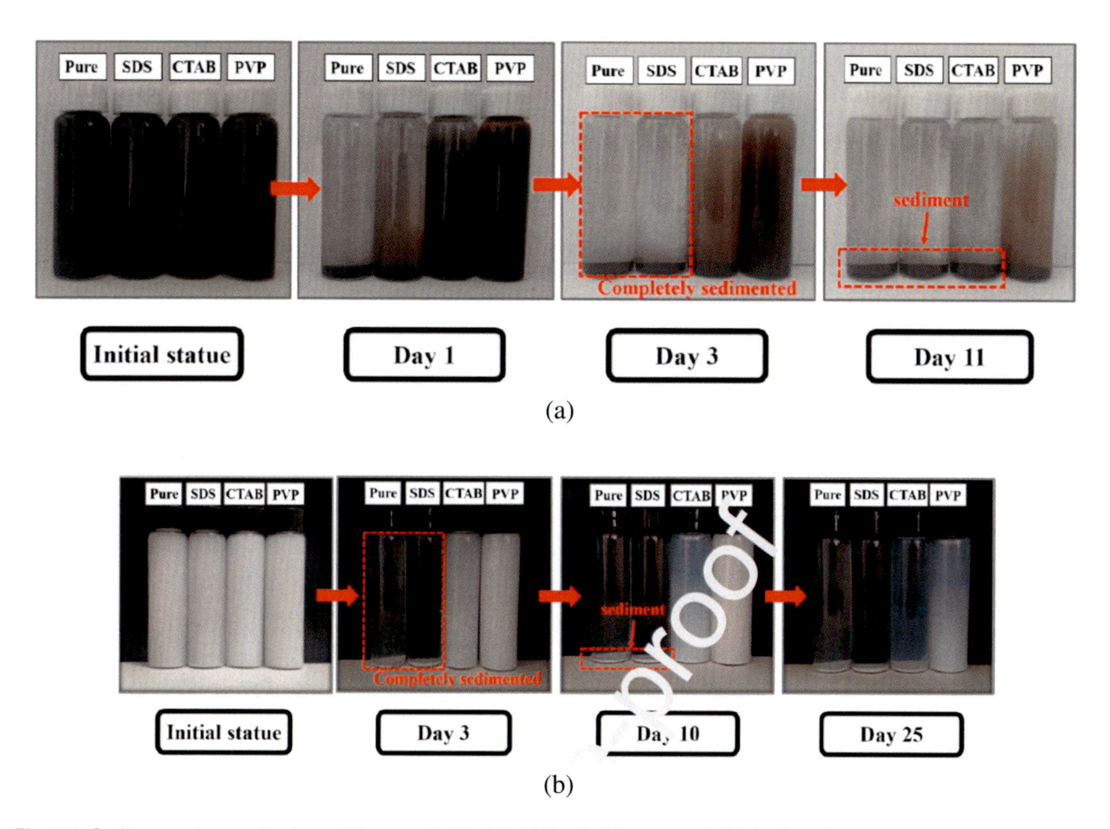

Fig. 2.4 Sedimentation method used for nanofluids (A) Al_2O_3-CuO/water and (B) Al_2O_3-TiO_2/water. Copyright Elsevier 2021 (Lic# 5000300424864). From M. Ma, Y. Zhai, P. Yao, et al., Effect of surfactant on the rheological behavior and thermophysical properties of hybrid nanofluids, Powder Technol. 379 (2021) 373–383.

inspection is carried out in this method. Singh and Raykar [20] analyzed silver/ethanol nanofluid stability with both sedimentation photograph and centrifugation methods. They observed that the former method took 1 month to find out nanofluid stability where only 10 h are taken by the latter method. Mehrali et al. [21] performed stability analysis for graphene/distilled water nanofluid with the centrifuge method. They reported that within 5–20 min, analysis of four concentrations of nanofluid was completed.

Zeta potential method

The stability of nanofluid is analyzed by observing the electrophoretic behavior of nanofluid. This is due to the free charge present in base fluid that gets attracted with the opposite charge present on the dispersed nanoparticles and forms a charged ions-loaded layer known as the stern layer. An additional layer of individual charge also surrounds the stern layer, called the diffuse layer. In the stern layer, positively charged ions are strongly bound, but negatively charged ions are loosely bound in the diffuse layer. In between the boundary of the diffuse layer, an imaginary line exists where a stable entity is formed. If particles with their charge move beyond this line but inside the diffuse layer, the electric charge potential of this line changes, called zeta potential. It can also be defined as a potential difference present between the free charge of the base fluid and the charge on the stern layer of dispersed nanoparticles [18, 22–24] (Fig. 2.5). It can be measured in millivolts (mV).

As the electrostatic repulsive forces increase between particles, zeta potential also increases; thus, the stability of nanofluid increases. The degree of stability varies according to the values of zeta potential recorded in experiments for different nanofluids (Fig. 2.6). Measurement of the zeta potential value in a nanofluid can be performed using a Zeta Sizer Nano (ZSN) device.

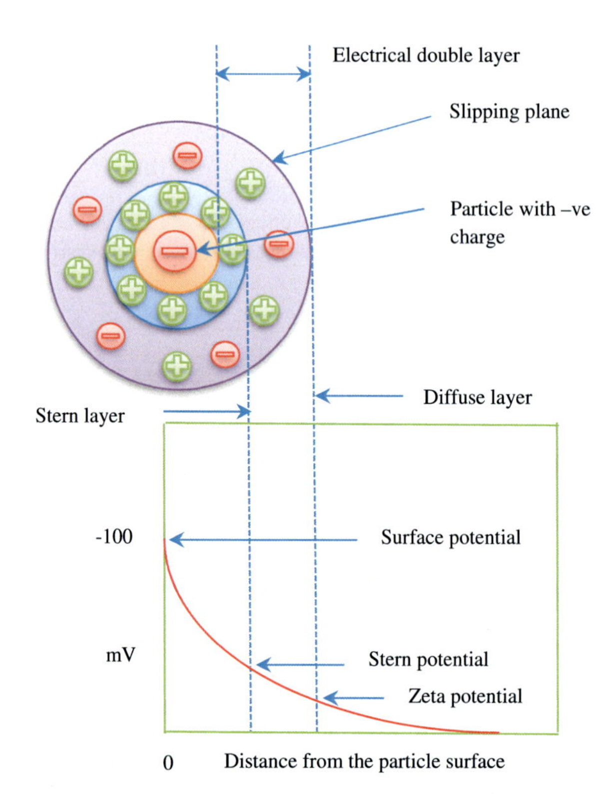

Fig. 2.5 Concept of zeta potential.

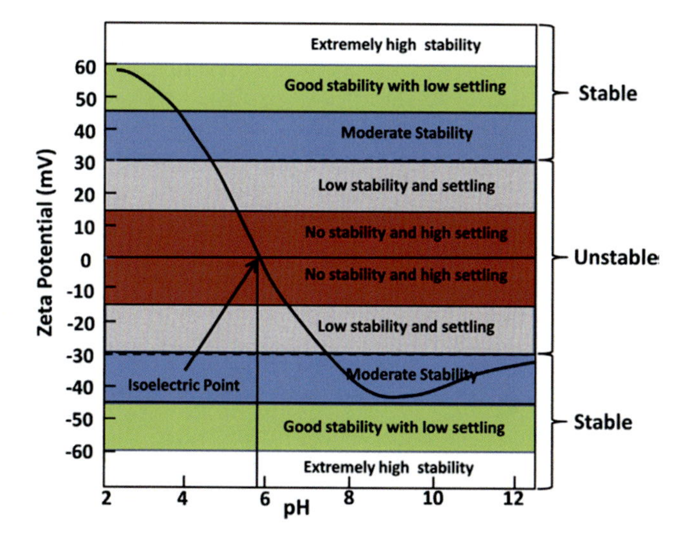

Fig. 2.6 Relationship between zetapotential values and conditions of stability [25]. Copyright Elsevier 2021 (Lic# 5000290015924). From S. Chakraborty, P.K. Panigrahi, Stability of nanofluid: a review, Appl. Thermal Eng. (2020) 115259.

It can be concluded that the higher the value of zeta potential (either in −ve or +ve), the higher the stability of nanofluid. Also, the lesser the zeta potential value, the solution tends to coagulate (or flocculate) and hence less stability. Various investigators measured the zeta potential values for checking the stability level for various nanofluids listed in Table 2.3.

Gallego et al. [31] showed positive zeta potential values between +20 and +30 mV for different concentrations of Al_2O_3 nanofluid. With the addition of anionic SDBS surfactant, these values converted to negative ones. The highest zeta potential value of -53 ± 3 mV has been observed for 0.5 wt.% of Al_2O_3 nanofluid and 0.32 wt.% of SDBS surfactant. This comparative high zeta potential value indicated good stability of nanofluid with the addition of surfactant.

There are also some limitations connected with the application of zeta potential:

- Cannot be used for higher conductive nanofluids because electrode polarization and degradation occur with the movement of highly conductive ions.
- Not used for low concentration of nanofluids because fewer particles are present for appropriate light scattering phenomenon.
- Also not used for a very high concentration of nanofluids because maximum light is absorbed by a number of particles, which reduces the intensity of scattered light.

Table 2.3 Experimental zeta potential values for various hybrid nanofluids.

Researchers	Nanofluid	Surfactant used	Zeta potential value (mV)	Duration of time (h)	Remarks on stability
Song et al. [26]	Stainless steel/ water	SDBS CTAB	−70 60.1	240 240	Excellent Excellent
Sundar et al. [27]	Nanodiamond/ Fe_3O_4-Water	None	−34.6	72	Good
Chakraborty et al. [28]	Cu-Zn-Al/LDH[a]-water	None	38.6	18	Moderate
Chakraborty et al. [29]	Cu-Zn-Al LDH/water Cu-Zn-Al LDH/water	SDS Tween 20	−50.6 24.3	>1 day 12	Good Low
Kumar et al. [15]	Al_2O_3-SiO_2/water (0.2, 0.4, and 0.6 wt.%)	None	−30.9 −42.7 −60.7	4 weeks	Moderate Moderate Extremely high
Cakmak et al. [30]	rGO-Fe_3O_4-TiO_2/EG (0.01, 0.05, 0.1, 0.15, 0.2, 0.25)	None	+63.45 +53.4 +53.3 +51.5 +53.15 +52.43	Just after preparation	Excellent
Xian et al. [16]	TiO_2-(COOH-GnP)/ water-EG	PVP, Triton X-100, CTAB, SDS, SDC, SDBS	Above 30 and highest −63.7	Just after preparation	Excellent

[a]LDH, layered double hydroxide.

Spectral absorbance analysis

This method is firstly adopted by Jiang et al. [32] in 2003 and applied only when nanoparticles have an absorption capacity of wavelength between 190 and 1100 nm. It is based on "Beer-Lambert law," which provides the linear relationship between the intensity of absorbing light and the concentration of nanoparticles present in nanofluids [33,34]. This can be written as

$$A = \alpha lc$$

where A is the absorbance, α is the absorption coefficient, l is the distance that light traveled throughout the material, and c is the concentration of absorbing elements in the material or nanoparticle concentration in nanofluids.

This law is used to measure the absorbing capacity of light in colloidal dispersions. Also, some instrumental and chemical factors divert the linearity of this law described below:

- Absorption coefficients get somewhat altered at high nanoparticle concentrations.
- Due to the presence of particles, scattering of light occurs.
- Refractive index may also vary at high nanoparticle fractions.
- Chemical equilibrium of dispersion can be shifted due to variation in nanoparticle loading and colloidal solution property.

A UV-vis spectrophotometer is used to measure the maximum absorbance of nanoparticles in different nanofluids at different wavelengths. This peak absorbance capacity is mainly dependent upon the concentration of nanoparticles that exist in nanofluids. In the first step of this spectroscope analysis, the peak absorbance of dispersed nanoparticles is measured by scanning the maximum wavelength, which acts as a reference point for further analysis. In the second step, at this fixed peak wavelength, the absorbance for different fractions of nanofluids is measured. It shows the linear relationship between absorbance and concentration of nanoparticles, which follows "Beer-Lambert law." In the third step, the same nanofluids' concentration is prepared by varying sonication time and repeats the whole steps after a certain interval of time [35].

One of the advantages of this method is applicable for all kinds of base fluids. On the other hand, disadvantages of this method are not used for high concentration or dark-colored nanofluids because these kinds of nanofluids absorb the high intensity of incident light and reduce the intensity of scattered light, which diminishes the quality of data [36,37].

Jana et al. [38] examined the stability of CNT-Au/water and CNT-Cu/water hybrid nanofluids using a UV-vis spectrophotometer. They observed that adding CNT nanoparticles in Cu/water and Au/water nanofluids increased the absorbance capability and reduced sedimentation time. Fig. 2.6A showed the absorbance capacity of different concentrations of Di-Ag nanofluid at different wavelengths. It was observed that with an increase in the concentration of nanofluid, absorption peaks are increased, which is also stated by Jana et al. [38]. Kumar et al. [39] performed UV-vis analysis for copper nanofluid and results showed the existence of copper nanoparticles with a peak at 422.15 nm wavelength (Fig. 2.7).

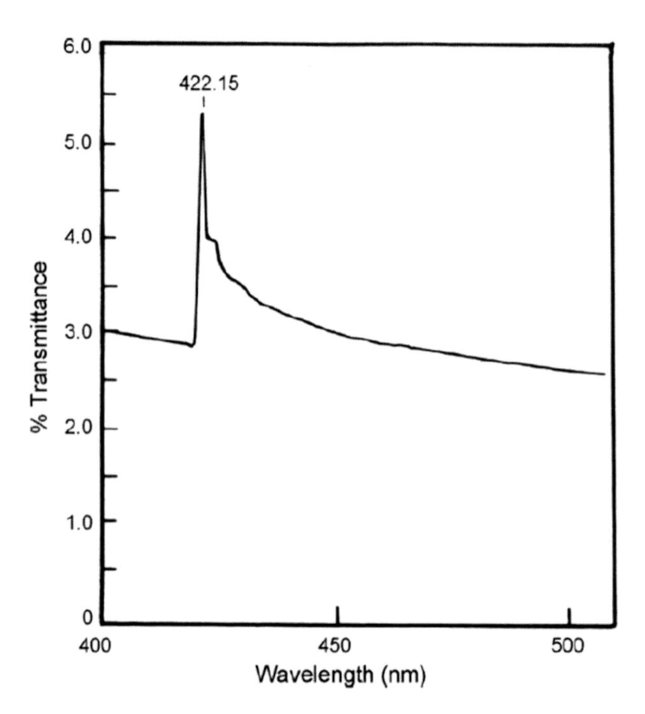

Fig. 2.7 Analysis of copper nanofluid using UV-vis method. Copyright Elsevier 2021 (Lic# 5000580432303). From S.A. Kumar, K.S. Meenakshi, B.R.V. Narashimhan, S. Srikanth, G. Arthanareeswaran, Synthesis and characterization of copper nanofluid by a novel one-step method, Mater. Chem. Phys. 113 (2009) 57–62.

Esfahani et al. [40] measured UV-vis spectrum for graphene oxide nanofluids at different weight concentrations (0.01, 0.05, 0.1, and 0.5 wt.%). They demonstrated that as particle loading increased, the absorbance of light also increased (Fig. 2.8A). A linear relationship was made between absorbance and volumetric concentration of nanofluid at fixed wavelength as shown in Fig. 2.8B.

Oliveira et al. [41] also obtained similar findings with Di-Ag nanofluid as obtained by a previous study [40] and depicted in Fig. 2.9.

Kumar et al. [15] measured the stability of Al_2O_3-SiO_2/water nanofluid at three different weight concentrations (0.2, 0.4, and 0.6 wt.%) by UV-vis spectroscopy. They observed excellent stability just after the preparation of nanofluids due to the sufficient amount of nanoparticles that were well-dispersed in a nanofluid. They also obtained that light absorption strength is wider just after nanofluid preparation compared to that of nanofluid after preparation of 4 weeks. After 4 weeks, fair stability was obtained for 0.2 wt.% due to less amount of particles that were uniformly

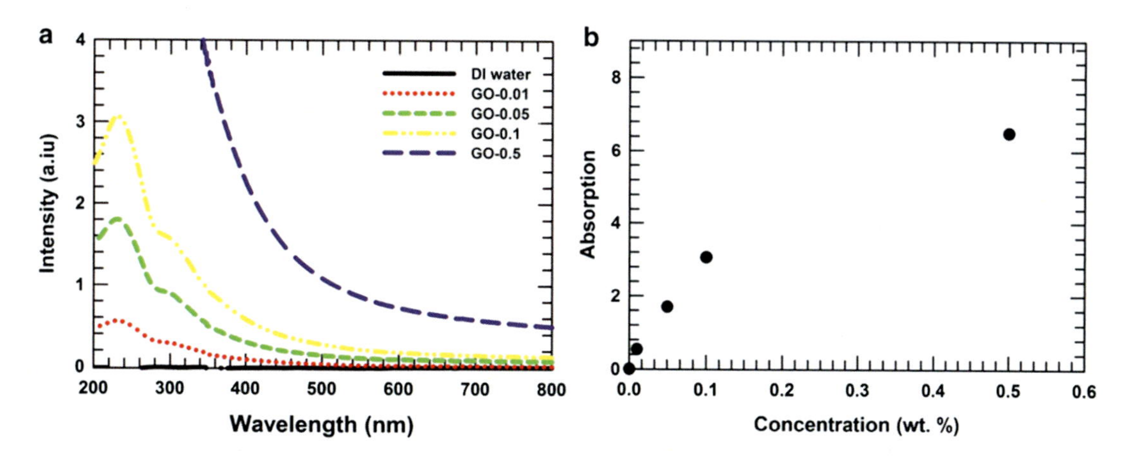

Fig. 2.8 (A) UV-vis absorption spectra for different concentrations of graphene oxide nanofluid; (B) linear relationship between absorbance and volume concentration of graphene oxide nanofluid. Copyright Elsevier 2021 (Lic# 5000620974263). From M.R. Esfahani, E.M. Languri, M.R. Nunna, Effect of particle size and viscosity on thermal conductivity enhancement of graphene oxide nanofluid, Int. Commun. Heat Mass Transfer 76 (2016) 308–315.

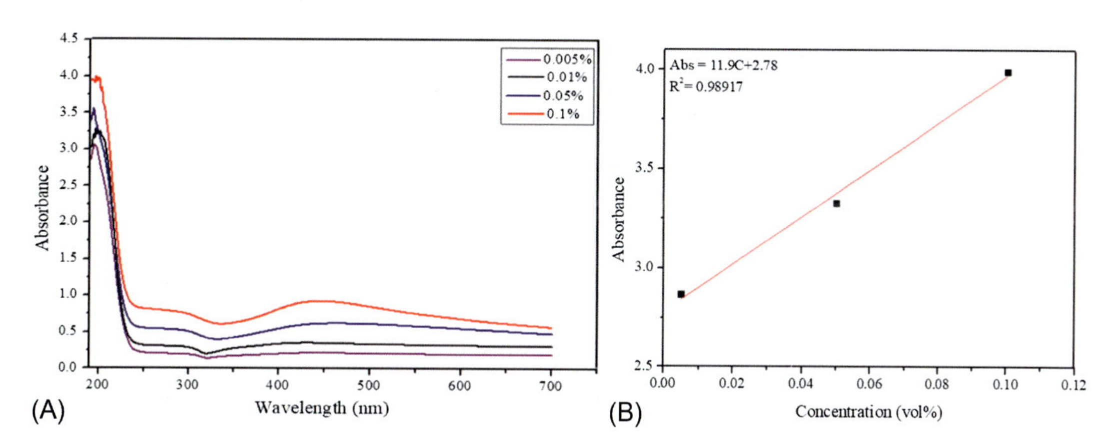

Fig. 2.9 (A) UV-vis absorption spectra for different concentrations of Di-Ag hybrid nanofluid; (B) linear relationship between absorbance and volume concentration of Di-Ag hybrid nanofluid. Copyright Elsevier 2021 (Lic# 5000590102493). From L.R. Oliveira, S.R.F.L. Ribeiro, M.H.M. Reis, V.L. Cardoso, E.P.B. Filho, Experimental study on the thermal conductivity and viscosity of ethylene glycol-based nanofluid containing diamond-silver hybrid material, Diam. Rel. Mater. 96 (2019) 216–230.

dispersed in nanofluid solution and a large amount of nanoparticles that settle down at the bottom of container (Fig. 2.10). Good stability of 0.4 and 0.6 wt.% of nanofluid was obtained after 4 weeks because a sufficient amount of nanoparticles are present in well-dispersed form; thus, a sufficient amount of light was absorbed by these suspended particles (Figs. 2.11 and 2.12).

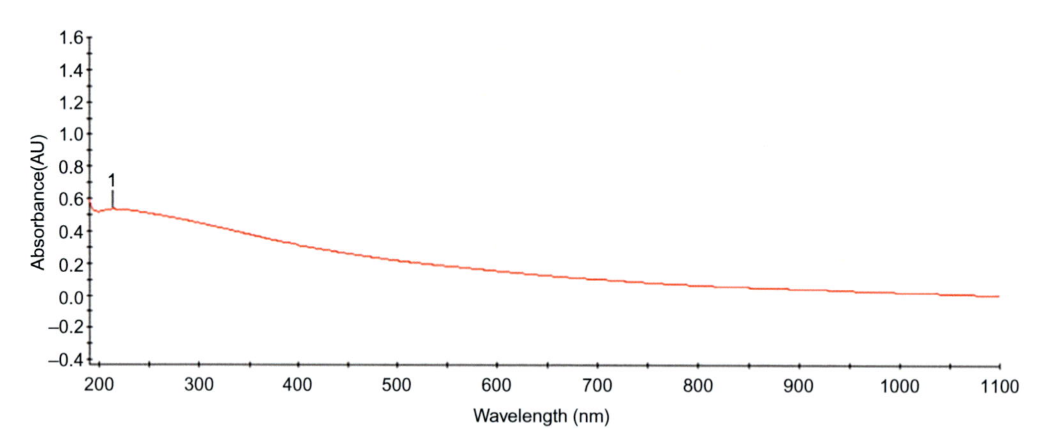

Fig. 2.10 UV-vis spectrum of 0.2 wt.% nanofluids after 4 weeks of preparation. Copyright Elsevier 2021 (Lic# 5000291061530). From P.C.M. Kumar, K. Palanisamy, V. Vijayan, Stability analysis of heat transfer hybrid/water nanofluids, Mater. Today: Proc. 21 (1) (2020) 708–712.

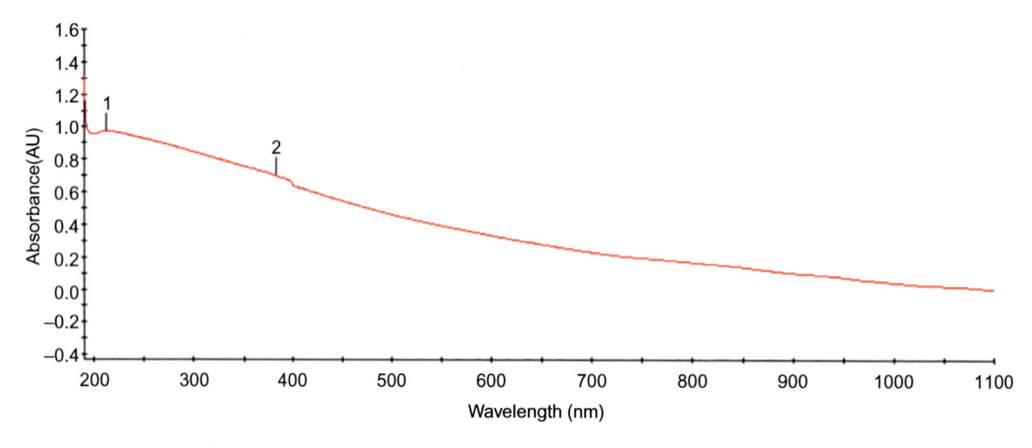

Fig. 2.11 UV-vis spectrum of 0.4 wt.% nanofluids after 4 weeks of preparation. Copyright Elsevier 2021 (Lic# 5000291061530). From P.C.M. Kumar, K. Palanisamy, V. Vijayan, Stability analysis of heat transfer hybrid/water nanofluids, Mater. Today: Proc. 21 (1) (2020) 708–712.

Gallego et al. [31] showed greater absorbance at high volume concentrations and low absorbance at low concentrations of Al_2O_3 nanofluids also with SDBS surfactant at different concentrations (Fig. 2.13). They measured absorbance patterns for different concentrations of both nanoparticles and surfactants over weeks using the UV method.

Xian et al. [16] used UV spectroscopy for finding the stability of TiO_2-(COOH-GnP)/water-EG hybrid nanofluid (Fig. 2.14A). A linear relationship was shown between absorbance values

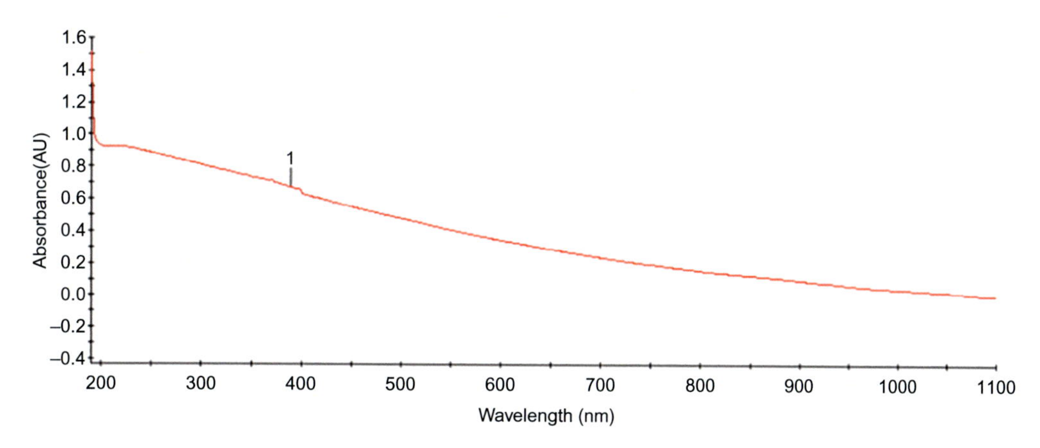

Fig. 2.12 UV-vis spectrum of 0.6 wt.% nanofluids after 4 weeks of preparation. Copyright Elsevier 2021 (Lic# 5000291061530). From P.C.M. Kumar, K. Palanisamy, V. Vijayan, Stability analysis of heat transfer hybrid/water nanofluids, Mater. Today: Proc. 21 (1) (2020) 708–712.

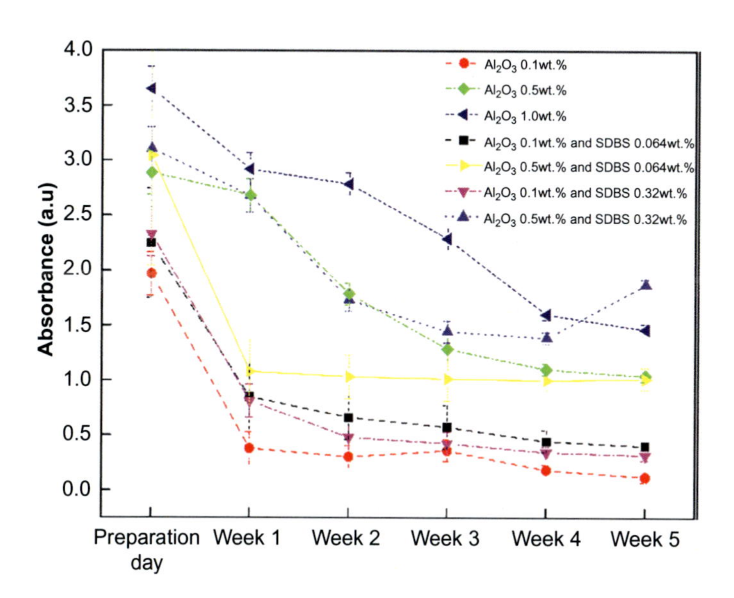

Fig. 2.13 Absorbance pattern analysis using the UV method for different Al_2O_3 concentrations and also with the addition of different surfactant concentrations. Copyright Elsevier 2021 (Lic# 5000590757831). From A. Gallego, K. Cacua, B. Herrera, D. Cabaleiro, M.M. Piñeiro, L. Lugo, Experimental evaluation of the effect in the stability and thermophysical properties of water-Al_2O_3 based nanofluids using SDBS as dispersant agent, Adv. Powder Technol. 31 (2) (2020) 560–570.

and the concentration of nanofluids (Fig. 2.14B). With the increase in nanoparticle concentration, the absorbance values also increased. They also used six different types of surfactants (PVP, Triton X-100, CTAB, SDS, SDC) to enhance the stability at different ultrasonication timings (30, 60, and 90 min). They found that longer ultrasonication timing gave better stability than less ultrasonication timing did. Their results indicated that all

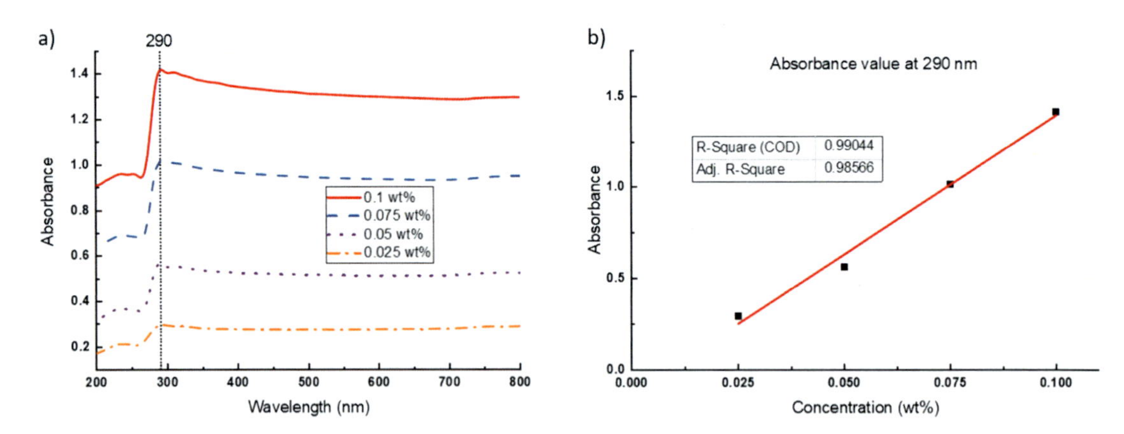

Fig. 2.14 (A) UV spectroscopy for finding the stability of TiO_2-(COOH-GnP)/water-EG hybrid nanofluid; (B) linear relationship shown between absorbance and concentration of nanofluids. Copyright Elsevier 2021 (Lic# 5000590447702). From H.W. Xian, N.A.C. Sidik, R. Saidur, Impact of different surfactants and ultrasonication time on the stability and thermophysical properties of hybrid nanofluids, Int. Commun. Heat Mass Transf. 110 (2020) 104389.

nanofluids with different surfactants showed good stability even after 40 days of preparation. They also mentioned that nanofluid with the addition of CTAB surfactant showed the best stability compared with the other nanofluids used in that particular study.

Thermal conductivity measurement

Measuring the thermal conductivity of nanofluids after a certain period of time is also a method of stability evaluation in nanofluids. It is the simple behavior of nanoparticles to form clusters/agglomerates in nanofluids after some time of preparation. Thus, the thermal conductivity of nanofluids decreases due to the formation of clusters. Hence, if there is a decrement observed in thermal conductivity values after some time, it means that sedimentation of nanoparticles occurs, which is the measure of instability. Various methods or devices used to measure the thermal conductivity of nanofluids are 3-ω, transient hot-wire method, KD2 Pro device, etc.

Electron microscopy

Several researchers have used this technique for measuring the stability of various nanofluids. The information regarding particle size, aggregation, and distribution is observed in this method. Scanning electron microscopy (SEM) and transmission electron microscopy (TEM) are the devices used to capture digital images of high resolution of 0.1 nm in size [42,43]. These images are known as electron micrographs [44]. If clusters of nanoparticles are found in obtained images, then it means that sedimentation occurs in dispersion and thus shows nanofluid instability [45]. Yu et al. [46] stated that both SEM and TEM are widely used to characterize matter at the nanoscale level. The digital images captured by software give information regarding particle size, distribution, and orientation. With the help of inter-particle distance spacing, the amount of particles' aggregation can easily be examined.

For efficient use of nanofluids in any thermal system, stability enhancement is the major concern for most of the researchers working in this field. Clogging of pipelines could occur with aggregation of nanoparticles; thus, uniform dispersion of nanoparticles is vital in nanofluids. Furthermore, instability in nanofluids causes depreciation in their thermophysical properties. Thus, different methods are discussed in further sections for the enhancement of stability in nanofluids.

2.2.2 Stability enhancement methods

It involves different physical and chemical treatments on nanoparticles' surface in order to improve the stability of nanofluids for the two-step method. Physical treatments involve ultrasonication and homogenization in which a high amount of energy forces are employed to break down nanoparticle clusters. The addition of surfactants, surface modifications, and pH change are some of the chemical treatments applied in nanofluids for improving their stability.

Ultrasonication

In this technique, breaking down or rupturing particle clusters occurs by applying ultrasonic waves in the fluid. This method has shown great potential for enhancing the stability of the suspensions [47]. The ultrasonication process has different purposes of applications [48]. It includes a dispersion of nanoparticles in base fluids, de-agglomeration of particles, and reduction of particle size.

The ultrasonication process can be performed by using probe-type or bath-type ultrasonic devices. Both sonicators have significant differences in terms of their capabilities and efficiencies [45,49]. Desirable effects of ultrasonication include dispersion, homogenization, de-agglomeration, and sonochemical effects that are caused by "cavitation." This cavitation process occurs uncontrollably distributed in ultrasonic bath devices. In this device, the ultrasonication effect is of less intensity and unevenly distributed. On the other side, in probe devices, more intense and focused ultrasonic effects are produced under the probe. This process is fully controlled, and intensity is also evenly distributed. The probe device is more effective and strong compared with the bath device [50]. A bath device can produce 20–40 W/L ultrasonic waves in fluids and in a nonuniform distribution manner. On the other hand, the probe device can produce ultrasonication of 20,000 W/L to the fluid and in a uniform manner. Thus, the efficiency of the probe device is 1000 times greater than that of the bath device. In general, increasing the sonication time and power reduces cluster size and improves the stability of suspension. Though this statement is not true for very high power of sonication and for large time intervals, it should be necessary to find out the optimum period and power up to which sonication shows results assisting the stability of nanofluids [47]. In general, with increasing sonication time, the thermal conductivity of nanofluids increased. But in some cases, thermal conductivity decreased after some optimum sonication time because of various reasons [51]. Nasiri et al. [52] prepared nanofluids with both probe and bath type sonicators for a fixed ultrasonication time of 40 min. They reported the better stability of nanofluids prepared with the probe sonicator than with the bath sonicator. Kole and Dey [53] stated that as the ultrasonication time increases, the thermal conductivity of ZnO/EG nanofluid also increases. They reported the highest thermal conductivity of nanofluid at 60 h of sonication period. Shahsavar et al. [54] performed an experimental study to find out the optimum sonication time in order to get the highest thermal conductivity of used nanofluids (Fe_3O_4-water nanofluid (ferrofluid) and Fe_3O_4/CNT-water hybrid nanofluid). They measured thermal conductivity at different probe sonication timings (2.5, 5, 7.5, and 10 min) and found the highest thermal conductivity of 0.85 W/mK at 5 min of sonication timings. This is due to the reason that, below 5 min of sonication time, there was no sufficient time available to attach CNTs with magnetic particles, thus lowering thermal conductivity. Above 5 min of sonication time, there was a reduction in aspect ratio and lowered the quality of the 3-D structure of CNTs.

Tiwari et al. [55] prepared 0.75 vol.% of CeO_2 + MWCNT (80:20)/water nanofluid with a sonication period of 30, 60, 90, 120, 150, and 180 min with an ultrasonic processor. They observed the highest zeta potential values at 90 min (63 mV) of the sonication period of 90 min; before and after the sonication period, less zeta potential value (55 mV) has been observed, which validates good stability at 90 min of sonication period.

Addition of surfactants

Preparation of nanofluids consists of mainly two components: nanoparticles and base fluids. The stability of nanofluids depends only on these two mentioned components. Nanoparticles are hydrophobic and hydrophilic in nature, and base fluids are polar and nonpolar [56]. The hydrophilic nature of nanoparticles such as metal oxides was easily dispersed in polar nature-based fluids such as water. On the other side, the hydrophobic nature of nanoparticles such as CNT can be easily dispersed in nonpolar nature of base fluids such as oils [2]. But, when the hydrophilic nature of nanoparticles are dispersed in nonpolar nature of base fluid or the hydrophobic nature of nanoparticles are dispersed in the polar nature of base fluids, then in both cases the third component is required called surfactants in order to stabilize the resultant nanofluid [57,58]. Surfactants act as a bridge between nanoparticles and base fluids to form the continuity (wettability) between them by reducing the surface tension of base fluids and improving the dispersion process of nanoparticles [59].

Surfactant is a substance that in the aqueous solution tries to congregate at the boundary surfaces. This tendency to accumulate surfactant at boundary surfaces depends upon its structure and the nature of two meeting surfaces. Surfactant has hydrophilic nature polar head group and hydrophobic nature tail portion. Based on charge present on polar head groups, the surfactant can be classified into four categories (Fig. 2.15):

- Anionic: A negative charge is present on polar head groups such as carboxylate, sulfate, sulfonate, and phosphate. Examples include SDBS, SDS, potassium and ammonium lauryl sulfate, sodium stearate, etc. [60].
- Cationic: A positive charge is present on polar head groups such as primary, secondary, and tertiary nitrogen (amine ethoxylates, amineoxidesandfatty amines). Examples include DDC (distearyl dimethylammonium chloride), CTAB, cetrimoniumchloride, benzalkonium chloride, etc. [61].
- Nonionic: No charge is present on polar head groups such as polyether and polyhydroxyl unit. Examples include gum arabic, Rokacet O7, Rokanol K7, polyoxyethylene (10) nonyl phenyl ether, PVP, stearyl alcohol, oleic acid, etc. [45].
- Zwitterionic or amphoteric: Both positive and negative charges are present on polar head groups. Ammonium is most commonly used as a source in positively charged and carboxylates are used as sources in negatively charged. Examples include sodium lauroamphoacetate, lecithin, cocamidopropylbetaine, Rokacet O7, hydroxysultaine, etc. [14].

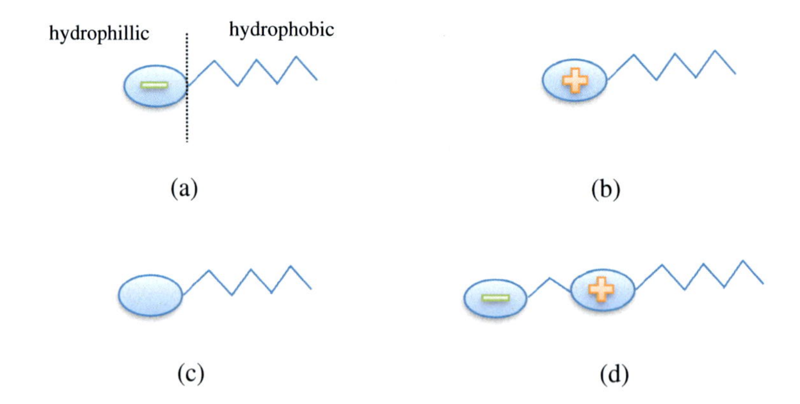

Fig. 2.15 Different kinds of surfactants: (A) anionic, (B) cationic, (C) nonionic, and (D) zwitterionic or amphoteric.

Ma et al. [17] measured the stability of Al_2O_3-CuO/water nanofluid and Al_2O_3-TiO_2/water nanofluid with or without adding different surfactants. They showed that the addition of PVP surfactant showed better stability even after 25 days of preparation for Al_2O_3-TiO_2/water nanofluid and 11 days for Al_2O_3-CuO/water nanofluid. After that, the addition of the CTAB surfactant showed good stability and then the addition of the SDS surfactant in both nanofluids (PVP $>$ CTAB $>$ SDS). No surfactant-containing nanofluids showed the least stability and sediment nanoparticles even after 1 day of preparation for Al_2O_3-CuO/water nanofluids and 3 days of preparation for Al_2O_3-TiO_2/water nanofluids.

Tiwari et al. [55] reported stability of 0.75 vol.% of CeO_2 + MWCNT (80:20)/water nanofluids using two anionic (SDS, SDBS), two cationic (CTAB, DDC), and two nonionic (PVP,GA) surfactants with different nanoparticle-to-surfactant mixing ratios (5,0, 4:1, 3:2, 2:3, and 1:4). They reported that the CTAB surfactant showed the highest zeta potential values (50–60 mV) just after preparation, 15th and 30th days, and after that, the SDBS surfactant showed the highest zeta potential values (50–60 mV) after 45th, 60th, and 90th days.

Various surfactants were used to enhance the stability of various nanofluids reported by various researchers depicted in Table 2.4.

Some of the disadvantages are also reported in the literature regarding the addition of surfactants in nanofluids. With heating and cooling of nanofluids, foam formation occurs in nanofluids with surfactants, which lower the efficiency of nanofluids [52,72]. Somewhere, the purity of nanofluids must affect the addition of surfactant, and the reduction in thermal

Table 2.4 Combination of surfactants used with nanofluids.

Researchers	Nanofluids	Surfactants
Wang et al. [62]	Al_2O_3/water and Cu/water	SDBS
Agarwal et al. [63]	Al_2O_3/kerosene	Hydrochloric acid (HCl)
Khairul et al. [64]	CuO/water	SDBS
Li et al. [65]	MWCHT-NH_2	β-Cyclodextrin
Das et al. [66]	TiO_2/water	CTAB, acetic acid, oleic acid, and SDS
Sarsam et al. [67]	Graphene/water	CTAB, SDS, and SDBS
Wei et al. [68]	TiO_2/diathermic oil	Oleic acid
Asadi et al. [69]	$Mg(OH)_2$/water	CTAB, oleic acid, and SDS
Gallego et al. [31]	Al_2O_3/water	SDBS
Shah et al. [70]	Reduced graphene oxide (rGO)/EG	SDS, SDBS, and CTAB
Xian [16]	Mono (COOH-GnP)/water and hybrid TiO_2-(COOH-GnP)/water	PVP, Triton X-100, CTAB, SDS, SDC, and SDBS
Ilyas et al. [71]	GnP/saline water	SDS

conductivity has been observed. Enhancement in viscosity has also been noticed, which is not desirable.

Surface modification of nanoparticles

Modifying the surface of nanoparticles via functionalization or directly using surface-modified (or functionalized) nanoparticles is one of the promising approaches to attain long-term stability in nanofluids. No surfactant is used in this technique. The stability of suspension mainly depends upon two kinds of repulsion forces present in it, such as steric repulsion and electrostatic (charge) repulsion. For stabilizing steric repulsion forces, polymers such as polymethacrylic acid (PMAA) and PVP are mostly used; they adsorb on the surface of particles and produce additional steric repulsion forces. Thus, particles do not adhere to each other, aggregation of particles does not occur, and it hinders sedimentation. This process of stabilizing dispersions is also called polymeric stabilization. High concentration of nanofluids can also stabilize with this method [73–76]. In electrostatic stabilization, the same charge is developed on approaching particles so that repelling effect can be produced. Different ways of developing surface charge are listed.

- By adsorption of ions
- With dissociating surface charged species in suspensions
- By substituting ions with other ions
- By depleting electrons onto the surface

The purpose of this stabilization technique is to make a charge barrier between particles. They do not come close to each other and not form clusters. Thus, sedimentation can be prevented with this technique, and also stability can be improved.

Esumi et al. [77] enhanced CNT nanoparticle-based nanofluids' stability by chemically treated with HNO_3 and H_2SO_4 in a ratio of 1:3 for better dispersion. Xie et al. [78] also used HNO_3 and H_2SO_4 for modifying the surface of CNT nanoparticles. They noticed that sedimentation of particles was not happened for about 2 months. Yang and Liu [79] showed the stability of SiO_2/deionized water nanofluid up to 12 months by functionalizing with silanes of (3-glycidoxylpropyl) trimethoxysilane. Kayani et al. [80] used functionalized TiO_2 by chemical treatment to prepare TiO_2/water nanofluid. First, TiO_2 nanopowder was mixed with hexamethyldisilazane ($C_6H_{19}NSi_2$) in 2:1 mass fraction and then ultrasonicated for 1 hour at 30°C and soaked nanoparticles were obtained. Furthermore, these soaked nanoparticles were dried with evaporation process and then these functionalized nanoparticles were used for preparing nanofluids. Fabricated nanofluids were observed to be stable for few more days than nanofluids fabricated with untreated TiO_2 nanoparticles. Chougule and Sahu [81] used functionalized and surface-treated CNT nanoparticles in water. Functionalized CNT nanoparticle-based nanofluids showed better heat transfer performance in radiator than surface-treated CNT nanoparticle nanofluids. Yarmand et al. [82] conducted acid treatment to make functionalized graphene nanoparticles. Functionalization is done to introduce functional groups (hydroxyl and carboxyl) on the surface of nanoparticles so that graphene nanoparticles are converted into hydrophilic nature and easily dispersed in polar solvents like water. Jadar et al. [83] showed an enormous increase in thermal conductivity and heat transfer coefficient using functionalized MWCNT. Also, the efficiency of radiators in which functionalized nanofluids are used as coolants is enhanced.

pH change

pH can be understood as the hydrogen ion concentration of any solution. If hydrogen ions are more in suspension, it has a low value of pH and the suspension is acidic in nature. If hydrogen ions are less in suspension, it has a high pH value and the suspension is alkaline in nature. The pH of the solution lies around 7 so that the solution is not of acidic or alkaline nature; otherwise, it will damage or corrode the surfaces of heat transfer devices [84]. pH should make 7 by adding some nonreactive acidic or alkaline solution in nanofluids [85]. Relation between the pH and stability of nanofluids is given in terms of IEP (isoelectric point) [13,59]. IEP is that it is a particular value of pH of dispersion at which no net electric charge is present. At this point, the zeta potential value becomes zero, and thus from previous studies, it has been concluded that stability of dispersion becomes poor because no repulsive forces are present, and coagulation occurs, which is not desirable. Thus, for stable suspension, some electric charge should be present in the solution and the value of IEP should not be kept zero in any case. pH at which nanofluid becomes stable is called effective pH.

Witharana et al. [86] showed the effect of pH on the stability of Al_2O_3/water nanofluid. They showed that nanofluid having pH 6.3 was 30 min more stable than nanofluid having pH 7.8. Song et al. [26] examined the stability of stainless steel in water using SDS and SDBS surfactants by different stability measurement methods. They demonstrated that nanofluid with pH 11 was not sedimented until 10 days, nanofluid with pH 10 for 3 days, and nanofluids with lower pH values less than a day. Also, nanofluids with a high pH of 12.6 contain a great amount of OH^- ions and quickly settle down. Cacua et al. [87] inspected the stability of Al_2O_3/water nanofluids by zeta potential analysis using two different surfactants (CTAB and SDBS) at varying pH values (2–12). They found IEP at pH of 8.6 for this particular nanofluid without adding any surfactant. The pH value could be varied with the addition of HCl and NaOH. At a pH value of 6, nanofluid is examined to be stable, but lower pH nanofluid could corrode the thermal system.

2.3 Challenges and outlook

Hybrid nanofluids are becoming an interesting topic of research due to comparatively better thermophysical properties than mono nanofluids. However, there are a lot of discrepancies as one can see in the information depicted in this chapter. Some of the major challenges in research studies regarding hybrid nanofluids are briefly discussed.

- Selection of proper pairs of nanomaterials for preparing hybrid nanofluids is an open question for researchers. The emergence of hybrid nanofluids came into existence two decades ago; yet, the best combination of suitable pairs is not yet affirmed.

- A big contradiction among the outcomes of experimental, theoretical, and numerical studies occurs. There are differences in the results found by these studies concerning measuring the stability or thermophysical properties or heat transfer and flow characteristics of hybrid nanofluids.
- There is a lack of quality information to find out the optimal ultrasonication time for the fabrication of hybrid nanofluids so that the prepared nanofluids have good thermophysical properties and they do not settle down early.
- There is insufficient knowledge about the use of surfactants and their quantity to enhance nanofluids' stability.
- A deep understanding of the mechanisms behind the heat transfer enhancement through hybrid nanofluids is still unknown.
- Cost estimation is also a matter of concern for producing hybrid nanofluids.

All these issues are yet to be resolved by the team of coordinating researchers. Consequently, hybrid nanofluids could be used in real-world applications and are not limited to laboratory purposes.

2.4 Summary

A lot of research work has been garnered on different aspects regarding hybrid nanofluids to date. Different fabrication methods of hybrid nanofluids—one-step and two-step methods—have been described. Furthermore, the importance of stability of hybrid nanofluids has been discussed for the efficient working of thermal systems. Sedimentation, zeta potential, spectral absorbance, and electron microscopy are some of the stability evaluation methods that have been deliberated in detail. In practice, the sedimentation method is mostly used because researchers themselves visualize the sedimentation period of nanofluids. No other method is so real, but it is a long-time process. Zeta potential is also widely used after the sedimentation method due to its reliability and ease of access. UV-vis spectroscopy and electron microscopy are also used for determining the aggregation and precipitation behavior of nanoparticles in fluids. These stability evaluation methods gave only indicative information about the stability of nanofluids. Instability in nanofluids causes depreciation in thermophysical properties and also reduces the efficiency of the heat transfer system. For improving the stability of nanofluids, various stability enhancement techniques are investigated by various researchers. Ultrasonication, addition of surfactants, surface modification of nanoparticles, and change in pH are some of the methods discussed in this chapter for enhancement of stability. An optimum ultrasonication time period should be needed to investigate for every kind of nanofluid for obtaining maximum stability interval for nanofluids. Surfactants used in synthesizing nanofluids are mostly of organic nature (e.g., gum arabic, PVP, CTAB, SDS, and SDBS) and bio-degradation is not found yet. Also, nanofluids with surfactants used at a higher temperature cause foam formation and fouling of inner walls of pipes to occur. Thus, long-time use of nanofluids containing surfactants can cause deterioration of thermal devices, which should be addressed. pH is also another important factor for the stability of nanofluids. Very low or high pH can cause acidity or alkalinity in nanofluids, which damage the heat transferring equipment and also restrict the use of nanofluids in practical applications.

One of the major drawbacks of using hybrid nanofluids is an increment in pressure losses caused by the increasing viscosity of nanofluids. More pumping power is required for transferring the nanofluid from one place to the other. Thus, proper selection of nanomaterials and base fluids is necessary according to their synergy level.

References

[1] J.A. Ranga Babu, K.K. Kumar, S.S. Srinivasa Rao, State-of-art review on hybrid nanofluids, Renew. Sust. Energ. Rev. 77 (2017) 551–565.

[2] D.D. Kumar, A.V. Arasu, A comprehensive review of preparation, characterization, properties and stability of hybrid nanofluids, Renew. Sust. Energ. Rev. 81 (2018) 1669–1689.

[3] M. Gupta, V. Singh, S. Kumar, S. Kumar, N. Dilbaghi, Z. Said, Up to date review on the synthesis and thermophysical properties of hybrid nanofluids, J. Clean. Prod. 190 (2018) 169–192.

[4] S. Suresh, K.P. Venkitaraj, P. Selvakumar, M. Chandrasekar, Synthesis of Al_2O_3-Cu/water hybrid nanofluids using two step method and its thermo physical properties, Colloids Surf. A Physicochem. Eng. Asp. 388 (1-3) (2011) 41–48.

[5] L.S. Sundar, M.K. Singh, A.C. Sousa, Enhanced heat transfer and friction factor of MWCNT-Fe_3O_4/water hybrid nanofluids, Int. Commun. Heat Mass Transf. 52 (2014) 73–83.

[6] D. Madhesh, R. Parameshwaran, S. Kalaiselvam, Experimental investigation on convective heat transfer and rheological characteristics of Cu-TiO_2 hybrid nanofluids, Exp. Thermal Fluid Sci. 52 (2014) 104–115.

[7] T.T. Baby, S. Ramaprabhu, Experimental investigation of the thermal transport properties of a carbon nanohybrid dispersed nanofluid, Nanoscale 3 (5) (2011) 2208–2214.

[8] L. Megatif, A. Ghozatloo, A. Arimi, M. Shariati-Niasar, Investigation of laminar convective heat transfer of a novel TiO_2-carbon nanotube hybrid water-based nanofluid, Exp. Heat Transfer 29 (1) (2016) 124–138.

[9] M. Batmunkh, M.R. Tanshen, M.J. Nine, M. Myekhlai, H. Choi, H. Chung, H. Jeong, Thermal conductivity of TiO_2 nanoparticles based aqueous nanofluids with an addition of a modified silver particle, Ind. Eng. Chem. Res. 53 (20) (2014) 8445–8451.
[10] S.M. Abbasi, A. Rashidi, A. Nemati, K. Arzani, The effect of functionalisation method on the stability and the thermal conductivity of nanofluid hybrids of carbon nanotubes/gamma alumina, Ceram. Int. 39 (4) (2013) 3885–3891.
[11] M.J. Nine, B. Munkhbayar, M.S. Rahman, H. Chung, H. Jeong, Highly productive synthesis process of well dispersed Cu_2O and Cu/Cu_2O nanoparticles and its thermal characterization, Mater. Chem. Phys. 141 (2-3) (2013) 636–642.
[12] L. Chen, W. Yu, H. Xie, Enhanced thermal conductivity of nanofluids containing Ag/MWNT composites, Powder Technol. 231 (2012) 18–20.
[13] N. Sezer, M.A. Atieh, M. Koç, A comprehensive review on synthesis, stability, thermophysical properties, and characterization of nanofluids, Powder Technol. 344 (2019) 404–431.
[14] Babita, S.K. Sharma, G.S. Mital, Preparation and evaluation of stable nanofluids for heat transfer application: a review, Exp. Thermal Fluid Sci. 79 (2016) 202–212.
[15] P.C.M. Kumar, K. Palanisamy, V. Vijayan, Stability analysis of heat transfer hybrid/water nanofluids, Mater. Today: Proc. 21 (1) (2020) 708–712.
[16] H.W. Xian, N.A.C. Sidik, R. Saidur, Impact of different surfactants and ultrasonication time on the stability and thermophysical properties of hybrid nanofluids, Int. Commun. Heat Mass Transf. 110 (2020) 104389.
[17] M. Ma, Y. Zhai, P. Yao, et al., Effect of surfactant on the rheological behavior and thermophysical properties of hybrid nanofluids, Powder Technol. 379 (2021) 373–383.
[18] H.T. Zhu, C.Y. Zhang, Y.M. Tang, J.X. Wang, Novel synthesis and thermal conductivity of CuO nanofluid, J. Phys. Chem. C 111 (4) (2007) 1646–1650.
[19] E.B. Haghighi, N. Nikkam, M. Saleemi, M. Behi, S.A. Mirmohammadi, H. Poth, R. Khodabandeh, M.S. Toprak, M. Muhammed, B. Palm, Shelf stability of nanofluids and its effect on thermal conductivity and viscosity, Meas. Sci. Technol. 24 (10) (2013) 105301.
[20] A.K. Singh, V.S. Raykar, Microwave synthesis of silver nanofluids with polyvinylpyrrolidone (PVP) and their transport properties, Colloid Polym. Sci. 286 (14–15) (2008) 1667–1673.
[21] M. Mehrali, E. Sadeghinezhad, M.A. Rosen, Heat transfer and entropy generation for laminar forced convection flow of graphene nanoplatelets nanofluids in a horizontal tube, Int. Commun. Heat Mass Transf. 66 (2015) 23–31.
[22] A. Ghadimi, R. Saidur, H.S.C. Metselaar, A review of nanofluid stability properties and characterization in stationary conditions, Int. J. Heat Mass Transf. 54 (17-18) (2011) 4051–4068.
[23] X. Li, D. Zhu, X. Wang, Evaluation on dispersion behavior of the aqueous copper nano-suspensions, J. Colloid Interface Sci. 310 (2) (2007) 456–463.
[24] D. Zhu, X. Li, N. Wang, X. Wang, J. Gao, H. Li, Dispersion behavior and thermal conductivity characteristics of Al_2O_3-H_2O nanofluids, Curr. Appl. Phys. 9 (1) (2009) 131–139.
[25] S. Chakraborty, P.K. Panigrahi, Stability of nanofluid: a review, Appl. Thermal Eng. (2020) 115259.
[26] Y.Y. Song, H.K.D.H. Bhadeshia, D.W. Suh, Stability of stainless-steel nanoparticle and water mixtures, Powder Technol. 272 (2015) 34–44.
[27] L.S. Sundar, E.V. Ramana, M.P.F. Graça, M.K. Singh, A.C. Sousa, Nanodiamond-Fe_3O_4 nanofluids: preparation and measurement of viscosity, electrical and thermal conductivities, Int. Commun. Heat Mass Transf. 73 (2016) 62–74.

[28] S. Chakraborty, I. Sarkar, A. Ashok, I. Sengupta, S.K. Pal, S. Chakraborty, Thermophysical properties of Cu-Zn-Al LDH nanofluid and its application in spray cooling, Appl. Therm. Eng. 141 (2018) 339–351.
[29] S. Chakraborty, I. Sengupta, I. Sarkar, S.K. Pal, S. Chakraborty, Effect of surfactant on thermo-physical properties and spray cooling heat transfer performance of Cu-Zn-Al LDH nanofluid, Appl. Clay Sci. 168 (2019) 43–55.
[30] N.K. Cakmak, Z. Said, L. Syam Sundar, Z.M. Ali, A.K. Tiwari, Preparation, characterization, stability, and thermal conductivity of rGO-Fe_3O_4-TiO_2 hybrid nanofluid: an experimental study, Powder Technol. 372 (2020) 235–245.
[31] A. Gallego, K. Cacua, B. Herrera, D. Cabaleiro, M.M. Piñeiro, L. Lugo, Experimental evaluation of the effect in the stability and thermophysical properties of water-Al_2O_3 based nanofluids using SDBS as dispersant agent, Adv. Powder Technol. 31 (2) (2020) 560–570.
[32] L. Jiang, L. Gao, J. Sun, Production of aqueous colloidal dispersions of carbon nanotubes, J. Colloid Interface Sci. 260 (1) (2003) 89–94.
[33] R.T. Peixoto, V.M.F. Paulinelli, H.H. Sander, M.D. Lanza, L.A. Cury, L.T.A. Poletto, Light transmission through porcelain, Dent. Mater. 23 (11) (2007) 1363–1368.
[34] J.G. Smith Jr., J.W. Connell, K.A. Watson, P.M. Danehy, Optical and thermo-optical properties of space durable polymer/carbon nanotube films: experimental results and empirical equations, Polymer 46 (7) (2005) 2276–2284.
[35] M.Z. Sharif, W.H. Azmi, A.A.M. Redhwan, N.N.M. Zawawi, R. Mamat, Improvement of nanofluid stability using 4-step UV-vis spectral absorbency analysis, J. Mech. Eng. (2017) 233–247.
[36] R. Sadeghi, S.G. Etemad, E. Keshavarzi, M. Haghshenasfard, Investigation of alumina nanofluid stability by UV-vis spectrum, Microfluid. Nanofluid. 18 (5-6) (2015) 1023–1030.
[37] A. Ghadimi, R. Saidur, H.S.C. Metselaar, A review of nanofluid stability properties and characterization in stationary conditions, Int. J. Heat Mass Transf. 54 (17-18) (2011) 4051–4068.
[38] S. Jana, A. Saheli-Khojin, W.H. Zhong, Enhancement of fluid thermal conductivity by the addition of single and hybrid nano-additives, Thermochim. Acta 462 (2007) 45–55.
[39] S.A. Kumar, K.S. Meenakshi, B.R.V. Narashimhan, S. Srikanth, G. Arthanareeswaran, Synthesis and characterization of copper nanofluid by a novel one-step method, Mater. Chem. Phys. 113 (2009) 57–62.
[40] M.R. Esfahani, E.M. Languri, M.R. Nunna, Effect of particle size and viscosity on thermal conductivity enhancement of graphene oxide nanofluid, Int. Commun. Heat Mass Transf. 76 (2016) 308–315.
[41] L.R. Oliveira, S.R.F.L. Ribeiro, M.H.M. Reis, V.L. Cardoso, E.P.B. Filho, Experimental study on the thermal conductivity and viscosity of ethylene glycol-based nanofluid containing diamond-silver hybrid material, Diam. Relat. Mater. 96 (2019) 216–230.
[42] S. Mukherjee, S. Paria, Preparation and stability of nanofluids—a review, IOSR J. Mech. Civil Eng. 9 (2) (2013) 63–69.
[43] L. Kong, J. Sun, Y. Bao, Preparation, characterization and tribological mechanism of nanofluids, RSC Adv. 7 (21) (2017) 12599–12609.
[44] H. Zhu, C. Zhang, Y. Tang, J. Wang, B. Ren, Preparation and thermal conductivity of suspensions of graphite nanoparticles, Carbon (New York, NY) 45 (1) (2007) 226–228.
[45] N. Ali, J.A. Teixeira, A. Addali, A review on nanofluids: fabrication, stability, and thermophysical properties, J. Nanomater. 6978130 (2018).
[46] F. Yu, Y. Chen, X. Liang, J. Xua, C. Lee, Q. Liang, P. Tao, T. Deng, Dispersion stability of thermal nanofluids, Progr. Nat. Sci. Mater. Int. 27 (5) (2017) 531–542.

[47] I.M. Mahbubul, E.B. Elcioglu, R. Saidur, M.A. Amalina, Optimization of ultrasonication period for better dispersion and stability of TiO_2-water nanofluid, Ultrason. Sonochem. 37 (2017) 360–367.
[48] R. Gangadevi, B.K. Vinayagam, S. Senthilraja, Effects of sonication time and temperature on thermal conductivity of CuO/water and Al_2O_3/water nanofluids with and without surfactant, Mater. Today: Proc. 5 (2) (2018) 9004–9011.
[49] I.M. Mahbubul, R. Saidur, M.A. Amalina, E.B. Elcioglu, T. Okutucu-Ozyurt, Effective ultrasonication process for better colloidal dispersion of nanofluid, Ultrason. Sonochem. 26 (2015) 361–369.
[50] S.J. Chung, J.P. Leonard, I. Nettleship, J.K. Lee, Y. Soong, D.V. Martello, M.K. Chyu, Characterization of ZnO nanoparticle suspension in water: effectiveness of ultrasonic dispersion, Powder Technol. 194 (1-2) (2009) 75–80.
[51] B. Ruan, A.M. Jacobi, Ultrasonication effects on thermal and rheological properties of carbon nanotube suspensions, Nanoscale Res. Lett. 7 (1) (2012) 127.
[52] M. Nasiri, S.G. Etemad, R. Bagheri, Experimental heat transfer of nanofluid through an annular duct, Int. Commun. Heat Mass Transf. 38 (7) (2011) 958–963.
[53] M. Kole, T.K. Dey, Effect of prolonged ultrasonication on the thermal conductivity of ZnO-ethylene glycol nanofluids, Thermochim. Acta 535 (2012) 58–65, https://doi.org/10.1016/j.tca.2012.02.016.
[54] A. Shahsavar, M.R. Salimpour, M. Saghafian, M.B. Shafii, An experimental study on the effect of ultrasonication on thermal conductivity of ferrofluid loaded with carbon nanotubes, Thermochim. Acta 617 (2015) 102–110.
[55] A.K. Tiwari, N.S. Pandya, Z. Said, H.S. Öztop, N. Abu-Hamdeh, 4S consideration (synthesis, sonication, surfactant, stability) for the thermal conductivity of CeO_2 with MWCNT and water based hybrid nanofluid: an experimental assessment, Coll. Surf. A: Physicochem. Eng. Aspects 610 (2021) 125918.
[56] S. Mukherjee, S. Paria, Preparation and stability of nanofluids—a review, IOSR J. Mech. Civil Eng. 9 (2) (2013) 63–69.
[57] H. Yu, S. Hermann, S.E. Schulz, T. Gessner, Z. Dong, W.J. Li, Optimizing sonication parameters for dispersion of single-walled carbon nanotubes, Chem. Phys. 408 (2012) 11–16.
[58] A.M. Tiara, S. Chakraborty, I. Sarkar, A. Ashok, S.K. Pal, S. Chakraborty, Heat transfer enhancement using surfactant based alumina nanofluid jet from a hot steel plate, Exp. Thermal Fluid Sci. 89 (2017) 295–303.
[59] A.I. Khan, A.V. Arasu, A review of influence of nanoparticle synthesis and geometrical parameters on thermophysical properties and stability of nanofluids, Therm. Sci. Eng. Progr. 11 (2019) 334–364.
[60] B. Kronberg, K. Holmberg, B. Lindman, Types of surfactants, their synthesis, and applications, Surf. Chem. Surfact. Polym. (2014) 1–47.
[61] D.C. Cullum, Surfactant types; classification, identification, separation, in: Introduction to surfactant analysis, Springer, Dordrecht, 1994, pp. 17–41.
[62] X.J. Wang, X. Li, S. Yang, Influence of pH and SDBS on the stability and thermal conductivity of nanofluids, Energy Fuel 23 (5) (2009) 2684–2689.
[63] D.K. Agarwal, A. Vaidyanathan, S.S. Kumar, Synthesis and characterization of kerosene–alumina nanofluids, Appl. Therm. Eng. 60 (1-2) (2013) 275–284.
[64] M.A. Khairul, K. Shah, E. Doroodchi, R. Azizian, B. Moghtaderi, Effects of surfactant on stability and thermo-physical properties of metal oxide nanofluids, Int. J. Heat Mass Transf. 98 (2016) 778–787.
[65] X. Li, C. Zou, W. Chen, X. Lei, Experimental investigation of β-cyclodextrin modified carbon nanotubes nanofluids for solar energy systems: stability, optical properties and thermal conductivity, Sol. Energy Mater. Sol. Cells 157 (2016) 572–579.

[66] P.K. Das, A.K. Mallik, R. Ganguly, A.K. Santra, Synthesis and characterization of TiO_2-water nanofluids with different surfactants, Int. Commun. Heat Mass Transf. 75 (2016) 341–348.
[67] W.S. Sarsam, A. Amiri, S.N. Kazi, A. Badarudin, Stability and thermophysical properties of non-covalently functionalized graphene nanoplatelets nanofluids, Energy Convers. Manag. 116 (2016) 101–111.
[68] B. Wei, C. Zou, X. Li, Experimental investigation on stability and thermal conductivity of diathermic oil based TiO_2 nanofluids, Int. J. Heat Mass Transf. 104 (2017) 537–543.
[69] A. Asadi, M. Asadi, M. Siahmargoi, T. Asadi, M.G. Andarati, The effect of surfactant and sonication time on the stability and thermal conductivity of water-based nanofluid containing Mg (OH) 2 nanoparticles: an experimental investigation, Int. J. Heat Mass Transf. 108 (2017) 191–198.
[70] S.N.A. Shah, S. Shahabuddin, M.F.M. Sabri, M.F.M. Salleh, M.A. Ali, N. Hayat, N.Z.C. Sidik, M. Smykano, R. Saidur, Experimental investigation on stability, thermal conductivity and rheological properties of rGO/ethylene glycol based nanofluids, Int. J. Heat Mass Transf. 150 (2020) 118981.
[71] S.U. Ilyas, S. Ridha, F.A.A. Kareem, Dispersion stability and surface tension of SDS-Stabilized saline nanofluids with graphene nanoplatelets, Colloids Surf. A Physicochem. Eng. Asp. 592 (2020) 124584.
[72] S. Wu, D. Zhu, X. Li, H. Li, J. Lei, Thermal energy storage behavior of Al_2O_3-H_2O nanofluids, Thermochim. Acta 483 (1-2) (2009) 73–77.
[73] K.J. Park, D. Jung, S.E. Shim, Nucleate boiling heat transfer in aqueous solutions with carbon nanotubes up to critical heat fluxes, Int. J. Multiphase Flow 35 (6) (2009) 525–532.
[74] M. Shanbedi, S. Zeinali Heris, M. Baniadam, A. Amiri, The effect of multi-walled carbon nanotube/water nanofluid on thermal performance of a two-phase closed thermosyphon, Exp. Heat Transfer 26 (1) (2013) 26–40.
[75] A. Amiri, M. Shanbedi, H. Amiri, S.Z. Heris, S.N. Kazi, B.T. Chew, H. Eshghi, Pool boiling heat transfer of CNT/water nanofluids, Appl. Therm. Eng. 71 (1) (2014) 450–459.
[76] A. Amiri, M. Naraghi, G. Ahmadi, M. Soleymaniha, M. Shanbedi, A review on liquid-phase exfoliation for scalable production of pure graphene, wrinkled, crumpled and functionalized graphene and challenges, FlatChem 8 (2018) 40–71.
[77] K. Esumi, M. Ishigami, A. Nakajima, K. Sawada, H. Honda, Chemical treatment of carbon nanotubes, Carbon (New York, NY) 34 (2) (1996) 279–281.
[78] H. Xie, H. Lee, W. Youn, M. Choi, Nanofluids containing multiwalled carbon nanotubes and their enhanced thermal conductivities, J. Appl. Phys. 94 (8) (2003) 4967–4971.
[79] X.F. Yang, Z.H. Liu, Pool boiling heat transfer of functionalized nanofluid under sub-atmospheric pressures, Int. J. Therm. Sci. 50 (12) (2011) 2402–2412.
[80] M.H. Kayhani, H. Soltanzadeh, M.M. Heyhat, M. Nazari, F. Kowsary, Experimental study of convective heat transfer and pressure drop of TiO_2/water nanofluid, Int. Commun. Heat Mass Transf. 39 (3) (2012) 456–462.
[81] S.S. Chougule, S.K. Sahu, Thermal performance of automobile radiator using carbon nanotube-water nanofluid—experimental study, J. Therm. Sci. Eng. Applicat. 6 (4) (2014). 041009-1.
[82] H. Yarmand, S. Gharehkhani, S.F.S. Shirazi, M. Goodarzi, A. Amiri, W.S. Sarsam, M.S. Alehashem, M. Dahari, S.N. Kazi, Study of synthesis, stability and thermo-physical properties of graphene nanoplatelet/platinum hybrid nanofluid, Int. Commun. Heat Mass Transf. 77 (2016) 15–21.
[83] R. Jadar, K.S. Shashishekar, *f*-MWCNT nanomaterial integrated automobile radiator, Mater. Today: Proc. 4 (2017) 11028–11033.

[84] D. Wen, Y. Ding, Experimental investigation into the pool boiling heat transfer of aqueous based γ-alumina nanofluids, J. Nanopart. Res. 7 (2-3) (2005) 265–274.

[85] R. Azizian, E. Doroodchi, B. Moghtaderi, Influence of controlled aggregation on thermal conductivity of nanofluids, J. Heat Transfer 138 (2) (2016) 021301.

[86] S. Witharana, C. Hodges, D. Xu, X. Lai, Y. Ding, Aggregation and settling in aqueous polydisperse alumina nanoparticle suspensions, J. Nanopart. Res. 14 (5) (2012) 851.

[87] K. Cacua, F. Ordoñez, C. Zapata, B. Herrera, E. Pabón, R. Buitrago-Sierra, Surfactant concentration and pH effects on the zeta potential values of alumina nanofluids to inspect stability, Colloids Surf. A Physicochem. Eng. Asp. 583 (2019) 123960.

3

Thermophysical, electrical, magnetic, and dielectric properties of hybrid nanofluids

E. Venkata Ramana[a], L. Syam Sundar[b], Zafar Said[c,d,e,*], and Antonio C.M. Sousa[b]

[a]I3N, Department of Physics, University of Aveiro, Aveiro, Portugal. [b]Centre for Mechanical Technology and Automation (TEMA-UA), Department of Mechanical Engineering, University of Aveiro, Aveiro, Portugal. [c]Department of Sustainable and Renewable Energy Engineering, University of Sharjah, Sharjah, United Arab Emirates. [d]Research Institute for Sciences and Engineering, University of Sharjah, Sharjah, United Arab Emirates. [e]U.S.-Pakistan Center for Advanced Studies in Energy (USPCAS-E), National University of Sciences and Technology (NUST), Islamabad, Pakistan

**Corresponding author: zsaid@sharjah.ac.ae, zaffar.ks@gmail.com*

Chapter outline

3.1 Thermophysical properties 65

3.1.1 Thermal conductivity 67

3.1.2 Viscosity of hybrid nanofluids 74

3.1.3 Specific heat and density of hybrid nanofluids 79

3.1.4 Magnetic property 86

3.1.5 Dielectric property 86

3.2 Conclusion 87

Acknowledgments 88

References 88

3.1 Thermophysical properties

There are different ways to prepare hybrid nanofluids [1–3]. First, synthesize the hybrid nanoparticles and then disperse them in the base fluid (water, ethylene glycol, propylene glycol, engine oil, etc.) called hybrid nanofluids [4, 5]. Fig. 3.1 shows the synthesis procedure of hybrid nanoparticles, and Fig. 3.2 illustrates the prepared hybrid nanofluids. Another procedure for preparation of

Hybrid Nanofluids: Preparation, Characterization and Applications. https://doi.org/10.1016/B978-0-323-85836-6.00003-X

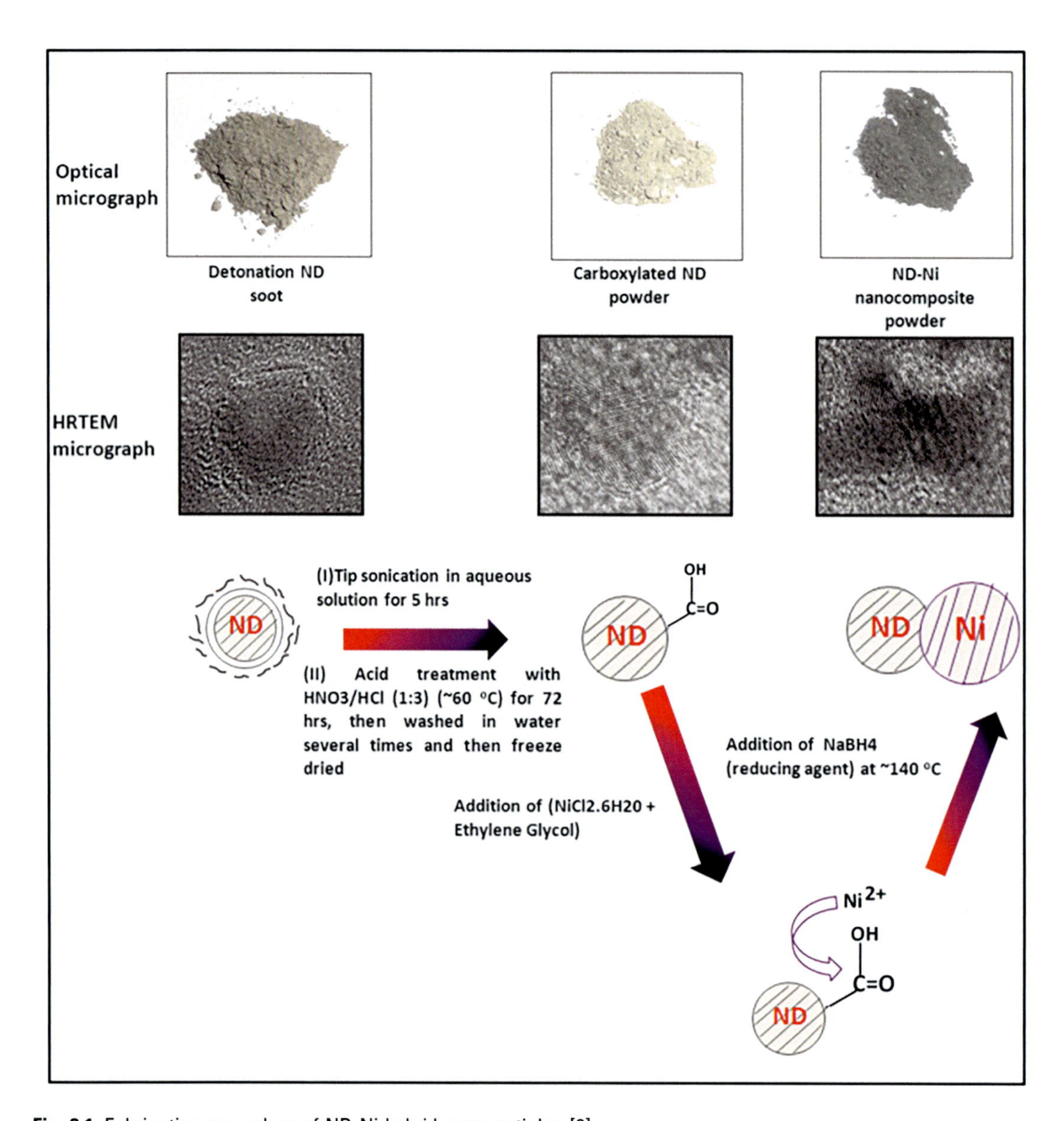

Fig. 3.1 Fabrication procedure of ND-Ni hybrid nanoparticles [6].

hybrid nanofluids is to prepare nanofluid 1 (single nanoparticle) and nanofluid 2 (single nanoparticle) and then mix them with an ultrasonic agitator. Fig. 3.3 displays the schematic representation of the preparation of hybrid nanofluids. Before using the hybrid nanofluids in any equipment, it is necessary to study their thermophysical properties [7, 8]. Various researchers analyze the thermophysical properties [7] such as thermal conductivity, viscosity, specific heat, and density regarding particle loadings and temperature, and these are discussed in what follows.

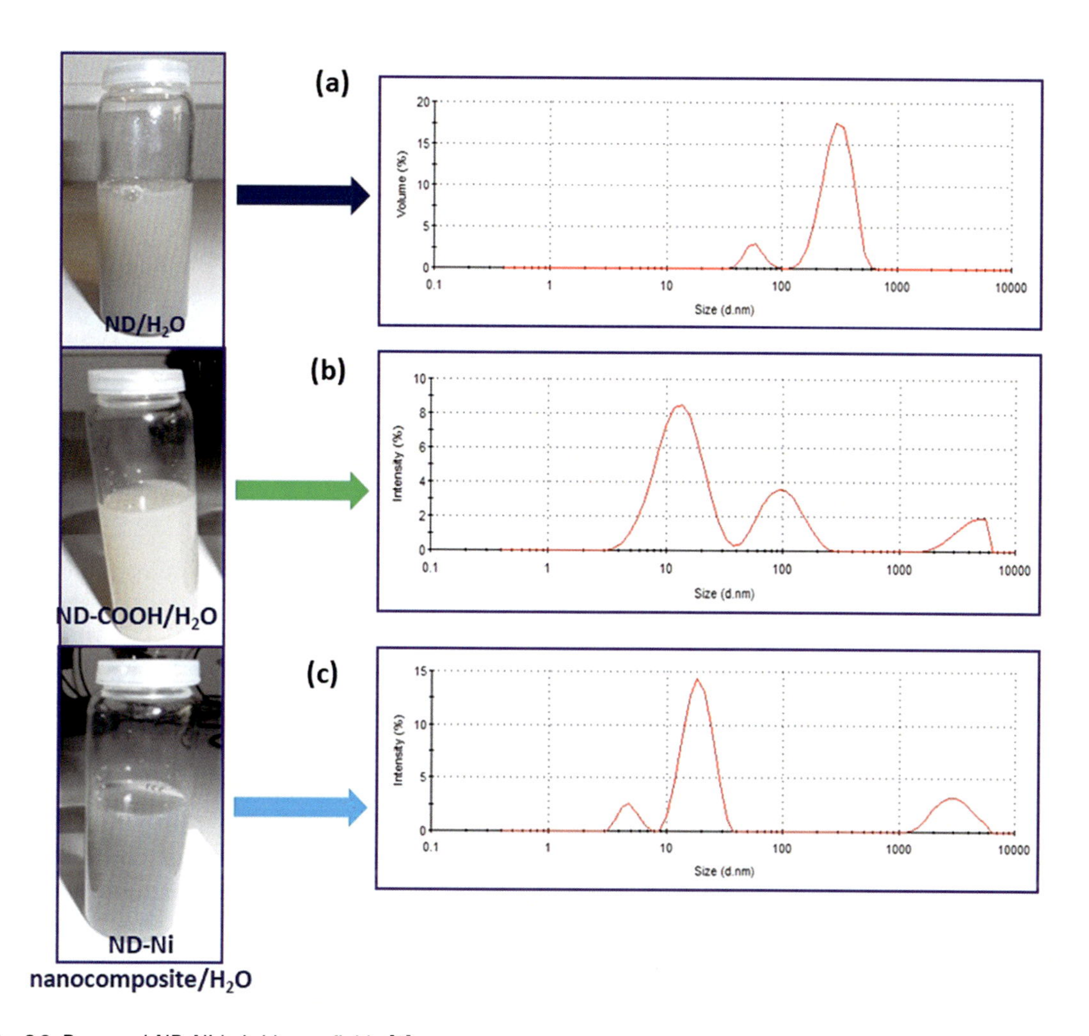

Fig. 3.2 Prepared ND-Ni hybrid nanofluids [6].

3.1.1 Thermal conductivity

The thermal conductivity of the hybrid nanofluid can be analyzed mainly by using KD 2 Pro thermal conductivity analyzer [9, 10]. Baghbanzadeh et al. [11] analyzed k_{nf} for silica/MWCNT nanofluid in the 0.1%, 0.55, and 1% weight concentration and at a temperature of 27°C and 40°C. They found thermal conductivity increment of nanofluid with an increase of weight concentration and temperature. Nine et al. [12] prepared 95:5 and 90:10 compositions of MWCNT-Al_2O_3/water hybrid nanofluid and studied

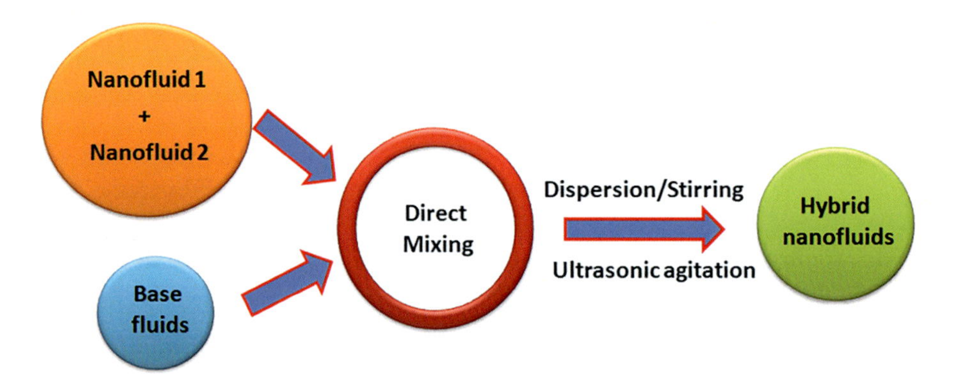

Fig. 3.3 Schematic for preparation of hybrid nanofluids.

thermal conductivity using the transient hot-wire method. They observed that thermal conductivity is enhanced with higher particle weight loadings and temperatures. Jana et al. [13] measured water-diluted Au-CNT and Cu-CNT nanofluid thermal conductivity using KD 2 pro instrument and found higher k_{nf} values for nanofluids than the individual mono nanofluids.

Suresh et al. [14] have observed k_{nf} enhancement of 12.11% at ϕ = 2% of Al_2O_3-Cu/water hybrid nanofluids. They also revealed the higher viscosity value of 115% with the same concentration and same hybrid nanofluid. Jyothirmayee Aravind and Ramaprabhu [15] conducted k_{nf} experiments for water- and EG-based graphene MWNT (GC) hybrid nanofluids. They reported that k_{nf} at ϕ = 0.04% is raised to 11.3% and 13.7% for water-based and 13.7% and 24% for EG-based nanofluids at 25°C and 50°C, respectively. Aravind and Ramaprabhu [16] achieved k_{nf} augmentation of 9.2% and 10.5% at ϕ = 0.04% for graphene nanofluid.

Baby and Ramaprabhu [17] have presented higher k_{nf} for Ag/MWNT–HEG ethylene glycol nanofluid. Baby and Ramaprabhu [18] presented augmented k_{nf} of 2% at ϕ = 0.08% of f-MWNT/f-HEG nanofluid and also k_{nf} of 1% at ϕ = 0.08% of f-HEG nanofluid. Sundar et al. [19] presented augmented k_{nf} and μ_{nf} values for water/MWCNT-Fe_3O_4 nanofluid at ϕ values of 0.1% and 0.3% and at temperature range from 20°C to 60°C. Fig. 3.4 shows the k_{nf} increment of 13.88% and 28.46% at ϕ = 0.3% at 20°C and 60°C compared with the base fluid. Esfahani et al. [20] have noticed k_{nf} raised from 0.663 to 0.788 W/m K at 25°C and 50°C for ϕ = 2% of water-diluted ZnO-Ag nanofluid. Esfe et al. [21] have shown k_{nf} rise with Ag-MgO/W hybrid nanofluid compared with the mono nanofluid.

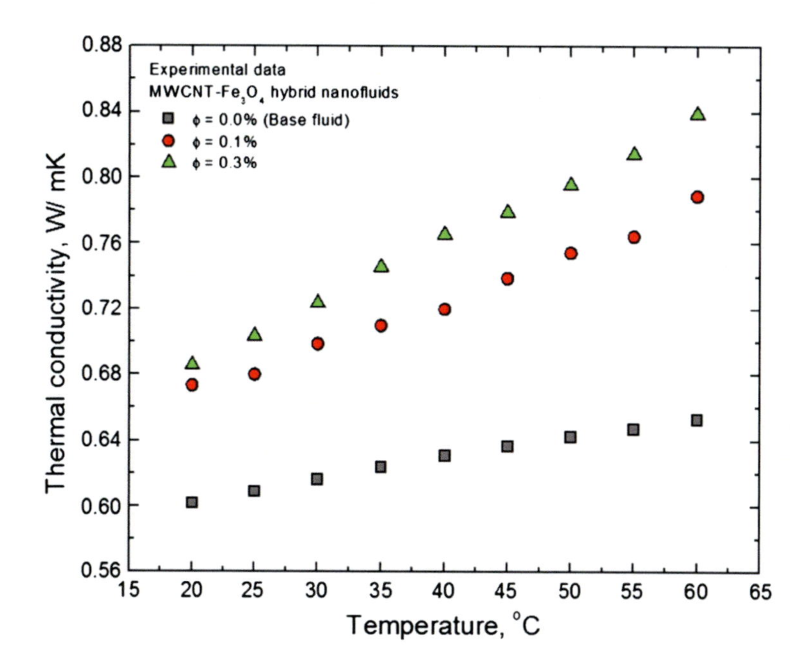

Fig. 3.4 Thermal conductivity of water-based MWCNT-Fe_3O_4 hybrid nanofluid. Copyright Elsevier 2014 (Lic# 5087501488977) (L.S. Sundar, M.K. Singh, A.C. Sousa, Enhanced heat transfer and friction factor of MWCNT–Fe3O4/water hybrid nanofluids, Int. Commun. Heat Mass Transfer 52 (2014) 73–83).

Esfe et al. [22] with ANNs (artificial neural networks) have shown the augmented k_{nf} values for the case of water-EG mixture Cu/TiO_2 hybrid nanofluid. Esfe et al. [23] have found k_{nf} augmentation to 24.9% and 32% for EG-diluted ZnO-DWCNT nanofluid and W/EG-diluted ZnO-DWCNT nanofluid. Chen et al. [24] have observed k_{nf} augmentation of 23% for Ag/MWNT hybrid nanofluids, but k_{nf} augmentation of 10% for MWNT nanofluids. Batmunkh et al. [25] have studied the thermal conductivity of Ag/TiO_2-H_2O nanofluid applying transient hot-wire technique and observed enhanced thermal conductivity with increased particle concentration. Yarmand et al. [26] have presented k_{nf} increase of 16.94% and 22.22% at 0.1 wt% of GNP-Ag nanofluid at 20°C and 40°C, respectively, whereas μ_{nf} is raised to 30% under the same conditions. Jha and Ramaprabhu [27] presented that k_{nf} increased to 35.2% and 10.1% at $\phi = 0.03\%$ for water- and EG-based MWCNTs/Cu nanofluids.

Fig. 3.5 indicates the experimental k_{nf} values of ND-Fe_3O_4 nanofluids of Sundar et al. [28]. Their outcomes of k_{nf} for water-based nanofluids are raised to 9.15% and 17.8%; k_{nf} for 20:80% EG/W-based nanofluids is raised to 6.6% and 13.4%; k_{nf} for 40:60% EG/W-based nanofluids is raised to 5% and 13.6%; and k_{nf} for 60:40% EG/W-based nanofluids is raised to 4.1% and 14.6% at $\phi = 0.2\%$ and at 20°C and 60°C, respectively. Botha

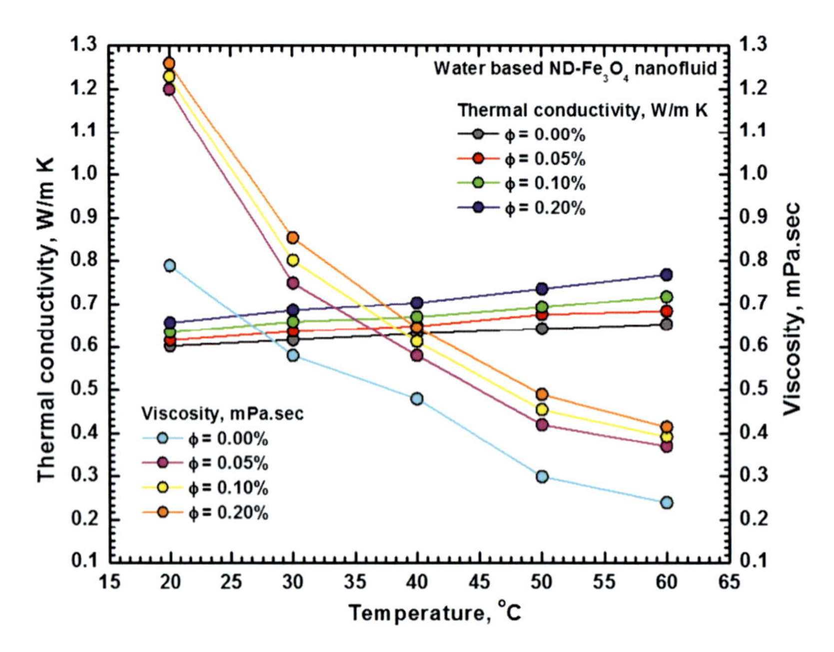

Fig. 3.5 Thermal conductivity and viscosity of water-based ND-Fe_3O_4 hybrid nanofluid. Copyright Elsevier 2021 (Lic# 5087510442690) (L.S. Sundar, et al., Nanodiamond-Fe3O4 nanofluids: preparation and measurement of viscosity, electrical and thermal conductivities, Int. Commun. Heat Mass Transfer 73 (2016) 62–74).

et al. [29] have noticed k_{nf} augmentation of 15% for 0.6 wt% of Ag–0.07 wt% silica transformer oil-based nanofluid.

Farbod et al. [30] have revealed maximum k_{nf} increase of 20% for 4 and 1 wt% of Ag-MWCNT nanofluid at 40°C. Sundar et al. [6] with H_2O- and EG-based nanodiamond-Ni nanofluids have found k_{nf} and μ_{nf} enhancements. The k_{nf} increase of water-based ND-Ni nanofluid is presented in Fig. 3.6A, and that of EG-based ND-Ni nanofluid is portrayed in Fig. B3.6b. Figure shows k_{nf} increase of 21% and μ_{nf} increase of 14.6% for 60EG and 40W ND-Ni nanofluid at 60°C. Figs. 3.7 and 3.8 show the water- and EG-based GO/Co_3O_4 nanofluids of Sundar et al. [31]. The k_{nf} increase is 7.64% and 19.14%, whereas μ_{nf} increase is 1.075-fold and 1.66-fold at $\phi = 0.05\%$ and at 7.64% and 19.14% and 20°C and 60°C.

Figs. 3.9 and 3.10 show the thermal conductivity experimental data of Sundar et al. [31] for H_2O- and EG-based nanodiamond-Co_3O_4 hybrid nanofluids, and they observed an enhancement of 16%, 9%, 14%, 11%, and 10% for H_2O and EG 20:80%, 40:60%, and 60:40% of EG/W-based ND-Co_3O_4 nanofluid and observed viscosity increment of 1.45-times, 1.46-times, 1.15-times, 1.19-times, and 1.51-times for H_2O and EG 20:80%, 40:60%, and

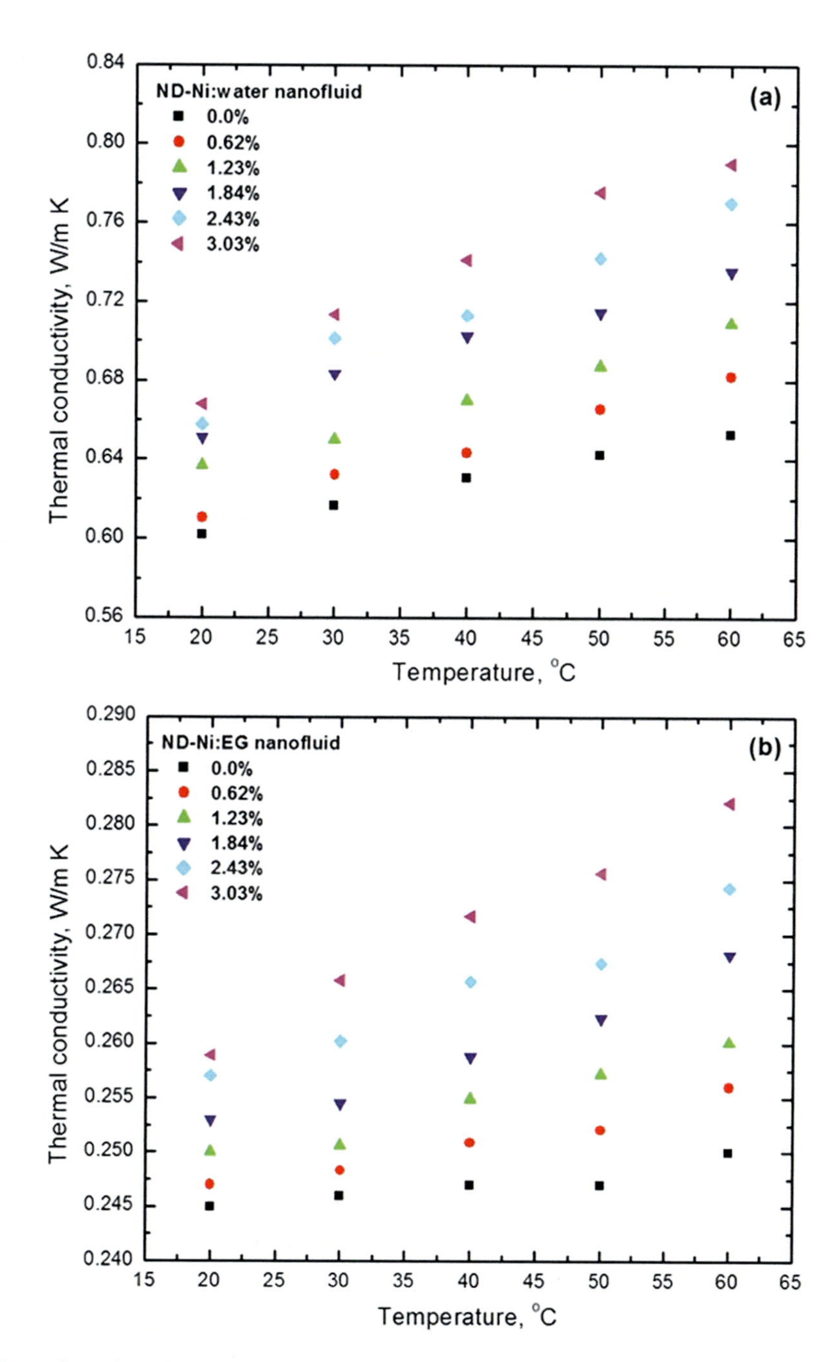

Fig. 3.6 (A) Thermal conductivity of water-based ND-Ni nanofluid and (B) ethylene glycol-based ND-Ni hybrid nanofluid [6].

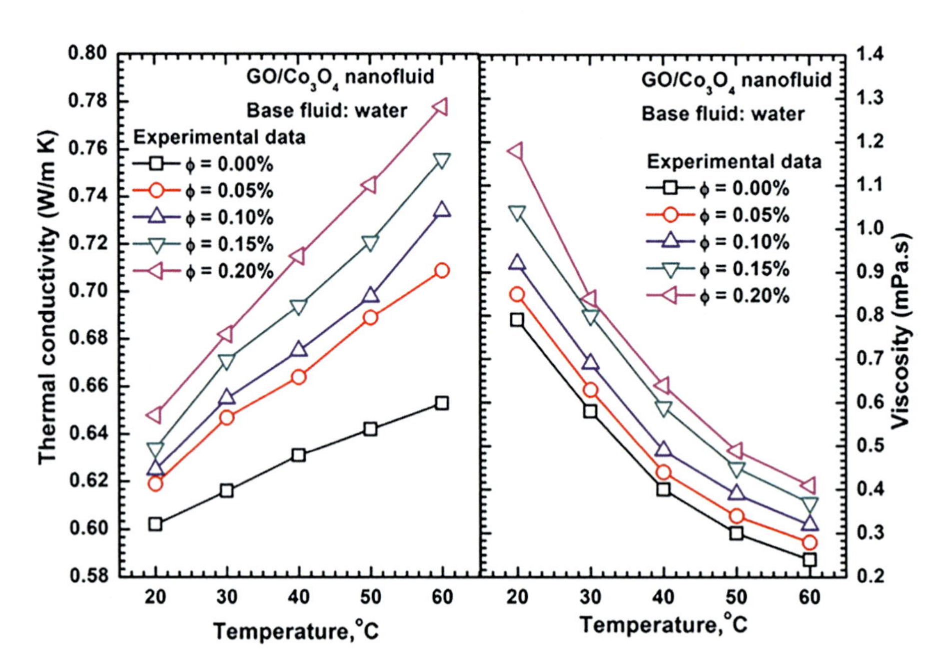

Fig. 3.7 Thermal conductivity and viscosity of GO + Co_3O_4-H_2O hybrid nanofluid. Copyright Elsevier 2021 (Lic# 5087510751269) (L.S. Sundar, et al., Experimental investigation of the thermal transport properties of graphene oxide/Co3O4 hybrid nanofluids, Int. Commun. Heat Mass Transfer 84 (2017) 1–10).

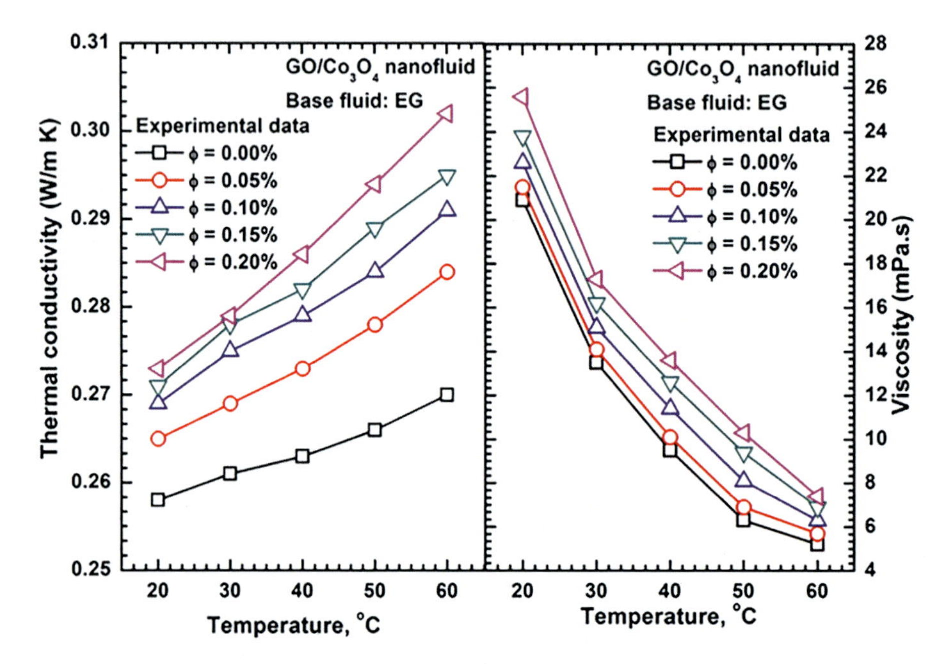

Fig. 3.8 Thermal conductivity and viscosity of GO + Co_3O_4-EG hybrid nanofluid. Copyright Elsevier 2021 (Lic# 5087510751269) (L.S. Sundar, et al., Experimental investigation of the thermal transport properties of graphene oxide/Co3O4 hybrid nanofluids, Int. Commun. Heat Mass Transfer 84 (2017) 1–10).

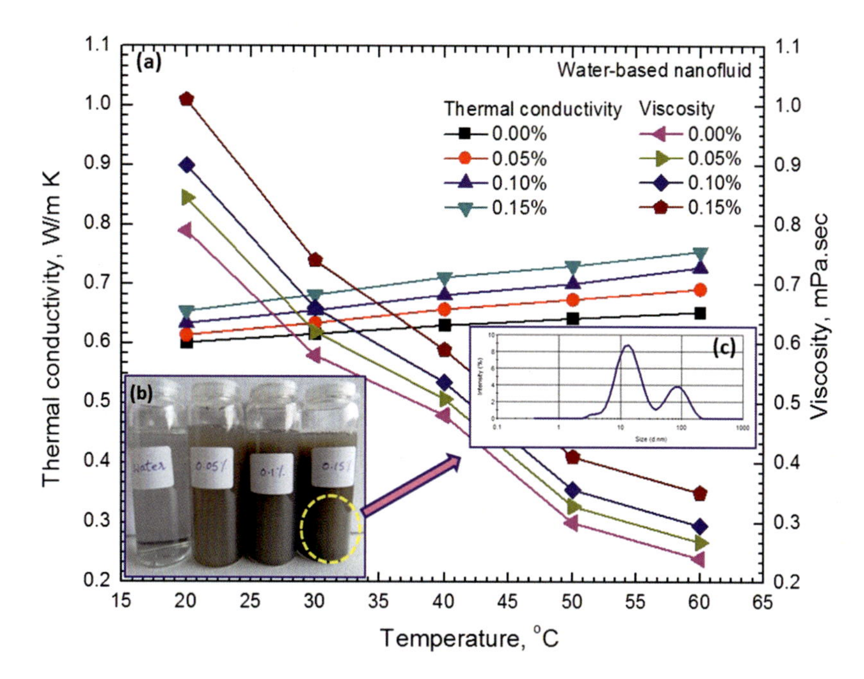

Fig. 3.9 Thermal conductivity and viscosity of ND + Co_3O_4-H_2O hybrid nanofluid. Copyright Elsevier 2021 (Lic# 5087510884804) (L.S. Sundar, et al., Thermal conductivity and viscosity of hybrid nanofluids prepared with magnetic nanodiamond-cobalt oxide (ND-Co3O4) nanocomposite, Case Stud. Therm. Eng. 7 (2016) 66–77).

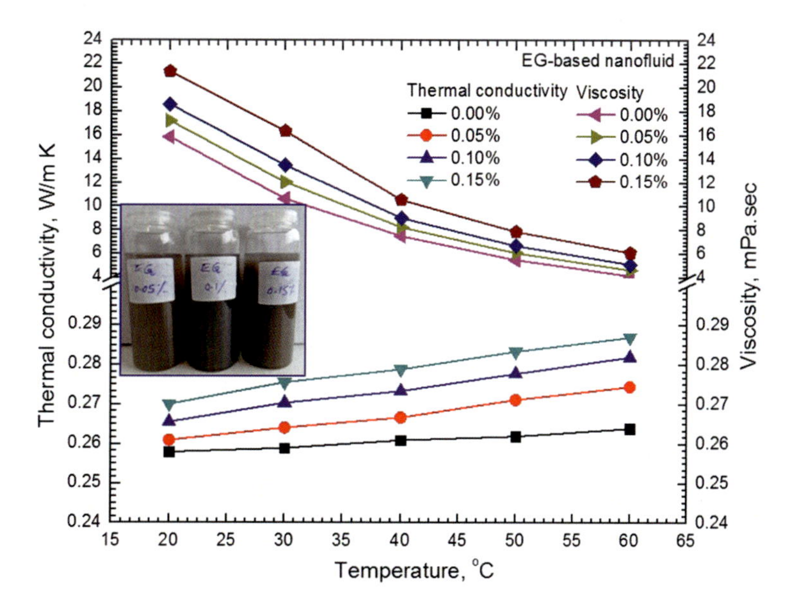

Fig. 3.10 Thermal conductivity and viscosity of EG-based ND-Co_3O_4 hybrid nanofluid. Copyright Elsevier 2021 (Lic# 5087510884804) (L.S. Sundar, et al., Thermal conductivity and viscosity of hybrid nanofluids prepared with magnetic nanodiamond-cobalt oxide (ND-Co3O4) nanocomposite, Case Stud. Therm. Eng. 7 (2016) 66–77).

60:40% EG/W-based nanofluid at 0.15 wt.% and 60°C. Minea et al. [32] have noticed k_{nf} increase of 12% based on the numerical analysis. Harandi et al. [33] revealed k_{nf} augmentation of 30% at $\phi = 2.3\%$ for EG-based f-MWCNTs-Fe_3O_4 nanofluid at 50°C.

Akilu et al. [34] measured the thermal conductivity of 80:20 wt.% of SiO_2-CuO/C@60:40 wt.% glycerol and EG-based hybrid nanofluid and observed thermal conductivity increment of 6.9% at 2 vol.% of nanofluid at 353.15K. Fig. 3.11 is the thermal conductivity of SiO_2-CuO/C@60:40 hybrid nanofluid [35]. Fig 3.12 is the thermal conductivity data of Oliveira et al. [36] for diamond-Ag/EG-based hybrid nanofluid and then observed thermal conductivity increment of 6.92% at 0.1 vol.%. Recent advancements on the thermal conductivity of hybrid nanofluids is summarized in Table 3.1.

3.1.2 Viscosity of hybrid nanofluids

When diluting the nanoparticles in the single-phase fluid, it will affect the fluid's viscosity. Various instruments are used to measure the viscosity of nanofluids [50, 51]. The viscosity of a fluid is a measure of its resistance to deformation at a given rate. In general, the fluids are classified as Newtonian and Non-Newtonian [51]. When the shear stress is proportional to the deformation rate at a constant pressure and temperature is called a Newtonian fluid. A non-Newtonian fluid is said to have constant viscosity independent of stress. The increased viscosity of the hybrid nanofluids affects the fluid pressure drop while they flow in a system [52]. When using the hybrid nanofluids in various

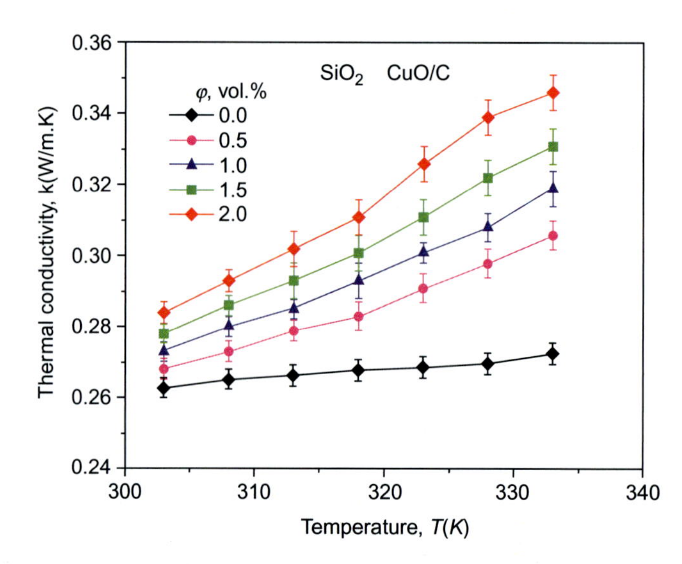

Fig. 3.11 Thermal conductivity of SiO_2-CuO/C@60:40 hybrid nanofluid. Copyright Elsevier 2021 (Lic# 5087511120990) (S. Akilu, et al., Properties of glycerol and ethylene glycol mixture based SiO2-CuO/C hybrid nanofluid for enhanced solar energy transport, Solar Energy Mater. Solar Cells 179 (2018) 118–128).

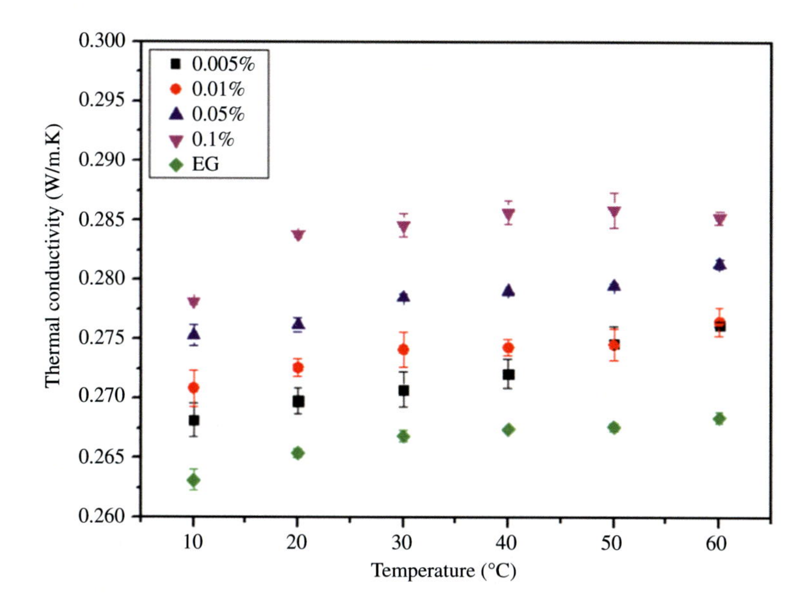

Fig. 3.12 Thermal conductivity of EG-based Di-Ag hybrid nanofluid. Copyright Elsevier 2021 (Lic# 5087511241674) (L.R. de Oliveira, et al., Experimental study on the thermal conductivity and viscosity of ethylene glycol-based nanofluid containing diamond-silver hybrid material, Diamond Relat. Mater. 96 (2019) 216–230).

Table 3.1 The k_{nf} summary of hybrid nanofluids and the recent advancements.

Researchers	Hybrid nanoparticles	Base fluid	Parameters	Remarks
Wu et al. [37]	Ag-CNTs Cu-CNTs	Water and EG	0.01–0.05 wt.%; 25–65° C	• Cu-CNT- and Ag-CNT-based hybrid nanofluids show remarkable thermal conductivity • The highest improvement in thermal conductivity for H_2O and EG Ag-CNT nanofluids is reported to be 52.37% and 40.42%, respectively, at 65°C and 0.05 wt.%, compared to the base fluid • Metal nanoparticle CNT composites can enhance the nanofluids' thermal conductivity and stimulate the advancement of novel thermal conductive medium

Continued

Table 3.1 The k_{nf} summary of hybrid nanofluids and the recent advancements—cont'd

Researchers	Hybrid nanoparticles	Base fluid	Parameters	Remarks
Kamkari et al. [38]	TiO_2-graphene	Water	0.005–0.5 vol.%; 25–75°C	• Increment in thermal conductivity is observed with increasing volume fraction and temperature • The effect of volume concentration was more significant than the temperature • Maximum thermal conductivity was reported at 0.5 vol.% and 75°C increased by 27.84%, compared to the base fluid
Aghaei et al. [39]	Cu-SiO_2-MWCNT	Water	1–3 vol.%; 15–65°C	• Excellent thermal conductivity was observed • Thermal conductivity depends on the volume concentration and temperature
Said et al. [40]	CeO_2-MWCNT	Water	Surfactants: SDS, SDBS, CTAB, SDS, DDC, gum arabic, PVP with ratio (5:0, 4:1, 3:2, 2:3, and 1:4)	• Stabilization approach has a significant relation with thermal conductivity • The addition of a further quantity of surfactant (more than 3:2 mixing ratio) has a reduction in thermal conductivity
Said et al. [41]	rGO-Co_3O_4	Distilled water	0.05, 0.1, and 0.2 wt.%; 20–60°C	• The maximum increment in thermal conductivity was reported to be 70.83% with 0.2 wt.% at 60°C
Sunil et al. [42]	Graphene-NiO	Coconut oil	Graphene sheet thickness (1–4 nm)	• It was reported that the thickness of nanoparticle base fluid interface and aggregated path of nanomaterials performs a substantial role to improve thermal conductivity

Table 3.1 The k_{nf} summary of hybrid nanofluids and the recent advancements—cont'd

Researchers	Hybrid nanoparticles	Base fluid	Parameters	Remarks
Zhai et al. [43]	Al_2O_3-CuO Al_2O_3-TiO_2	Water	Surfactants: SDS, PVP, and CTAB	• At the optimum PVP surfactant 0.005 and 0.01 wt.% concentrations, the highest thermal conductivity is reported to be 12% and 14% at 60°C for Al_2O_3-CuO/H_2O and Al_2O_3-TiO_2/H_2O hybrid nanofluids • Adding a slight quantity of surfactant can enhance the stability and thermal conductivity of hybrid nanofluids
Wanatasanapan et al. [44]	Al_2O_3-TiO_2	Water	1 vol.%; 30–70°C	• Maximum thermal conductivity of 1.134 W/m K is reported with a 50:50 nanoparticle ratio at 70°C • The thermal conductivity for all mixing ratios of hybrid nanofluid increases with increasing temperature • The Al_2O_3-TiO_2 hybrid nanofluid enhances the thermal conductivity to a maximum of 71% as compared to the base fluid
Behbahani et al. [45]	Mesoporous silica-Cu	Water	0.019–0.075 wt.%; 25–50°C	• Thermal conductivity ratio of hybrid nanofluids was strengthened by increasing concentration and temperature • Maximum increment of 24.24% in thermal conductivity for 0.075 wt.% and at 50°C
Karimipour et al. [46]	WO_3-MWCNTs	Engine oil	0.05–0.6 vol.%; 20–60°C	• Volume concentration has a more significant impact on thermal conductivity as compared to temperature • The k_{nf} is increased to 19.85% at $T = 60°C$ and $\phi = 0.6\%$ compared with the base fluid

Continued

Table 3.1 The k_{nf} summary of hybrid nanofluids and the recent advancements—cont'd

Researchers	Hybrid nanoparticles	Base fluid	Parameters	Remarks
Aparna et al. [47]	Al_2O_3-Ag	Water	0.005–0.1 vol.%	• It was reported that Al_2O_3-Ag nanofluids display a remarkable thermal conductivity increment than Al_2O_3 nanofluids • With increasing temperature, the thermal conductivity of Ag-Al_2O_3 and Al_2O_3-Ag nanofluids increases significantly for larger volume fractions
Singh et al. [48]	GO-CuO	Distilled water	0.03, 0.1, 0.3 vol.%; 30, 45, and 60°C	• Thermal conductivity improves with volume fraction and temperature • Improvement of 30% was reported with 0.3 wt% and 60°C for hybrid nanofluids
Toghraie et al. [49]	MWCNTs-ZnO-titania	Water-EG (80:20)	0.1–0.4 vol.%; 25–50°C	• Increment in thermal conductivity with rising temperature and volume concentration

applications, the higher viscosity of hybrid nanofluids compared with the base fluid must be analyzed. Usually, viscosity is greater against its base fluid, and it also increases with increasing volume fractions. More researchers have evaluated the viscosity of the hybrid nanofluids with an effect of volume concentrations and temperatures.

Even the hybrid nanofluids offer higher viscosity, but many researchers have noticed higher heat transfer rates than the base fluid. When it is observed in the heat transfer process, hybrid nanofluids' higher thermal conductivity is the main reason for achieving augmented heat transfer coefficients. Akilu et al. [34] noticed a viscosity increase of 1.33-times with CuO-SiO_2+C- (G/EG(60:40)) hybrid nanofluid at 2 vol.% of nanofluid at 353.15 K. Soltani and Akbari [53] have noticed μ_{nf} augmentation of 168% at $\phi = 1\%$ for EG-based MgO-MWCNT nanofluid. Dynamic viscosity rise is occurred with dilution of nanoparticles in single phase fluids, and it decreases with augmented temperatures [54–58].

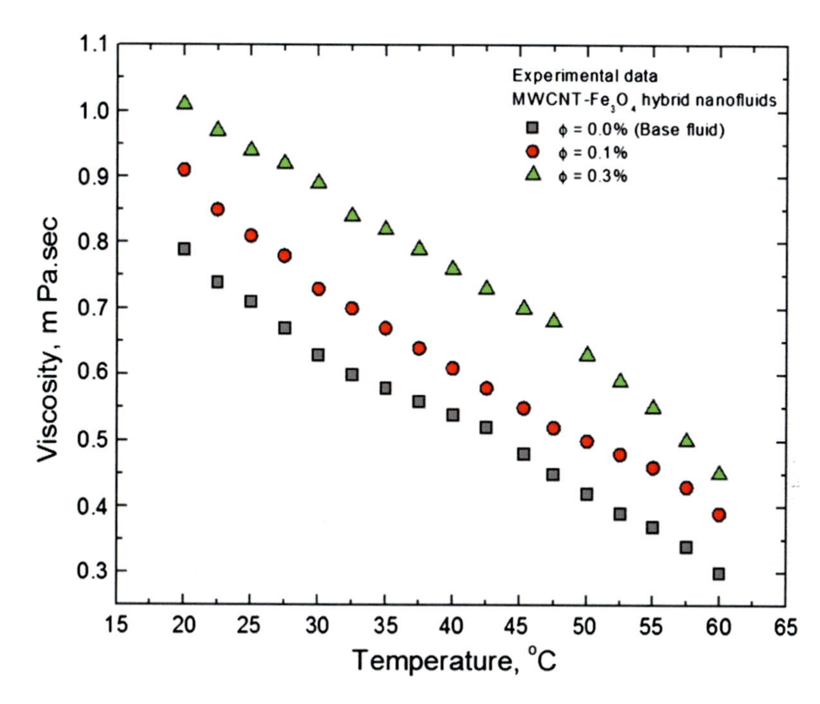

Fig. 3.13 Dynamic viscosity of water-based MWCNT-Fe_3O_4 hybrid nanofluid. Copyright Elsevier 2021 (Lic# 5087511376453) (L. S. Sundar, M.K. Singh, A.C. Sousa, Enhanced heat transfer and friction factor of MWCNT–Fe3O4/water hybrid nanofluids, Int. Commun. Heat Mass Transfer 52 (2014) 73–83).

Asadi and Asadi [59] found that μ_{nf} is increased to 45% at $\phi =$ 1% and at 55°C for engine oil-based MWCNT/ZnO nanofluid. Afrand et al. [60] have noticed Newtonian behavior of EG-based Fe_3O_4/Ag nanofluid in the maximum measured ϕ (1%) and temperatures (50°C). Fig. 3.13 indicates the dynamic viscosity of water-based MWCNT-Fe_3O_4 hybrid nanofluid [19]. Fig. 3.14 shows the viscosity of SiO_2-CuO/C@60:40 hybrid nanofluid [35]. Sundar et al. [28] noticed a maximum viscosity increase of 15.1, 29.8, 40.3, and 50.4% at 0.05, 0.1, 0.2, and 0.3 vol.% ND+Fe_3O_4 hybrid nanofluids at 60°C. Kazemi et al. [61] also obtained a similar trend of increased viscosity with increased particle loadings. Oliveira et al. [36] also observed similar nature for ND-Ag/EG hybrid nanofluids. Summary of the recent advancements on the μ_{nf} of nanofluids is presented in Table 3.2.

3.1.3 Specific heat and density of hybrid nanofluids

The specific heat of hybrid nanofluids is an important parameter that must be evaluated. The modified differential scanning calorimeter determines the specific heat of nanofluids. The

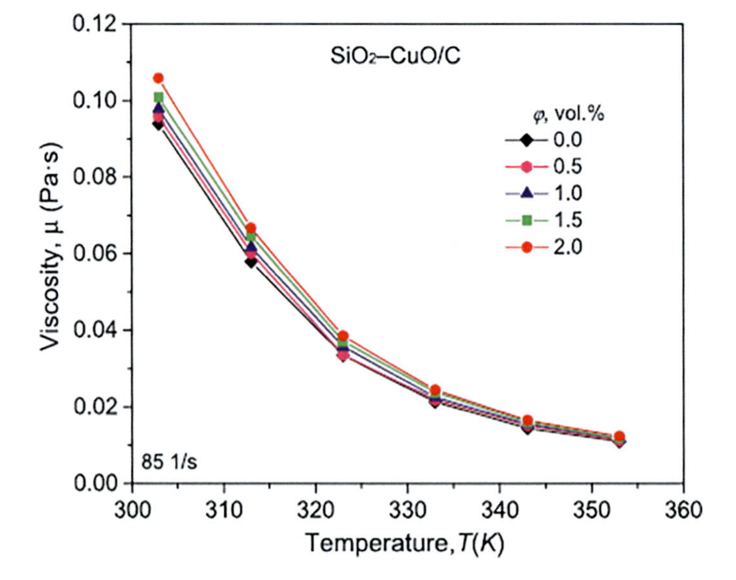

Fig. 3.14 Viscosity of SiO_2-CuO/C@60:40 hybrid nanofluid. Copyright Elsevier 2021 (Lic# 5087511511090) (S. Akilu, A.T. Baheta, K. Sharma, Experimental measurements of thermal conductivity and viscosity of ethylene glycol-based hybrid nanofluid with TiO2-CuO/C inclusions, J. Mol. Liq. 246 (2017) 396–405).

particle size is an influencing parameter on the specific heat, and thinner size nanoparticles have a higher specific heat than larger size particles [69]. With the addition of hybrid nanoparticles in the base fluid, interestingly, its specific heat is decreased. When the particle size is decreasing, the surface atom size is bigger than the large size particles.

Fig. 3.15 indicates the specific heat of CuO-SiO_2+C@60:40 hybrid nanofluid [35]. It is observed that with the increase of particle loadings, the specific heat is decreased. Wole-Osho et al. [70] also observed decreased specificity of Al_2O_3-ZnO/water hybrid nanofluids. Tiwari et al. [71] conducted specific heat experiments for CuO+MWCNT, MgO+MWCNT, and SnO_2+MWCNT and DI water and CTAB surfactant at 3:2 ratio and studied the effect of particle size d_p (20–50 nm) at 25–50°C and 0.25–1.5 vol.%. They observed the greatest reduction in specific heat value, which is about 15.09% greater than that of the base fluid at 25°C and 1.50 vol.% for MgO+MWCNT (80:20)/water nanofluids. Okonkwo et al. [72] studied the thermophysical properties Al_2O_3-Fe nanofluids in temperature ranges from 25°C to 65°C at three nanoparticle concentrations of 0.05%, 0.1%, and 0.2%. They observed 14% thermal conductivity enhancement with a 0.2% concentration for the hybrid nanofluids, while an enhancement of 7.8% and 7.5% was recorded using Al_2O_3-Fe nanofluids and Al_2O_3 nanofluids, respectively, at 0.1% concentrations.

Çolak et al. [73] studied the specific heat of Cu-Al_2O_3/H_2O hybrid nanofluid using an artificial neural network model in the

Table 3.2 Summary of the recent advancements on the μ_{nf} of nanofluids.

Researchers	Hybrid nanoparticles	Base fluid	Parameters	Remarks
Huminic et al. [62]	Fe-Si	Water	0.25–1 wt.%; 20–50°C	• Maximum dynamic viscosity was observed at 1.0 wt% and 25°C
Zhang et al. [37]	Cu-CNT Ag-CNT	Water EG	0.01–0.05 wt.%	• Lower viscosity was observed than functionalized CNT nanofluids
Aghaei et al. [39]	Cu-SiO_2-MWCNT	Water	1–3 vol.%; 15–65°C	• Dynamic viscosity is a function of temperature and particle loading • It exhibited Newtonian behavior
Toghraie et al. [63]	WO_3-MWCNT	Water+EG	0.1–0.6 vol.%; 25–50°C	• As volume concentration increases at a constant temperature, the dynamic viscosity increases • As temperature increases at constant concentration, the dynamic viscosity reduces
Said et al. [64]	MWCNT-CeO_2	Water, EG, silicone oil, and therminol VP-I	55–80°C; surfactants: BAC; CTAC; ALS; PLS, Brij 700 and Span 80	• It is reported that with increasing sonication duration, the viscosity of hybrid nanofluids reduces up to a specific period, then it improves gradually • CeO_2-MWCNT/water-based hybrid nanofluids show as a non-Newtonian fluid as there is no linear relationship reported between shear stress and shear rate
Said et al. [41]	rGO-Co_3O_4	Water	0.05, 0.10, and 0.20 wt.%; 20–60°C	• Maximum increment in viscosity was reported to be 70.83% at 0.2% particle and 60°C
Zhai et al. [65]	Cu-Al_2O_3	Water	0.3–1.5 vol.%; Cu = 50 nm, Al_2O_3 = 20, 30, and 50 nm	• At constant 0.3% concentration, viscosity increases with increasing nanoparticle size • Minimal viscosity was observed with 20 nm Al_2O_3-Cu/water at room temperature • At lower shear rates (<100 s^{-1}), the 20-nm Al_2O_3-Cu/W hybrid nanofluid exhibits non-Newtonian behavior

Continued

Table 3.2 Summary of the recent advancements on the μ_{nf} of nanofluids—cont'd

Researchers	Hybrid nanoparticles	Base fluid	Parameters	Remarks
Kalbasi et al. [66]	MWCNT-TiO_2	5W40	0.05–1 vol.%; 20–60°C	• It exhibited non-Newtonian fluid • The viscosity increases with decreased shear rate as well as temperature and increasing volume concentration
Li et al. [67]	SiC-MWCNTs	EG	0.04–0.4 vol.%; 30–50°C	• The SiC-MWCNT nanofluids exhibited Newtonian fluid • Viscosity rises with particle loadings
Vidhya et al. [68]	ZnO-MgO	Water and EG	0–0.1 vol.%; 30–80°C	• The viscosity is decreased due to the reduced van der Waals forces between the molecules • It is increased 67.4% higher than the base fluid for 0.1 vol.% at 80°C

temperature range from 20°C to 65°C. The value was taken into three main portions, with the addition of 901 (70%) for instruction, 257 (20%) for the test, and 129 (10%) for the confirmation. They observed that experimental specific heat values predict 0.99994 average relative errors compared with model values. Çolak et al. [74] measured experimentally specific heat values of Al_2O_3/water, Cu/water nanofluids, and Al_2O_3-Cu/water hybrid nanofluids using the DTA procedure and compared the results with those frequently used in the literature and observed that the specific heat of nanofluid is decreased compared with water.

The density of hybrid with nanofluid is also a necessary property.

With the augmented dilution of nanoparticles in individual phases, fluid caused the enhanced density of the fluid, which is also further increased due to the more dilution of particle loadings. In another way, the fluid density is degraded with the augmented temperature values. Wole-Osho et al. [70] also observed decreased specificity of Al_2O_3-ZnO/water hybrid nanofluids. Ramalingam et al. [75] reported the increased density of 50:50% W/EG-based Al_2O_3-SiC hybrid nanofluids with an increase of particle loadings, as shown in Fig. 3.16. Summary of the recent advancements on the specific heat of hybrid nanofluids is presented in Table 3.3. Summary of the recent advancements on the density of hybrid nanofluids is presented in Table 3.4.

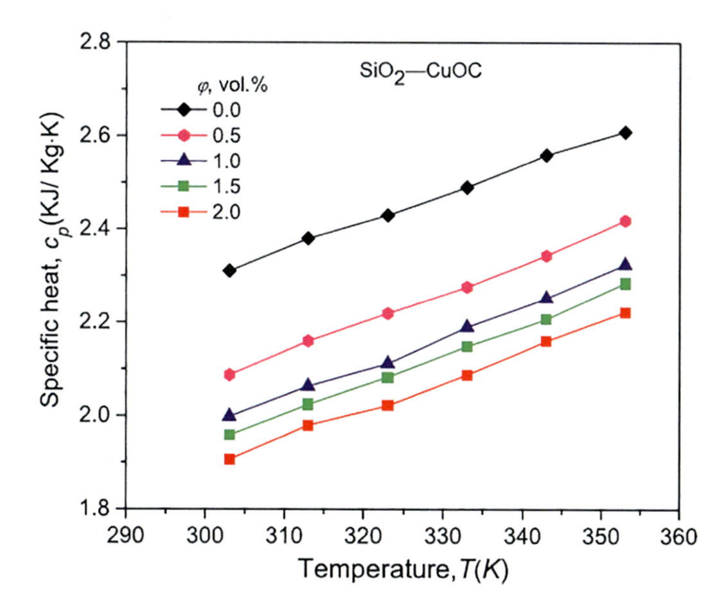

Fig. 3.15 Specific heat of SiO_2-CuO/C@60:40 hybrid nanofluid. Copyright Elsevier 2021 (Lic#5087520150540) (S. Akilu, et al., Properties of glycerol and ethylene glycol mixture based SiO2-CuO/C hybrid nanofluid for enhanced solar energy transport, Solar Energy Mater. Solar Cells 179 (2018) 118–128).

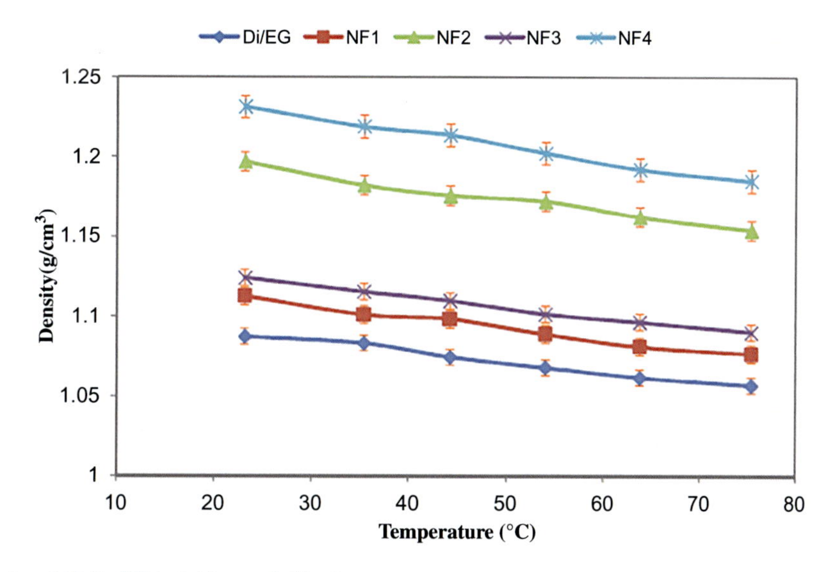

Fig. 3.16 Density of Al_2O_3-SiC hybrid nanofluids. Copyright Elsevier 2021 (Lic#5087520614075) (S. Ramalingam, R. Dhairiyasamy, M. Govindasamy, Assessment of heat transfer characteristics and system physiognomies using hybrid nanofluids in an automotive radiator, Chem. Eng. Process. Process Intens. 150 (2020) 107886).

Table 3.3 Summary of the recent advancements on the specific heat of hybrid nanofluids.

Researchers	Hybrid nanoparticles	Base fluid	Parameters	Remarks
Wole-Osho et al. [70]	Al_2O_3-ZnO	Water	0.33, 0.67, 1, 1.33, and 1.67 vol.%	• Specific heat reduces with increasing particle concentration • Maximum specific heat decrement of 30.12% at 1.67 vol.% and 25°C
Minea et al. [76]	Al_2O_3-TiO_2 Al_2O_3-SiO_2	Water	0.10, 0.015, 0.02, and 0.30 vol.%	• Specific heat declines with increased volume fraction • Al_2O_3-SiO_2 hybrid nanofluids show higher specific heat than Al_2O_3-TiO_2
Said et al. [41]	rGO-Co_3O_4	Water	0.05, 0.10, and 0.20 vol.%; 20–60°C	• At 0.2 wt.%, the maximum reduced specific heat was 0.192% and 0.194% at 20°C and 60°C
Rajan et al. [77]	ZnO-paraffin wax	Water-propylene glycol	ZnO: 0–2 vol.%; paraffin wax: 4–16 wt.%	• Specific heat was improved to a maximum of 18.7% as compared to the base fluid
Verma et al. [78]	CuO-MWCNTs MgO-MWCNTs	Water	0.25–2 vol.%	• Specific heat reduces with volume fraction and enhances with temperature • Specific heat decrement is reported maximum for MgO hybrid nanofluid of 10.51%
Kumar et al. [79]	Al_2O_3-MWCNT; TiO_2-MWCNT; ZnO-MWCNT; CeO_2-MWCNT	Water	0.25–2 vol.%; 35°C	• Specific heat reduces with an increment in volume concentration • Al_2O_3-MWCNT/H_2O hybrid nanofluids show the maximum value of specific heat • The specific heat reduces by 2.12%, 2.31%, 1.39%, and 2.88% for Al_2O_3-MWCNT, TiO_2-MWCNT, ZnO-MWCNT, and CeO_2-MWCNT H_2O-based hybrid nanofluids, respectively

Table 3.4 Summary of the recent advancements on the density of hybrid nanofluids.

Researchers	Hybrid nanoparticles	Base fluid	Parameters	Remarks
Said et al. [41]	rGO-Co_3O_4	Water	0.05, 0.10, and 0.20 wt. %; 20–60°C	• At 0.05, 0.1, and 0.2 wt.%, the density improved by 0.115%, 0.23%, and 0.451% at 20°C • It further improved by 0.117%, 0.235%, and 0.469% at 60°C
Kumar et al. [68]	ZnO-MgO	Water and EG	0–0.1 vol.%; 30–80°C	• The influence of particle loadings and temperatures. • Density is augmented with higher loadings and temperatures.
Kishore et al. [80]	Cu-graphene	Distilled water	0.01 and 0.02 vol.%; 45–100°C	• The density of hybrid nanofluid increased due to more particle dilution • The density for 0.02 vol.%, 70:30 Cu: GnP/H_2O hybrid nanofluid is increased by 12.5% as compared with base fluid water
Ghosh et al. [79]	Al_2O_3-MWCNT; TiO_2-MWCNT; ZnO-MWCNT; CeO_2-MWCNT	Water	0.25–2 vol.%; 35°C	• The CeO_2+MWCNT/H_2O hybrid nanofluid offers maximum density than the other hybrid nanofluids • The density increases with an increment in volume fraction • The density raised sequence for Al_2O_3+MWCNT is 3.19%, TiO_2+MWCNT is 2.77%, ZnO +MWCNT is 2.97%, and CeO_2+MWCNT is 2.76% water-based hybrid nanofluids
Verma et al. [78]	CuO-MWCNTs MgO-MWCNTs	Water	0.25–2 vol.%	• MgO-MWCNTs show better density but are higher than MgO mono nanofluids • It increases with increasing volume concentration

3.1.4 Magnetic property

The magnetic analysis of the hybrid nanofluids is very important when they are used in a specific application. For the controlled momentum and thermal boundary layer, the magnetic dipole plays an important role. The magnetic field is also another parameter for heat transfer and fluid flow. Some of the literature related to magnetic property is presented here. Gul et al. [81] studied a magnetic dipole moment of Cu/H_2O, $Cu\text{-}Al_2O_3/H_2O$, and $Cu\text{-}Al_2O_3/H_2O$ hybrid nanofluids under laminar flow based on the ordinary differential equation (ODE) by using Runge Kutta 4th order technique. The effect of physical restraints like volume concentration, viscous dissipation, and Prandtl number is analyzed. They observed that the hybrid nanofluid is more effective in thermal conduction because of its powerful thermal properties than mono nanofluid and noticed that the turbulence of fluid flow can be monitored during magnetic dipole.

Aminian et al. [82] have studied the MHD effects of Al_2O_3-CuO/water nanofluid inside a partition cylinder by using the Darcy-Brinkman-Forchheimer equation. Their outcomes are higher Hartmann and Darcy numbers at the augmented heat transfer coefficients. Alharbi [83] studied the MHD (CuO-Al_2O_3) hybrid nanofluid flow, laminar flow, boundary layer, and heat transfer by mixed convection nearby a vertical permeable plate with magnetic induction. The conduction-radiation with magnetic Prandtl number is also assessed. The expressions that govern the analysis are basic using a boundary layer technique and relevant dimensionless variables. They observed that the induced magnetic field has demonstrated declining behavior when Hartmann and magnetic Reynolds numbers improved. Li et al. [84] studied the convective heat transfer of Fe_3O_4-carbon nanotube (CNT) hybrid nanofluid in a horizontal small circular tube under the influence of annular magnets. The experiments were conducted in volume concentration values of 0.4%–1.2% and Reynolds number in the range of 476–996. They observed that, at 1.44 vol.% of Fe_3O_4-CNTs-water nanofluid, the local Nusselt number is increased by 61.54% at Reynolds number of 996.

3.1.5 Dielectric property

A dielectric is an electrical insulator that can be divided with an applied electric field. If the dielectric material is influenced by an electric field, the electric charge does not pass through the material, because they are an electrical conductor, but their equilibrium position is shifted due to the dielectric polarization. Dielectrics are vital to defining several features in electronics, optics, solid-state physics, and cell biophysics.

Qing et al. [85] prepared hybrid SiO_2-graphene nanofluid in naphthenic oil and investigated thermal conductivity and dielectric property at 0.01, 0.04, and 0.08 wt.% loadings and between 20° C and 100°C. They observed that with the availability of SiO_2 on graphene, k_{nf} is increased to 80% and electrical conductivity decreased from 557% to 97%.

Mansour et al. [86] have studied the degradation of dielectric properties of transformer oil-based metal oxide (MO) nanoparticles: titania, alumina, and silica. The thermal and dielectric properties were measured to investigate the behavior of nanoparticle hybridization on transformer oil characteristics. Hybrid nanofluids using titania nanoparticles provided a superior composition concerning either BDV or heat transfer coefficient. Asadikia et al. [87] investigated the thermal and electrical features of ethylene glycol/water-based multiwalled carbon nanotube (MWCNT) and Fe_2O_3 nanofluids and observed the highest electrical conductivity (545 μS/cm) at 0.69 (%, w/v) of MWCNT and 1.67 (%, w/v) of Fe_2O_3 nanoparticles dispersed in a 44:56 EG-W mixture. Fal et al. [88] experimentally investigated the dielectric behavior of silicon oxide lignin (SiO_2-L) particles dispersed with various particle loadings in EG, and the measurements were performed from 298.15 to 333.15 K. Their results indicated that the addition of even a small amount of SiO_2-L nanoparticles to ethylene glycol increases permittivity and alternating current (AC) conductivity as well as DC conductivity, while relaxation time reduces.

3.2 Conclusion

The thermophysical properties of hybrid nanofluids such as thermal conductivity, viscosity, specific heat, density, and electric and dielectric properties are important before the hybrid nanofluids are used in a particular device. The properties may be estimated experimentally and theoretically. Various instruments are available, and also various theories are available to study the thermophysical properties. With the dilution of hybrid nanoparticles in the base fluid, thermal conductivity, viscosity, and density are increased simultaneously; the specific heat is decreased with increasing concentration and temperature. With the above study, the properties mainly depend on particle concentration and temperature.

Acknowledgments

The authors L.S.S. (Ref. 045-88-ARH/2018; CEEC 2018 Individual) and E.V.R. (Ref. 032-88-ARH/2018) acknowledge Fundaç˜ao para a Ci ̂encia e a Tecnologia (Portugal) for its funding of their positions, and the work was funded by projects UIDB/00481/2020, UIDP/00481/2020, and infrastructure support of CENTRO-01-0145-FEDER-022083.

References

[1] P.K. Kanti, et al., Experimental investigation on thermal conductivity of fly ash nanofluid and fly ash-Cu hybrid nanofluid: prediction and optimization via ANN and MGGP model, Partic. Sci. Technol. (2021) 1–14.
[2] L. Sundar, et al., Heat transfer of rGO/Co3O4 hybrid nanomaterial based nanofluids and twisted tape configurations in a tube, J. Therm. Sci. Eng. Appl. (2020) 1–41.
[3] P.K. Kanti, et al., Experimental investigation on thermo-hydraulic performance of water-based fly ash–Cu hybrid nanofluid flow in a pipe at various inlet fluid temperatures, Int. Commun. Heat Mass Transfer 124 (2021) 105238.
[4] Z. Said, et al., Recent advances on nanofluids for low to medium temperature solar collectors: energy, exergy, economic analysis and environmental impact, Prog. Energy Combust. Sci. 84 (2021) 100898.
[5] L.S. Sundar, et al., Experimental investigation of thermo-physical properties, heat transfer, pumping power, entropy generation, and exergy efficiency of nanodiamond+ Fe3O4/60: 40% water-ethylene glycol hybrid nanofluid flow in a tube, Therm. Sci. Eng. Prog. 21 (2021) 100799.
[6] L.S. Sundar, et al., Enhanced thermal conductivity and viscosity of nanodiamond-nickel nanocomposite nanofluids, Sci. Rep. 4 (1) (2014) 1–14.
[7] Z. Said, et al., Heat transfer, entropy generation, economic and environmental analyses of linear Fresnel reflector using novel rGO-Co3O4 hybrid nanofluids, Renew. Energy 165 (2021) 420–437.
[8] A.K. Tiwar, et al., A review on the application of hybrid nanofluids for parabolic trough collector: recent progress and outlook, J. Clean. Prod. (2021) 126031.
[9] L.S. Sundar, et al., Combination of Co3O4 deposited rGO hybrid nanofluids and longitudinal strip inserts: thermal properties, heat transfer, friction factor, and thermal performance evaluations, Therm. Sci. Eng. Prog. 20 (2020) 100695.
[10] A.A. Hachicha, et al., On the thermal and thermodynamic analysis of parabolic trough collector technology using industrial-grade MWCNT based nanofluid, Renew. Energy 161 (2020) 1303–1317.
[11] M. Baghbanzadeh, et al., Synthesis of spherical silica/multiwall carbon nanotubes hybrid nanostructures and investigation of thermal conductivity of related nanofluids, Thermochim. Acta 549 (2012) 87–94.
[12] M.J. Nine, et al., Highly productive synthesis process of well dispersed Cu2O and Cu/Cu2O nanoparticles and its thermal characterization, Mater. Chem. Phys. 141 (2–3) (2013) 636–642.
[13] S. Jana, A. Salehi-Khojin, W.-H. Zhong, Enhancement of fluid thermal conductivity by the addition of single and hybrid nano-additives, Thermochim. Acta 462 (1-2) (2007) 45–55.
[14] S. Suresh, et al., Synthesis of Al2O3–Cu/water hybrid nanofluids using two step method and its thermo physical properties, Colloids Surf. A Physicochem. Eng. Asp. 388 (1–3) (2011) 41–48.

[15] S. Jyothirmayee Aravind, S. Ramaprabhu, Graphene wrapped multiwalled carbon nanotubes dispersed nanofluids for heat transfer applications, J. Appl. Phys. 112 (12) (2012), 124304.
[16] S.J. Aravind, S. Ramaprabhu, Graphene-multiwalled carbon nanotube-based nanofluids for improved heat dissipation, RSC Adv. 3 (13) (2013) 4199–4206.
[17] T. Theres Baby, R. Sundara, Synthesis of silver nanoparticle decorated multiwalled carbon nanotubes-graphene mixture and its heat transfer studies in nanofluid, AIP Adv. 3 (1) (2013), 012111.
[18] T.T. Baby, S. Ramaprabhu, Investigation of thermal and electrical conductivity of graphene based nanofluids, J. Appl. Phys. 108 (12) (2010) 124308.
[19] L.S. Sundar, M.K. Singh, A.C. Sousa, Enhanced heat transfer and friction factor of MWCNT–Fe3O4/water hybrid nanofluids, Int. Commun. Heat Mass Transfer 52 (2014) 73–83.
[20] N.N. Esfahani, D. Toghraie, M. Afrand, A new correlation for predicting the thermal conductivity of ZnO–Ag (50%–50%)/water hybrid nanofluid: an experimental study, Powder Technol. 323 (2018) 367–373.
[21] M.H. Esfe, et al., Experimental determination of thermal conductivity and dynamic viscosity of Ag–MgO/water hybrid nanofluid, Int. Commun. Heat Mass Transfer 66 (2015) 189–195.
[22] M.H. Esfe, et al., Thermal conductivity of Cu/TiO2–water/EG hybrid nanofluid: experimental data and modeling using artificial neural network and correlation, Int. Commun. Heat Mass Transfer 66 (2015) 100–104.
[23] M.H. Esfe, et al., Experimental study on thermal conductivity of DWCNT-ZnO/water-EG nanofluids, Int. Commun. Heat Mass Transfer 68 (2015) 248–251.
[24] L. Chen, W. Yu, H. Xie, Enhanced thermal conductivity of nanofluids containing Ag/MWNT composites, Powder Technol. 231 (2012) 18–20.
[25] M. Batmunkh, et al., Thermal conductivity of TiO2 nanoparticles based aqueous nanofluids with an addition of a modified silver particle, Indus. Eng. Chem. Res. 53 (20) (2014) 8445–8451.
[26] H. Yarmand, et al., Graphene nanoplatelets–silver hybrid nanofluids for enhanced heat transfer, Energ. Conver. Manage. 100 (2015) 419–428.
[27] N. Jha, S. Ramaprabhu, Synthesis and thermal conductivity of copper nanoparticle decorated multiwalled carbon nanotubes based nanofluids, J. Phys. Chem. C 112 (25) (2008) 9315–9319.
[28] L.S. Sundar, et al., Nanodiamond-Fe3O4 nanofluids: preparation and measurement of viscosity, electrical and thermal conductivities, Int. Commun. Heat Mass Transfer 73 (2016) 62–74.
[29] S.S. Botha, P. Ndungu, B.J. Bladergroen, Physicochemical properties of oil-based nanofluids containing hybrid structures of silver nanoparticles supported on silica, Indus. Eng. Chem. Res. 50 (6) (2011) 3071–3077.
[30] M. Farbod, A. Ahangarpour, Improved thermal conductivity of Ag decorated carbon nanotubes water based nanofluids, Phys. Lett. A 380 (48) (2016) 4044–4048.
[31] L.S. Sundar, et al., Thermal conductivity and viscosity of hybrid nanfluids prepared with magnetic nanodiamond-cobalt oxide (ND-Co3O4) nanocomposite, Case Stud. Therm. Eng. 7 (2016) 66–77.
[32] A.A. Minea, Hybrid nanofluids based on Al2O3, TiO2 and SiO2: numerical evaluation of different approaches, Int. J. Heat Mass Transfer 104 (2017) 852–860.
[33] S.S. Harandi, et al., An experimental study on thermal conductivity of F-MWCNTs–Fe3O4/EG hybrid nanofluid: effects of temperature and concentration, Int. Commun. Heat Mass Transfer 76 (2016) 171–177.
[34] S. Akilu, A.T. Baheta, K. Sharma, Experimental measurements of thermal conductivity and viscosity of ethylene glycol-based hybrid nanofluid with TiO2-CuO/C inclusions, J. Mol. Liq. 246 (2017) 396–405.

[35] S. Akilu, et al., Properties of glycerol and ethylene glycol mixture based SiO2-CuO/C hybrid nanofluid for enhanced solar energy transport, Solar Energy Mater. Solar Cells 179 (2018) 118–128.
[36] L.R. de Oliveira, et al., Experimental study on the thermal conductivity and viscosity of ethylene glycol-based nanofluid containing diamond-silver hybrid material, Diamond Relat. Mater. 96 (2019) 216–230.
[37] C. Jin, et al., Investigation on hybrid nanofluids based on carbon nanotubes filled with metal nanoparticles: stability, thermal conductivity, and viscosity, Powder Technol. 389 (2021) 1–10.
[38] R. Bakhtiari, et al., Preparation of stable TiO2-graphene/water hybrid nanofluids and development of a new correlation for thermal conductivity, Powder Technol. 385 (2021) 466–477.
[39] A. Dezfulizadeh, et al., An experimental study on dynamic viscosity and thermal conductivity of water-Cu-SiO2-MWCNT ternary hybrid nanofluid and the development of practical correlations, Powder Technol. 389 (2021) 215–234.
[40] A.K. Tiwari, et al., 4S consideration (synthesis, sonication, surfactant, stability) for the thermal conductivity of CeO2 with MWCNT and water based hybrid nanofluid: an experimental assessment, Colloids Surf. A Physicochem. Eng. Asp. 610 (2021) 125918.
[41] Z. Said, et al., Optimizing density, dynamic viscosity, thermal conductivity and specific heat of a hybrid nanofluid obtained experimentally via ANFIS-based model and modern optimization, J. Mol. Liq. 321 (2021) 114287.
[42] N. Senniangiri, et al., Effects of temperature and particles concentration on the thermal conductivity of graphene-NiO/coconut oil hybrid nanofuids, Mater. Today Proc. (2021).
[43] M. Ma, et al., Effect of surfactant on the rheological behavior and thermophysical properties of hybrid nanofluids, Powder Technol. 379 (2021) 373–383.
[44] V.V. Wanatasanapan, M.Z. Abdullah, P. Gunnasegaran, Effect of TiO2-Al2O3 nanoparticle mixing ratio on the thermal conductivity, rheological properties, and dynamic viscosity of water-based hybrid nanofluid, J. Mater. Res. Technol. 9 (6) (2020) 13781–13792.
[45] R. Pourrajab, et al., Investigation of thermal conductivity of a new hybrid nanofluids based on mesoporous silica modified with copper nanoparticles: synthesis, characterization and experimental study, J. Mol. Liq. 300 (2020) 112337.
[46] F. Soltani, D. Toghraie, A. Karimipour, Experimental measurements of thermal conductivity of engine oil-based hybrid and mono nanofluids with tungsten oxide (WO3) and MWCNTs inclusions, Powder Technol. 371 (2020) 37–44.
[47] Z. Aparna, et al., Thermal conductivity of aqueous Al2O3/Ag hybrid nanofluid at different temperatures and volume concentrations: an experimental investigation and development of new correlation function, Powder Technol. 343 (2019) 714–722.
[48] J. Singh, et al., Thermal conductivity analysis of GO-CuO/DW hybrid nanofluid, Mater. Today Proc. 28 (2020) 1714–1718.
[49] A. Boroomandpour, D. Toghraie, M. Hashemian, A comprehensive experimental investigation of thermal conductivity of a ternary hybrid nanofluid containing MWCNTs-titania-zinc oxide/water-ethylene glycol (80:20) as well as binary and mono nanofluids, Synth. Met. 268 (2020) 116501.
[50] N.K. Cakmak, et al., Preparation, characterization, stability, and thermal conductivity of rGO-Fe3O4-TiO2 hybrid nanofluid: an experimental study, Powder Technol. 372 (2020) 235–245.
[51] M. Gupta, V. Singh, Z. Said, Heat transfer analysis using zinc Ferrite/water (Hybrid) nanofluids in a circular tube: an experimental investigation and development of new correlations for thermophysical and heat transfer properties, Sustain. Energy Technol. Assess. 39 (2020) 100720.

[52] A.A. Hachicha, et al., A review study on the modeling of high-temperature solar thermal collector systems, Renew. Sustain. Energy Rev. 112 (2019) 280–298.
[53] O. Soltani, M. Akbari, Effects of temperature and particles concentration on the dynamic viscosity of MgO-MWCNT/ethylene glycol hybrid nanofluid: experimental study, Physica E 84 (2016) 564–570.
[54] M. Baghbanzadeh, et al., Investigating the rheological properties of nanofluids of water/hybrid nanostructure of spherical silica/MWCNT, Thermochim. Acta 578 (2014) 53–58.
[55] M. Bahrami, et al., An experimental study on rheological behavior of hybrid nanofluids made of iron and copper oxide in a binary mixture of water and ethylene glycol: non-Newtonian behavior, Exp. Therm. Fluid Sci. 79 (2016) 231–237.
[56] M. Afrand, K.N. Najafabadi, M. Akbari, Effects of temperature and solid volume fraction on viscosity of SiO2-MWCNTs/SAE40 hybrid nanofluid as a coolant and lubricant in heat engines, Appl. Therm. Eng. 102 (2016) 45–54.
[57] E. Dardan, M. Afrand, A.M. Isfahani, Effect of suspending hybrid nano-additives on rheological behavior of engine oil and pumping power, Appl. Therm. Eng. 109 (2016) 524–534.
[58] M.H. Esfe, et al., Effects of temperature and concentration on rheological behavior of MWCNTs/SiO2 (20–80)-SAE40 hybrid nano-lubricant, Int. Commun. Heat Mass Transfer 76 (2016) 133–138.
[59] M. Asadi, A. Asadi, Dynamic viscosity of MWCNT/ZnO–engine oil hybrid nanofluid: an experimental investigation and new correlation in different temperatures and solid concentrations, Int. Commun. Heat Mass Transfer 76 (2016) 41–45.
[60] M. Afrand, D. Toghraie, B. Ruhani, Effects of temperature and nanoparticles concentration on rheological behavior of Fe3O4–Ag/EG hybrid nanofluid: an experimental study, Exp. Therm. Fluid Sci. 77 (2016) 38–44.
[61] I. Kazemi, M. Sefid, M. Afrand, A novel comparative experimental study on rheological behavior of mono & hybrid nanofluids concerned graphene and silica nano-powders: characterization, stability and viscosity measurements, Powder Technol. 366 (2020) 216–229.
[62] G. Huminic, et al., Experimental study on viscosity of water based Fe–Si hybrid nanofluids, J. Mol. Liq. 321 (2021) 114938.
[63] Y. Zhu, et al., A comprehensive experimental investigation of dynamic viscosity of MWCNT-WO3/water-ethylene glycol antifreeze hybrid nanofluid, J. Mol. Liq. 333 (2021) 115986.
[64] A.K. Tiwari, et al., 3S (Sonication, surfactant, stability) impact on the viscosity of hybrid nanofluid with different base fluids: an experimental study, J. Mol. Liq. 329 (2021) 115455.
[65] M.-Y. Ma, et al., Particle size-dependent rheological behavior and mechanism of Al2O3-Cu/W hybrid nanofluids, J. Mol. Liq. 335 (2021) 116297.
[66] Y.-M. Chu, et al., Examining rheological behavior of MWCNT-TiO2/5W40 hybrid nanofluid based on experiments and RSM/ANN modeling, J. Mol. Liq. 333 (2021) 115969.
[67] X. Li, H. Wang, B. Luo, The thermophysical properties and enhanced heat transfer performance of SiC-MWCNTs hybrid nanofluids for car radiator system, Colloids Surf. A Physicochem. Eng. Asp. 612 (2021) 125968.
[68] R. Vidhya, T. Balakrishnan, B. Suresh Kumar, Investigation on thermophysical properties and heat transfer performance of heat pipe charged with binary mixture based ZnO-MgO hybrid nanofluids, Mater. Today Proc. 37 (2021) 3423–3433.
[69] M. Ehyaei, et al., Energy, exergy and economic analyses for the selection of working fluid and metal oxide nanofluids in a parabolic trough collector, Solar Energy 187 (2019) 175–184.

[70] I. Wole-Osho, et al., An experimental investigation into the effect of particle mixture ratio on specific heat capacity and dynamic viscosity of Al2O3-ZnO hybrid nanofluids, Powder Technol. 363 (2020) 699–716.
[71] A.K. Tiwari, et al., Experimental comparison of specific heat capacity of three different metal oxides with MWCNT/water-based hybrid nanofluids: proposing a new correlation, Appl. Nanosci. (2020) 1–11.
[72] E.C. Okonkwo, et al., Comparison of experimental and theoretical methods of obtaining the thermal properties of alumina/iron mono and hybrid nanofluids, J. Mol. Liq. 292 (2019) 111377.
[73] A.B. Çolak, et al., Experimental study for predicting the specific heat of water based Cu-Al2O3 hybrid nanofluid using artificial neural network and proposing new correlation, Int. J. Energy Res. 44 (9) (2020) 7198–7215.
[74] A.B. Çolak, et al., Experimental study on the specific heat capacity measurement of water-based Al2O3-Cu hybrid nanofluid by using differential thermal analysis method, Curr. Nanosci. 16 (6) (2020) 912–928.
[75] S. Ramalingam, R. Dhairiyasamy, M. Govindasamy, Assessment of heat transfer characteristics and system physiognomies using hybrid nanofluids in an automotive radiator, Chem. Eng. Process.-Process Intens. 150 (2020) 107886.
[76] G.M. Moldoveanu, A.A. Minea, Specific heat experimental tests of simple and hybrid oxide-water nanofluids: proposing new correlation, J. Mol. Liq. 279 (2019) 299–305.
[77] M.N. Chandran, et al., Novel hybrid nanofluid with tunable specific heat and thermal conductivity: characterization and performance assessment for energy related applications, Energy 140 (2017) 27–39.
[78] S.K. Verma, et al., Performance analysis of hybrid nanofluids in flat plate solar collector as an advanced working fluid, Solar Energy 167 (2018) 231–241.
[79] V. Kumar, A.K. Tiwari, S.K. Ghosh, Exergy analysis of hybrid nanofluids with optimum concentration in a plate heat exchanger, Mater. Res. Exp. 5 (6) (2018), 065022.
[80] P.S. Kishore, et al., Preparation, characterization and thermo-physical properties of Cu-graphene nanoplatelets hybrid nanofluids, Mater. Today Proc. 27 (2020) 610–614.
[81] T. Gul, et al., Magnetic dipole impact on the hybrid nanofluid flow over an extending surface, Sci. Rep. 10 (1) (2020) 1–13.
[82] E. Aminian, H. Moghadasi, H. Saffari, Magnetic field effects on forced convection flow of a hybrid nanofluid in a cylinder filled with porous media: a numerical study, J. Therm. Anal. Calorim. (2020) 1–13.
[83] S.O. Alharbi, Impact of hybrid nanoparticles on transport mechanism in magnetohydrodynamic fluid flow exposed to induced magnetic field, Ain Shams Eng. J. 12 (1) (2021) 995–1000.
[84] G. Li, et al., Improvement of cooling performance of hybrid nanofluids in a heated pipe applying annular magnets, J. Therm. Anal. Calorim. (2021) 1–19.
[85] S.H. Qing, et al., Thermal conductivity and electrical properties of hybrid SiO2-graphene naphthenic mineral oil nanofluid as potential transformer oil, Mater. Res. Exp. 4 (1) (2017), 015504.
[86] D.-E.A. Mansour, et al., Multiple nanoparticles for improvement of thermal and dielectric properties of oil nanofluids, IET Sci. Meas. Technol. 13 (7) (2019) 968–974.
[87] A. Asadikia, et al., Characterization of thermal and electrical properties of hybrid nanofluids prepared with multi-walled carbon nanotubes and Fe2O3 nanoparticles, Int. Commun. Heat Mass Transfer 117 (2020) 104603.
[88] J. Fal, et al., Electrical conductivity and dielectric properties of ethylene glycol-based nanofluids containing silicon oxide–lignin hybrid particles, Nanomaterials 9 (7) (2019) 1008.

4

Hydrothermal properties of hybrid nanofluids

L. Syam Sundar[a], E. Venkata Ramana[b], Zafar Said[c,d,e,*], and Antonio C.M. Sousa[a]

[a]Centre for Mechanical Technology and Automation (TEMA–UA), Department of Mechanical Engineering, University of Aveiro, Aveiro, Portugal. [b]I3N, Department of Physics, University of Aveiro, Aveiro, Portugal. [c]Department of Sustainable and Renewable Energy Engineering, University of Sharjah, Sharjah, United Arab Emirates. [d]Research Institute for Sciences and Engineering, University of Sharjah, Sharjah, United Arab Emirates. [e]U.S.-Pakistan Center for Advanced Studies in Energy (USPCAS-E), National University of Sciences and Technology (NUST), Islamabad, Pakistan

**Corresponding author: zsaid@sharjah.ac.ae, zaffar.ks@gmail.com*

Chapter outline

4.1 Introduction 93
4.2 Surface tension 94
4.3 Friction factor 96
4.4 Pressure drop 99
4.5 Pumping power 101
4.6 Fouling factor of nanofluid 101
4.7 Conclusions and challenges 105
Acknowledgments 106
References 106

4.1 Introduction

Hybrid nanofluids are a novel type of nanofluids, synthesized by suspending two or more different types of nanoparticles in single-phase fluid [1–3]. Homogeneous properties can be expected with hybrid nanoparticles [4, 5] and they also possess exceptional physicochemical properties [6]. Hybrid nanofluids are believed to replace mono nanofluids due to their wide absorption range, lower extinction, remarkable thermal conductivity, minimal pressure drop, lower frictional losses, and pumping

Hybrid Nanofluids: Preparation, Characterization and Applications. https://doi.org/10.1016/B978-0-323-85836-6.00004-1

power compared with mono nanofluids [7, 8]. The hybrid nanoparticles consist of Al_2O_3-Cu, Al_2O_3-Ni, MgO-Fe, CNT-Al, ND-Ni, Al_2O_3-CNT, Al_2O_3-TiO_2, CNT-Fe_3O_4, MWCNT-Fe_3O_4, and so on [9].

The synthesis of hybrid nanofluids is the crucial phase in using nanoparticles to enhance the thermal conductivity of standard heat transfer fluids [10, 11]. Literature investigations worked on properties like thermal conductivity, viscosity, specific heat, and density of hybrid nanofluids, but studies on hydrothermal properties are very limited [12, 13]. The objective of employing hybrid nanofluids is to further enhance heat transfer and pressure drop characteristics by compromising between merits and demerits of suspension, better aspect ratio, superior thermal system, and synergetic force of nanomaterials. The prepared hybrid nanofluids [14, 15] possess augmented k_{nf} values due to their synergistic characteristics. Hybrid nanofluids contribute enhancement in heat transfer, which results in reduced cost and increased efficiency [16]. With the employment of hybrid nanofluids in a heat transfer, components will influence the decrease in cost, size, and material [17]. The pumping power of the system with the use of nanofluid in a thermal component is an influencing factor of the system [18, 19].

The increased friction factor outside the specific limit may need pumping power that will balance the advantages of improved heat transfer [20]. Hybrid nanofluids should guarantee a slight increment or declined friction factor and an improved heat transfer coefficient [21]. These properties depend on the volume concentration of nanofluids and Reynolds number. Increasing volume concentration will have a significant impact on the pumping power, friction factor, and pressure drop [22]. It is highly desirable to possess minimal pressure drop losses and pumping power requirements to achieve better thermal systems' performance [23].

The synergistic hydrothermal properties of hybrid nanofluids are deciding factor for future investigations [15, 24]. There is a wide range of applications with hybrid nanofluids especially in engine cooling, power industries, and electronic cooling [25–27]. The hybrid nanofluids are aspirations to execute these applications with exceptional performance. The hybrid nanofluids are quite a novel class of nanofluids and are still in the research and development phase. The increased friction factor for hybrid nanofluids may be important for the feasibility of commercial applications.

4.2 Surface tension

The definition of surface tension says that it is the tendency of the liquid surfaces at rest to shrink, which is larger than water, and it is a salient factor in the capillarity phenomenon, and it has a unit of N/m. Studies on the surface tension property of nanofluids are also important and described in this section. Tanvir and Qiao [28] have estimated the surface tension of ethanol and n-decane-based Al, Al_2O_3 boron, and MWCNT nanofluids based on the Young-Laplace equation. They observed that surface tension improves with particles' volume concentration and size due to Van der Waals force between particles at the liquid-gas interface and improves free energy and surface tension. However, nanoparticles have minimal impact on surface tension due to the considerable space between particles at low concentrations.

Chinnam et al. [29] measured the surface tension of nanofluids containing aluminum oxide (Al_2O_3), zinc oxide (ZnO), titanium dioxide (TiO_2), and silicon dioxide (SiO_2) nanoparticles in a 60:40 PG/H_2O base fluid at 30–70°C and from 0 to 6 vol% and 15–50 nm particle size. Their results indicate that the surface tension of nanofluids reduced with increasing temperature. Surface tension reduces with increasing volume fraction at a constant temperature. At constant concentration and temperature, the surface tension for smaller particle size is lower except for ZnO nanofluid.

Huminic et al. [30] have measured the surface tension of the FeC/H_2O nanofluids in the temperature range from 10°C to 70°C and for 0.1, 0.5, and 1.0 wt%. They observe enhanced surface tension of the nanofluid by increased particle loadings. Bhuiyan et al. [31] studied the surface tension of water-based Al_2O_3(13 and 50 nm), TiO_2 (21 nm), and SiO_2 (5–15 nm and 10–20 nm) nanofluids using the traditional Du-Noüy ring method. They observed that the surface tension of the nanofluids enhances from 2.62% to 4.82% in comparison with the base fluids for concentration variation of 0.05% to 0.25% at 25°C. Khaleduzzaman et al. [32] have noticed that the surface tension of nanofluids reduces with the increment in surfactant concentration and temperature. Still, there are a few contrary outcomes on the impact of particle concentrations and surfactants on the surface tension of nanofluids.

Michał Wanic et al. [33] have focused on the estimation of surface tension of ethylene glycol-based aluminum nitride, silicon nitride, and titanium nitride nanofluid in the mass concentration between 1% and 5%, and they measured at 298.15 K based on the du Noüy ring and pendant drop technique. They observed that the surface tension of the nanofluids enhances with an increase of particle loadings. Kumar and Milanova [34] explained that the heat transfer of single-walled carbon nanotube (SWCNT) nanofluids in a hot atmosphere can encompass the saturated boiling management and they observed that critical heat flux is approximately four times of 1:5 surfactant-to-CNT concentration ratio and they also observed that the material burnout is a strong function of the relaxation of the nanofluid surface tension with the base fluid.

Lu et al. [35] have examined the surface tension of water-based Ag nanofluids based on the molecular dynamics simulations and found that wettability is the reason for higher surface tension. Zhou et al. [36] have studied the surface tension of water-based Al_2O_3 nanofluid and found that the decreased surface tension of nanofluid with higher temperatures and lower particle loadings.

4.3 Friction factor

There are two ways to understand the fluid friction factor (f): (i) Darcy-Weisbach equation and (ii) Fanning friction factor. It is defined as the ratio between the local shear stress and the local flow kinetic energy density. This friction factor is one-fourth of the Darcy friction factor, so observation must be required from the "friction factor" chart or equation used. The friction factor can be estimated based on the below formula:

$$\text{Friction factor}, f = \frac{(\Delta P)}{\left(\frac{L}{d_i}\right)\left(\frac{\rho v^2}{2}\right)} \tag{4.1}$$

where mass flow rate,

$$\dot{m} = \rho A v \Rightarrow v = \frac{\dot{m}}{\rho A} \tag{4.2}$$

Aravind and Ramaprabhu [37] estimated the heat transfer coefficient and friction factor of sG-f-MWCNT hybrid nanofluids' flow in a tube and observed higher friction factors using nanofluids in a tube. Suresh et al. [6] studied the convective heat transfer and friction factor of Al_2O_3-Cu/H_2O hybrid nanofluid in a tube under laminar flow and noticed a higher friction factor for hybrid nanofluids and they also proposed a friction factor correction:

$$f = 26.44\, Re^{-0.8737}(1+\phi)^{156.23} \tag{4.3}$$

Sundar et al. [38] observed friction factor enhancement of 1.18-times for $\phi = 0.3\%$ of water-based MWCNT+Fe_3O_4 hybrid nanofluid flow in a tube at a Reynolds number of 22,000, and they also developed a friction factor correlation, which is shown below:

$$f = 0.3108\, Re^{-0.245}(1+\phi)^{0.42} \tag{4.4}$$

$$3000 < Re < 22000; 0 < \phi < 0.3\%$$

Yarmand et al. [39] have conducted heat transfer and friction factor experiments for GNP-Ag/water nanofluid in a tube under constant heat flux conditions in a turbulent flow. They observed a friction factor loss of 8% at $\phi = 0.1\%$ of GNP-Ag/H_2O nanofluid in a tube under the *Re* number = 17,500 as in comparison with water data:

$$f = 0.567322\,Re^{-0.285869}\phi^{0.02716} \tag{4.5}$$

Ahammed et al. [40] observed friction factor enhancement of 20.35% at 0.1 vol% of Al_2O_3-graphene/water nanofluid at 1000 *Re* number as compared to water. Madesh et al. [41] examined friction factor for Cu-TiO_2/water nanofluids in the Reynolds number range from 4000 to 8000, volume concentration range from 0.1% to 2.0%, and reported friction factor increment of 14.9% at 2.0 vol% of nanofluid. Maghadassi et al. [42] observed friction factor improvement of 15.53% at 0.1 vol% of Al_2O_3-Cu/water nanofluid at 2300 Reynolds number. Takabi and Shokouhmand [43] noticed friction factor loss of 13.76% at 2 vol% of Cu + Al_2O_3/H_2O nanofluid at a Reynolds number of 2300, compared with the water values.

Sundar et al. [44] estimated the friction factor of water-based nanodiamond-Ni nanofluids that circulate in a tube under turbulent flow and obtained *f* penalty of 1.07-times and 1.12-times for 0.1% and 0.3% volume concentrations at a Reynolds number of 22,000 compared with water. They also proposed a friction factor correlation, which is given below:

$$f = 0.295\,Re^{-0.241}(1+\phi)^{0.3097} \tag{4.6}$$

$$3000 < Re < 22000; 0 < \phi < 0.3\%$$

Ramadhan et al. [45] have estimated the friction factor of TiO_2 + -SiO_2-W/EG in the plain tube under fixed heat flux condition numerically and experimentally at 1.0, 2.0, and 3.0 vol% and the Reynolds number (*Re*) of 2900–11,200 and observed friction factor enhancement of 4.1, 3.8, and 3.5% at 1.0, 2.0, and 3.0 vol%, respectively.

Sundar et al. [5] have analyzed friction factor for ND-Fe_3O_4/60:40% water-EG hybrid nanofluid (Fig. 4.1) and noticed a maximum *f* penalty of 6.63% and 12.80% for 0.2 vol% and at $Re = 2105$ and 7502 over base fluid and also they proposed a friction factor correlation.

$$f = 0.2432\,Re^{-0.2169}(1+\phi)^{0.3874} \tag{4.7}$$

$$2000 < Re < 8500; 0 < \phi < 0.2\%$$

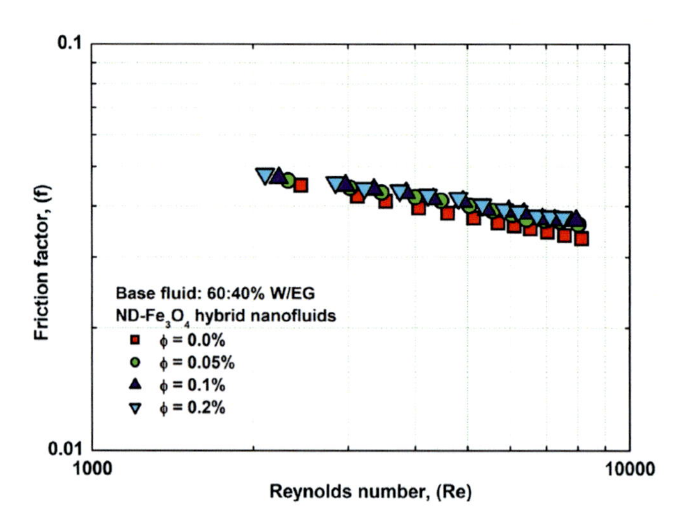

Fig. 4.1 Friction factor of 60 W and 40EG ND-Fe_3O_4 nanofluid. From *L.S. Sundar, et al., Experimental investigation of thermo-physical properties, heat transfer, pumping power, entropy generation, and exergy efficiency of nanodiamond+ Fe3O4/60: 40% water-ethylene glycol hybrid nanofluid flow in a tube, Therm. Sci. Eng. Prog. 21 (2021) 100799. Copyright Elsevier 2021 (Lic#5087580504436).*

Sundar et al. [12] analyzed f of GO/Co_3O_4@H_2O hybrid nanofluid in a tube (Fig. 4.2) and found a maximum f increase of 8.1% and 11.1% for observed for ϕ =0.2% and at Re equal to 1949 and 13,921 than the water data.

Saleh and Sundar [46] analyzed the friction factor of 60EG and 40 W based nanodiamond + Fe_3O_4 hybrid nanofluid and observed increased friction factors of 4.15% 7.01%, and 9.66% at 0.05%, 0.1%, and 0.2% vol. concentrations and at Reynolds numbers of 2204, 2105, and 2047. They also observed further friction factor rise of 9.44%, 12.42%, and 15.11% at 0.05%, 0.1%, and 0.2% vol. concentrations and at Re of 7621, 7421, and 7218 over the water.

$$f = 0.246\,Re^{-0.2178}(1+\phi)^{0.4926} \tag{4.8}$$

$$2000 < Re < 7925; 0 < \phi < 0.2\%$$

Nabil et al. [47] conducted friction factor experiments for 60 W and 40EG TiO_2-SiO_2 nanofluid and observed that at 3.0% vol. concentration, the friction factor increase was about 7%–8% over the base fluid. Sundar et al. [48] have found the friction factor increase of 4.77%, 7.94%, and 9.34% at 0.05%, 0.1%, and 0.2% weight concentration of rGO-Fe_3O_4-TiO_2/EG nanofluid in a tube

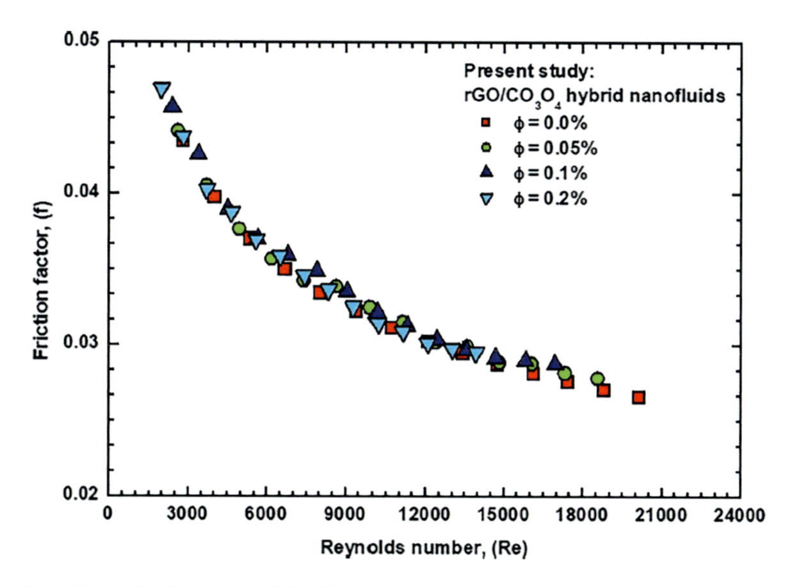

Fig. 4.2 Friction factor of GO/Co_3O_4@water hybrid nanofluid. From *L.S. Sundar, et al., Combination of Co3O4 deposited rGO hybrid nanofluids and longitudinal strip inserts: thermal properties, heat transfer, friction factor, and thermal performance evaluations, Therm. Sci. Eng. Prog. 20 (2020) 100695. Copyright Elsevier 2021 (Lic# 5087580805498).*

at *Re* of 1823, 1651, and 1548 over water, and they also proposed a friction factor correlation:

$$f = 69.42\, Re^{-1.018}(1+\phi)^{-1.367} \tag{4.9}$$

$$221.1 < Re < 2200; 0 < \phi < 0.2\,\text{wt.}\%$$

4.4 Pressure drop

When we use the hybrid nanofluids in a mechanical component tube, it is necessary to know their pressure drop. The below expression is generally used to estimate the pressure drop:

$$\text{Pressure drop, } \Delta p = \rho_{nf} g H \tag{4.9}$$

Gupta et al. [15] have found the pressure drop rise of 72%, 70.85%, 64.8%, and 52% at the 0.5 weight percentage of $ZnFe_2O_4$/water nanofluid at Reynolds numbers of 2200, 1875, 1500, and 1275, respectively, compared with water. Kumar and Sarkar [49] studied the pressure drop characteristics of Al_2O_3-TiO_2 hybrid nanofluids at 20°C, 30°C, and 40°C, and they observed the highest pressure drop of 306.2 N/m^2 for Al_2O_3/H_2O nanofluid at 0.5 LPM and an increased pressure drop of 10.86% than the base fluid for TiO_2 nanofluids at 20°C for 0.5 LPM. Kaska et al. [50] analyzed pressure drop of aluminum nitride (AlN) and alumina (Al_2O_3) nanofluid using the CFD analysis and observed a pressure drop increase of 14%.

Sundar et al. [5] analyzed pressure drop of 60 W and 40EG ND-Fe_3O_4 nanofluid (Fig. 4.3) and observed the highest pressure drop increment of 104.41% and 151.06% at $\phi = 0.2\%$ and *Re* equal to 2105 and 7520 over the water. Hussien et al. [51] have found the enhanced pumping power with the use of Al_2O_3 + graphene hybrid nanofluid in three sizes of tubes, such as 2.1, 1.1, and 0.8 mm over Al_2O_3/water.

Sundar et al. [48] analyzed the pressure drop of rGO-Fe_3O_4-TiO_2/EG and observed 10%, 12%, and 13.6% rise in pressure drop at 0.05%, 0.1%, and 0.2 wt% at *Re* equal to 260.5, 235.9, and 221.1 compared with the base fluid. They also observed further pressure drop increase of 4.11%, 6.87%, and 7.11% at 0.05, 0.1, and 0.2 wt% at *Re* equal to 1823, 1651, and 1548, respectively.

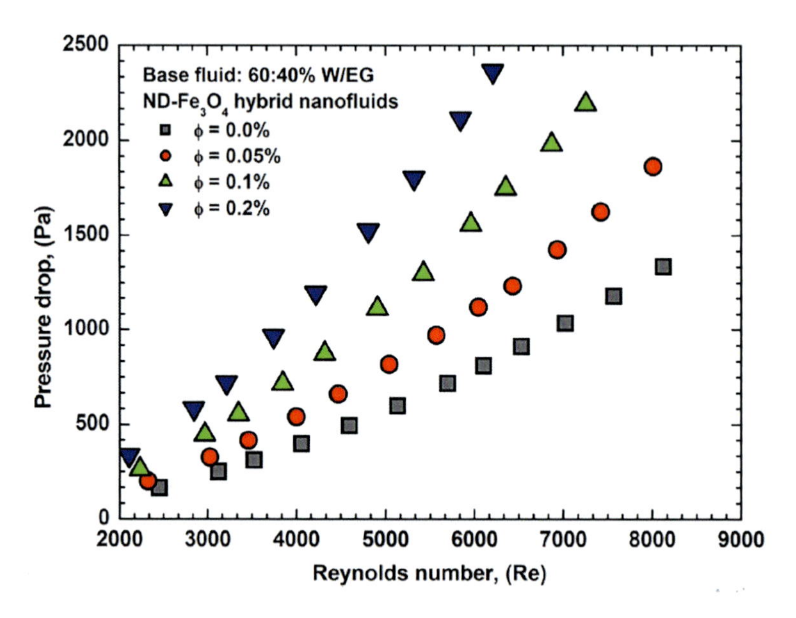

Fig. 4.3 Pressure drop of 60 W and 40EG ND-Fe_3O_4 hybrid nanofluid. From *L.S. Sundar, et al., Experimental investigation of thermo-physical properties, heat transfer, pumping power, entropy generation, and exergy efficiency of nanodiamond+ Fe3O4/60: 40% water-ethylene glycol hybrid nanofluid flow in a tube, Therm. Sci. Eng. Prog. 21 (2021) 100799. Copyright Elsevier 2021 (Lic# 5087580656920).*

Saleh and Sundar [46] have analyzed the friction factor of nano-diamond+Fe_3O_4/60EG and 40 W mixture-based hybrid nanofluid and they observed a maximum Δp rise of 73.3% and 125.25% at $\phi = 0.2\%$ and *Re* of 2047 and 7218 compared with the water.

4.5 Pumping power

When differentiating with single-phase fluid, the pressure drop is higher for hybrid nanofluids because of the addition of nano-particles, which leads to enhance the pumping power. That means, the power required to pump the hybrid nanofluid is more, which affects the system's overall performance. While designing a component to work with hybrid nanofluids, it is required to know the pumping power. The pumping power can be analyzed from the below equation:

$$\text{Pumping power}, Pp = \left(\frac{\dot{m}}{\rho_{nf}}\right)\Delta p \tag{4.10}$$

Sundar et al. [5] conducted pumping power analysis of 60 W and 40EG ND-Fe_3O_4 nanofluid (Fig. 4.4) and found a maximum of 0.035 W and 1.27 W increment in pumping power at $\phi = 0.2\%$ at *Re* equal to 2105 and 7520 compared with the water; in terms of augmentation, it is 3.72-fold.

Saleh and Sundar [46] have analyzed the friction factor of 60EG and 40 W nanodiamond+Fe_3O_4 hybrid nanofluid. They observed a maximum pumping power 0.38 W and 13 W at $\phi = 0.2\%$ and at *Re* of 2047 and 7218, compared with water. Sundar et al. [48] analyzed the pumping power of rGO-Fe_3O_4-TiO_2/EG nanofluid and reported that the pumping power needed is 0.07 W for EG at *Re* equal to 321, and it is 0.08 W for 0.2 wt% of nanofluid at $Re = 221.1$. Similarly, the pumping power expected to pump the base fluid at $Re = 2250$ is 3.6 W, and it is 3.77 W at 0.2 wt% and $Re = 1548$. Table 4.1. indicates the summary of the studies on hydrothermal properties of hybrid nanofluids.

4.6 Fouling factor of nanofluid

The fouling factor shows the theoretic resistance to heat flow because of dust deposition or other fouling substances on heat exchanger surfaces, but the end-user frequently overstates them in an effort to reduce the cleaning frequency. The below-mentioned papers are related to the nanofluid fouling in a thermal component.

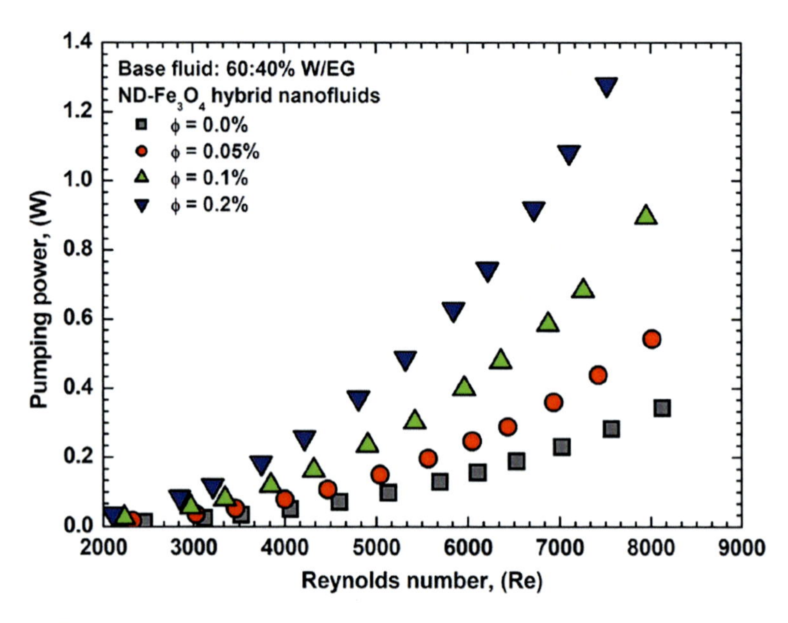

Fig. 4.4 Pumping power of 60 W and 40EG ND-Fe_3O_4 nanofluid. From *L.S. Sundar, et al., Experimental investigation of thermo-physical properties, heat transfer, pumping power, entropy generation, and exergy efficiency of nanodiamond+ Fe3O4/60: 40% water-ethylene glycol hybrid nanofluid flow in a tube, Therm. Sci. Eng. Prog. 21 (2021) 100799. Copyright Elsevier 2021 (Lic# 5087580656920).*

Table 4.1 Summary of the studies on hydrothermal properties of hybrid nanofluids.

Researchers	Hybrid nanofluids	Parameters	Findings
Nabil et al. [52]	$TiO_2 + SiO_2$-EG + H_2O	2.0, 2.5, 3 vol%; *Re* = 2000–10,000	• Friction factor reduces with increasing Reynolds number and increases with increasing volume concentration.
Yang et al. [53]	$Al_2O_3 + TiO_2$-H_2O Al_2O_3 + ZnO-H_2O	0.02, 0.04, and 0.06 vol%	• $Al_2O_3 + TiO_2$-H_2O hybrid nanofluids show a minimal friction factor compared with Al_2O_3 + ZnO-H_2O based nanofluids.
Madhesh et al. [41]	Cu + TiO2	0.1 vol%–2 vol%; *Re* = 3500–7500	• Increment in friction factor of 1.7% and pressure drop of 14.9% at 2 vol%.
Azmi et al. [45]	$TiO_2 + SiO_2$-H_2O + EG	1, 2, and 3 vol%; *Re* = 2900–11,200 0.1–0.3 vol%	• At 3 vol%, maximum friction factor was reported and followed by 2 and 1 vol%.

Table 4.1 Summary of the studies on hydrothermal properties of hybrid nanofluids—cont'd

Researchers	Hybrid nanofluids	Parameters	Findings
Sundar et al. [38]	Fe_3O_4 + MWCNT-H_2O		• The enhancement in friction factor for 0.3% vol %Fe_3O_4 + MWCNT nanofluid is 1.11-times and 1.18-times at the Reynolds number of 3000 and 22,000 Reynolds number, respectively. • Friction factor increment is insignificant.
Sundar et al. [54]	Nanodiamond + nickel-H_2O	0.1–0.3 vol%; *Re* = 3000–22,000	• The friction factor loss of 1.12-times and 1.248-times without inserts and with longitudinal strip insert, respectively, at 0.3 vol% and Reynolds number value of 22,000 as compared to the base fluid. • The friction factor increases with the increasing volume concentration and reduces with the increasing Reynolds numbers. • The friction factor of nanodiamond-Ni is less than MWCNT-Fe_3O_4 nanofluids because of nanoparticles' shape and size.
Mousavi et al. [55]	CuO + MgO + TiO_2-H_2O	0.1–0.5 vol%; 15–60°C	• Surface tension of ternary hybrid nanofluids increases with the increasing volume concentration. • For 0.1 vol%, surface tension is less than water.
Houshfar et al. [56]	MWCNT +Fe_3O_4-H_2O	0.1 vol% and 0.3 vol %; *Re* = 2000, 2500, and 3000	• The pumping power ratio less than 1 can be considered suitable for energy consumption. • By increasing the volume concentration, there is an increment in the pumping power ratio.
Minea et al. [57]	Al_2O_3 + TiO_2-H_2O Al_2O_3 + SiO_2-H_2O	1, 1.5, and 2 vol%	• Pumping power depends on nanoparticle type and volume fraction. • Al_2O_3 + SiO_2-H_2O hybrid nanofluid shows no pumping power penalty and reported an increment of 12.71% in heat transfer in laminar flow. • Pressure drop increases with increasing volume concentration.
Datta et al. [58]	Al_2O_3 + Ag-H_2O Al_2O_3 + SiC-H_2O	1, 1.5 and 2 vol%	• Addition of nanoparticles in the base fluid gives rise to pressure losses. • Pumping power increases with increasing volume concentration. • Greater pumping power is mainly due to increased pressure drop.

Continued

Table 4.1 Summary of the studies on hydrothermal properties of hybrid nanofluids—cont'd

Researchers	Hybrid nanofluids	Parameters	Findings
			• Increment of 17% in pumping power along with an enhancement of 42.2% in the convective average heat transfer coefficient for Al_2O_3 + SiC-H_2O-based hybrid nanofluids.
Sahoo et al. [59]	Al_2O_3 + Ag-EG Al_2O_3 + Cu-EG Al_2O_3 + SiC-EG Al_2O_3 + CuO-EG Al_2O_3 + TiO_2-EG	1 vol%	• SiC+Al_2O_3 hybrid nanofluids exhibit minimal pressure drop. • Ag + Al_2O_3 exhibits maximum pumping power of 2.5% higher than other hybrid nanofluids. • Pressure drop for Ag-based hybrid nanofluid increases by 2% followed by Cu, CuO, and TiO_2 at 1 vol%.
Selvakumar et al. [60]	Al_2O_3 + Cu-H_2O	0.1 vol%; 0.79–2.45 LPM	• Pressure drop increment of 14.25% for 0.1 vol % nanofluid flow in a copper heat sink at 1.347 L/min flow rate.
Garud et al. [61]	Al_2O_3 + Cu-H_2O	0.5, 1.0, and 2 vol%; 2000–12,000	• Pressure drop and friction factor increase with volume concentration. • Pressure drop of hybrid nanofluids increases with increasing *Re* number. • Pressure drop of hybrid nanofluids is higher than water due to the increased viscosity. • The friction factor increases up to the Reynolds number of 6000 and reduces beyond this value.
Yarmand et al. [39]	GNP + Ag-H_2O	0.02, 0.06, and 0.1 vol%; *Re* = 5000–17,500	• The enhancement in friction factor for 0.1 vol% GNP-Ag hybrid nanofluid is 8% at the Reynolds number of 17,500.
Saleh et al. [62]	MWCNT+Fe_3O_4	0.05–0.3 vol%; 0.1 to 0.75 L/min	• Friction factor is improved with increasing *Re* and volume concentration. • The friction factor is enhanced by 7.65%, 11.8%, 14.46%, and 18.91% for 0.05%, 0.1%, 0.2%, and 0.3 vol% and *Re* of 1774, 1674, 1528, and 1413, respectively, as compared to the base fluid.
Saleh et al. [46]	Nanodiamond+ Fe_3O_4-EG + H_2O	0.05–0.2 vol%; *Re* = 2000–8000	• With the increment in particle fraction and *Re* values, the pressure drop is increased. • The pressure drop, pumping power, and friction factor improved by 125%, 213%, and 15.11% at 0.2 vol% and Reynolds xnumber of 7218.

Sarafraz and Hormozi [63] studied the thermal performance, pressure drop characteristics, and fouling factor of water-based multiwalled carbon nanotube nanofluids in a heat exchanger in the Reynolds number range from 700 to 25,000, 0.5% to 1.5% vol% concentration, and inlet temperature from 50 to 70°C, and they observed a nonlinear fouling behavior over the extended time for nanofluids and a decrease of heat transfer coefficient as well. Sarafraz et al. [64] studied the fouling factor of TiO_2-water nanofluids under distinct operating conditions such as heat flux, mass loadings, and thermosyphon inclination and observed that the fouling rate can be improved by increasing the concentration.

Sarafraz et al. [65] studied the thermal behavior of a chevron-type flat plate heat exchanger working with CuO/water nanofluid with the fouling formation of nanoparticles in a heat exchanger and observed that nanoparticles' fouling could be mitigated via low-frequency vibration and thermal resistance reduces by introducing vibration in the system. Nikkhah et al. [66] experimentally investigated the convective boiling of dilute CuO-water nanofluids in an upward flow inside the conventional heat exchanger and developed particulate fouling correlated for nanofluids based on the data.

4.7 Conclusions and challenges

This chapter explains the hydrothermal properties of hybrid nanofluids, which are friction factor, pumping power, pressure drop, and fouling factor. In contrast to the base fluid, the pressure drop is high for hybrid nanofluids, leading to a higher friction factor. Once the pressure drop and friction factor are increased, the power required to pump the hybrid nanofluid is higher. The nanofluids prepared at higher particle concentrations required more pumping power than the lesser particle concentrations. Then, it is suggested that lower particle concentrations of nanofluids are preferable fluids in a thermal component with negligible pumping power. The fouling factor is also one of the parameters; the researcher has to focus on it. An increasing fouling factor is noticed with augmented particle loadings. Increased fouling factor reduces heat transfer coefficient with the use of nanofluids.

Developing mathematical models for the fouling factor of nanofluids with particle loadings and temperature is absolutely necessary. Reaching the long-term stable hybrid nanofluids with higher k_{nf} and μ_{nf} values is the focusing area. Hence, the stability and establishing a new technique for the preparation of hybrid nanofluids with lower price instruments are key objectives in this area. The applicability of hybrid nanofluids is also an important factor because it contains two or more varieties of nanoparticles. More research is required to understand various combinations of nanoparticles on hybrid nanofluid stability, properties, heat transfer, less friction factor, pumping power, and pressure drop characteristics.

Acknowledgments

The authors L.S.S. (Ref. 045-88-ARH/2018; CEEC 2018 Individual) and E.V.R. (Ref. 032-88-ARH/2018) acknowledge Fundação para a Ciência e a Tecnologia (Portugal) for its funding of their positions and the work was funded by projects UIDB/00481/2020, UIDP/00481/2020, and infrastructure support of CENTRO-01-0145-FEDER-022083.

References

[1] P.K. Kanti, et al., Experimental investigation on thermal conductivity of fly ash nanofluid and fly ash-Cu hybrid nanofluid: prediction and optimization via ANN and MGGP model, Part. Sci. Technol. (2021) 1–14.
[2] L. Sundar, et al., Heat Transfer of rGO/Co3O4 hybrid nanomaterial based nanofluids and twisted tape configurations in a tube, J. Therm. Sci. Eng. Appl. (2020) 1–41.
[3] P.K. Kanti, et al., Experimental investigation on thermo-hydraulic performance of water-based fly ash–cu hybrid nanofluid flow in a pipe at various inlet fluid temperatures, Int. Commun. Heat Mass Transf. 124 (2021) 105238.
[4] Z. Said, et al., Recent advances on nanofluids for low to medium temperature solar collectors: energy, exergy, economic analysis and environmental impact, Prog. Energy Combust. Sci. 84 (2021) 100898.
[5] L.S. Sundar, et al., Experimental investigation of thermo-physical properties, heat transfer, pumping power, entropy generation, and exergy efficiency of nanodiamond+ Fe3O4/60: 40% water-ethylene glycol hybrid nanofluid flow in a tube, Therm. Sci. Eng. Prog. 21 (2021) 100799.
[6] S. Suresh, et al., Effect of Al2O3–cu/water hybrid nanofluid in heat transfer, Exp. Thermal Fluid Sci. 38 (2012) 54–60.
[7] Z. Said, et al., Heat transfer, entropy generation, economic and environmental analyses of linear Fresnel reflector using novel rGO-Co3O4 hybrid nanofluids, Renew. Energy 165 (2021) 420–437.
[8] A.K. Tiwar, et al., A review on the application of hybrid nanofluids for parabolic trough collector: recent progress and outlook, J. Clean. Prod. (2021) 126031.
[9] L.S. Sundar, et al., Hybrid nanofluids preparation, thermal properties, heat transfer and friction factor—a review, Renew. Sust. Energ. Rev. 68 (2017) 185–198.
[10] Z. Said, et al., Acid-functionalized carbon nanofibers for high stability, thermoelectrical and electrochemical properties of nanofluids, J. Colloid Interface Sci. 520 (2018) 50–57.
[11] Z. Said, S. Arora, E. Bellos, A review on performance and environmental effects of conventional and nanofluid-based thermal photovoltaics, Renew. Sust. Energ. Rev. 94 (2018) 302–316.

[12] L.S. Sundar, et al., Combination of Co3O4 deposited rGO hybrid nanofluids and longitudinal strip inserts: thermal properties, heat transfer, friction factor, and thermal performance evaluations, Therm. Sci. Eng. Prog. 20 (2020) 100695.
[13] A.A. Hachicha, et al., On the thermal and thermodynamic analysis of parabolic trough collector technology using industrial-grade MWCNT based nanofluid, Renew. Energy 161 (2020) 1303–1317.
[14] N.K. Cakmak, et al., Preparation, characterization, stability, and thermal conductivity of rGO-Fe3O4-TiO2 hybrid nanofluid: an experimental study, Powder Technol. 372 (2020) 235–245.
[15] M. Gupta, V. Singh, Z. Said, Heat transfer analysis using zinc ferrite/water (hybrid) nanofluids in a circular tube: an experimental investigation and development of new correlations for thermophysical and heat transfer properties, Sustainable Energy Technol. Assess. 39 (2020) 100720.
[16] M. Ehyaei, et al., Energy, exergy and economic analyses for the selection of working fluid and metal oxide nanofluids in a parabolic trough collector, Sol. Energy 187 (2019) 175–184.
[17] M. Jamei, et al., On the specific heat capacity estimation of metal oxide-based nanofluid for energy perspective–a comprehensive assessment of data analysis techniques, Int. Commun. Heat Mass Transfer 123 (2021) 105217.
[18] M. Sheikholeslami, et al., Recent progress on flat plate solar collectors and photovoltaic systems in the presence of nanofluid: a review, J. Clean. Prod. (2021) 126119.
[19] L.S. Sundar, et al., Energy, efficiency, economic impact, and heat transfer aspects of solar flat plate collector with Al2O3 nanofluids and wire coil with core rod inserts, Sustain. Energy Technol. Assess. 40 (2020) 100772.
[20] Z. Said, et al., Heat transfer enhancement and life cycle analysis of a Shell-and-tube heat exchanger using stable CuO/water nanofluid, Sustain. Energy Technol. Assess. 31 (2019) 306–317.
[21] X. Wang, et al., Vegetable oil-based nanofluid minimum quantity lubrication turning: academic review and perspectives, J. Manuf. Process. 59 (2020) 76–97.
[22] L.S. Sundar, et al., Properties, heat transfer, energy efficiency and environmental emissions analysis of flat plate solar collector using nanodiamond nanofluids, Diam. Relat. Mater. 110 (2020) 108115.
[23] T. Gao, et al., Mechanics analysis and predictive force models for the single-diamond grain grinding of carbon fiber reinforced polymers using CNT nano-lubricant, J. Mater. Process. Technol. 290 (2021) 116976.
[24] A. Kumar, Z. Said, E. Bellos, An up-to-date review on evacuated tube solar collectors, J. Therm. Anal. Calorim. (2020) 1–17.
[25] Z. Said, et al., A comprehensive review on minimum quantity lubrication (MQL) in machining processes using nano-cutting fluids, Int. J. Adv. Manuf. Technol. 105 (5) (2019) 2057–2086.
[26] S. Rahman, et al., Performance enhancement of a solar powered air conditioning system using passive techniques and SWCNT/R-407c nano refrigerant, Case Stud. Therm. Eng. 16 (2019) 100565.
[27] F.I. Alhamadi, et al., Performance of Nano-fluids as coolants/moderator in APR1400–Neutronics case study, Transactions 120 (1) (2019) 524–527.
[28] S. Tanvir, L. Qiao, Surface tension of nanofluid-type fuels containing suspended nanomaterials, Nanoscale Res. Lett. 7 (1) (2012) 1–10.
[29] J. Chinnam, et al., Measurements of the surface tension of nanofluids and development of a new correlation, Int. J. Therm. Sci. 98 (2015) 68–80.
[30] A. Huminic, et al., Thermal conductivity, viscosity and surface tension of nanofluids based on FeC nanoparticles, Powder Technol. 284 (2015) 78–84.

[31] M. Bhuiyan, et al., Effect of nanoparticles concentration and their sizes on surface tension of nanofluids, Procedia Eng. 105 (2015) 431–437.
[32] S. Khaleduzzaman, et al., Effect of particle concentration, temperature and surfactant on surface tension of nanofluids, Int. Commun. Heat Mass Transfer 49 (2013) 110–114.
[33] M. Wanic, et al., Surface tension of ethylene glycol-based nanofluids containing various types of nitrides, J. Therm. Anal. Calorim. 139 (2) (2020) 799–806.
[34] R. Kumar, D. Milanova, Effect of surface tension on nanotube nanofluids, Appl. Phys. Lett. 94 (7) (2009), 073107.
[35] G. Lu, Y.-Y. Duan, X.-D. Wang, Surface tension, viscosity, and rheology of water-based nanofluids: a microscopic interpretation on the molecular level, J. Nanopart. Res. 16 (9) (2014) 1–11.
[36] Z.Y. Zhou, et al., Experimental study on the surface tension of Al2O3-H2O nanofluid, in: Materials Science Forum, Trans Tech Publ, 2016.
[37] T. Theres Baby, R. Sundara, Synthesis of silver nanoparticle decorated multiwalled carbon nanotubes-graphene mixture and its heat transfer studies in nanofluid, AIP Adv 3 (1) (2013) 012111.
[38] L.S. Sundar, M.K. Singh, A.C. Sousa, Enhanced heat transfer and friction factor of MWCNT–Fe3O4/water hybrid nanofluids, Int. Commun. Heat Mass Transfer 52 (2014) 73–83.
[39] H. Yarmand, et al., Graphene nanoplatelets–silver hybrid nanofluids for enhanced heat transfer, Energy Convers. Manag. 100 (2015) 419–428.
[40] N. Ahammed, L.G. Asirvatham, S. Wongwises, Entropy generation analysis of graphene–alumina hybrid nanofluid in multiport minichannel heat exchanger coupled with thermoelectric cooler, Int. J. Heat Mass Transf. 103 (2016) 1084–1097.
[41] D. Madhesh, R. Parameshwaran, S. Kalaiselvam, Experimental investigation on convective heat transfer and rheological characteristics of cu–TiO2 hybrid nanofluids, Exp. Thermal Fluid Sci. 52 (2014) 104–115.
[42] A. Moghadassi, E. Ghomi, F. Parvizian, A numerical study of water based Al2O3 and Al2O3–cu hybrid nanofluid effect on forced convective heat transfer, Int. J. Therm. Sci. 92 (2015) 50–57.
[43] B. Takabi, H. Shokouhmand, Effects of Al 2 O 3–cu/water hybrid nanofluid on heat transfer and flow characteristics in turbulent regime, Int. J. Mod. Phys. C 26 (04) (2015) 1550047.
[44] L.S. Sundar, M.K. Singh, A.C. Sousa, Turbulent heat transfer and friction factor of nanodiamond-nickel hybrid nanofluids flow in a tube: an experimental study, Int. J. Heat Mass Transf. 117 (2018) 223–234.
[45] A. Ramadhan, et al., Experimental and numerical study of heat transfer and friction factor of plain tube with hybrid nanofluids, Case Stud. Therm. Eng. 22 (2020) 100782.
[46] B. Saleh, L.S. Sundar, Entropy generation and exergy efficiency analysis of ethylene glycol-water based nanodiamond+ Fe3O4 hybrid nanofluids in a circular tube, Powder Technol. 380 (2021) 430–442.
[47] M. Nabil, et al., Heat transfer and friction factor of composite TiO2–SiO2 nanofluids in water-ethylene glycol (60: 40) mixture, in: IOP Conference Series: Materials Science and Engineering, IOP Publishing, 2017.
[48] L.S. Sundar, et al., Heat Transfer and second law analysis of ethylene glycol-based ternary hybrid Nanofluid under laminar flow, J. Therm. Sci. Eng. Appl. 13 (5) (2021), 051021.
[49] V. Kumar, J. Sarkar, Numerical and experimental investigations on heat transfer and pressure drop characteristics of Al2O3-TiO2 hybrid nanofluid in minichannel heat sink with different mixture ratio, Powder Technol. 345 (2019) 717–727.

[50] S.A. Kaska, R.A. Khalefa, A.M. Hussein, Hybrid nanofluid to enhance heat transfer under turbulent flow in a flat tube, Case Stud. Therm. Eng. 13 (2019) 100398.
[51] A.A. Hussien, et al., Numerical study of heat transfer enhancement using Al 2 O 3–graphene/water hybrid nanofluid flow in mini tubes, Iran. J. Sci. Technol. Trans. A: Sci. 43 (4) (2019) 1989–2000.
[52] M.F. Nabil, et al., Heat transfer and friction factor of composite TiO2–SiO2 nanofluids in water-ethylene glycol (60:40) mixture, IOP Conf. Ser.: Mater. Sci. Eng. 257 (2017), 012066.
[53] C. Yang, et al., Heat transfer performance assessment of hybrid nanofluids in a parallel channel under identical pumping power, Chem. Eng. Sci. 168 (2017) 67–77.
[54] L. Syam Sundar, M.K. Singh, A.C.M. Sousa, Heat transfer and friction factor of nanodiamond-nickel hybrid nanofluids flow in a tube with longitudinal strip inserts, Int. J. Heat Mass Transf. 121 (2018) 390–401.
[55] S. Mousavi, F. Esmaeilzadeh, X. Wang, Effects of temperature and particles volume concentration on the thermophysical properties and the rheological behavior of CuO/MgO/TiO 2 aqueous ternary hybrid nanofluid, J. Therm. Anal. Calorim. 137 (3) (2019) 879–901.
[56] M. Irandoost Shahrestani, et al., Convective heat transfer and pumping power analysis of MWCNT + Fe3O4/water hybrid nanofluid in a helical coiled heat exchanger with orthogonal rib turbulators, Front. Energy Res. 9 (12) (2021).
[57] A.A. Minea, Pumping power and heat transfer efficiency evaluation on Al2O3, TiO2 and SiO2 single and hybrid water-based nanofluids for energy application, J. Therm. Anal. Calorim. 139 (2) (2020) 1171–1181.
[58] A. Datta, P. Halder, Thermal efficiency and hydraulic performance evaluation on Ag–Al2O3 and SiC–Al2O3 hybrid nanofluid for circular jet impingement, Arch. Thermodyn. 42 (1) (2021).
[59] R.R. Sahoo, J. Sarkar, Heat transfer performance characteristics of hybrid nanofluids as coolant in louvered fin automotive radiator, Heat Mass Transf. 53 (2017) 1923.
[60] P. Selvakumar, S. Sivan, Use of/water hybrid Nanofluid in an electronic heat sink, IEEE Trans. Compon. Packag. Manuf. Technol. 2 (2012) 1600–1607.
[61] K.S. Garud, M.-Y. Lee, Numerical investigations on heat Transfer characteristics of single particle and hybrid Nanofluids in uniformly heated tube, Symmetry 13 (5) (2021) 876.
[62] B. Saleh, L.S. Sundar, Thermal efficiency, heat Transfer, and friction factor analyses of MWCNT + Fe3O4/water hybrid nanofluids in a solar flat plate collector under thermosyphon condition, Processes 9 (1) (2021) 180.
[63] M. Sarafraz, F. Hormozi, Heat transfer, pressure drop and fouling studies of multi-walled carbon nanotube nano-fluids inside a plate heat exchanger, Exp. Thermal Fluid Sci. 72 (2016) 1–11.
[64] M. Sarafraz, F. Hormozi, S. Peyghambarzadeh, Role of nanofluid fouling on thermal performance of a thermosyphon: are nanofluids reliable working fluid? Appl. Therm. Eng. 82 (2015) 212–224.
[65] M. Sarafraz, et al., Low-frequency vibration for fouling mitigation and intensification of thermal performance of a plate heat exchanger working with CuO/water nanofluid, Appl. Therm. Eng. 121 (2017) 388–399.
[66] V. Nikkhah, et al., Particulate fouling of CuO–water nanofluid at isothermal diffusive condition inside the conventional heat exchanger-experimental and modeling, Exp. Thermal Fluid Sci. 60 (2015) 83–95.

5

Rheological behavior of hybrid nanofluids

Abdulla Ahmad Alshehhi[a], Zafar Said[b,c,d,*], and Maham Aslam Sohail[b]

[a]Space Missions Department, UAE Space Agency, Abu Dhabi, United Arab Emirates. [b]Department of Sustainable and Renewable Energy Engineering, University of Sharjah, Sharjah, United Arab Emirates. [c]Research Institute for Sciences and Engineering, University of Sharjah, Sharjah, United Arab Emirates. [d]U.S.-Pakistan Center for Advanced Studies in Energy (USPCAS-E), National University of Sciences and Technology (NUST), Islamabad, Pakistan

**Corresponding author: zsaid@sharjah.ac.ae, zaffar.ks@gmail.com*

Chapter outline

5.1 Introduction 111
5.2 Experimental and numerical studies on rheology 113
5.3 Effects of various parameters on the rheology of hybrid nanofluids 117
5.3.1 Temperature 119
5.3.2 Particle size and shape 120
5.3.3 Volume concentration 121
5.3.4 Other factors 122
5.4 Conclusion and future outlook 122
References 123

5.1 Introduction

With the advancement of technologies and miniaturizing of components and systems, the demand for higher capabilities within thermal and heat transfer is essential. A considerable amount of the intensified consumption cost in the energy area is because of the utilization of former technologies in heat transfer. One of the novel techniques of heat transfer advancement is the use of nanotechnology. Nanoparticles have the excellent potential to enhance heat transfer capabilities and have drawn researchers' attention in recent years. One novel technique can be achieved by mixing nanoparticles (in nanometers) in the

Hybrid Nanofluids: Preparation, Characterization and Applications. https://doi.org/10.1016/B978-0-323-85836-6.00005-3

conventional heat transfer fluids termed "nanofluids" that significantly augment the thermophysical properties [1, 2]. These nanofluids have superior features such as remarkable thermo-physical properties, stability [3], and less pressure drop over conventional fluid [4, 5]. Nanofluids are the colloidal suspension of nanograde (10^{-9}) solid nanoparticles into a base fluid, introduced by Choi and his co-workers [6], and are promising for solar collectors [7–9], heat exchangers [10, 11], electronics coolant, and many other fluid-related properties [12, 13]. Nanofluids' thermal conductivity and viscosity are the two key parameters for heat and mass transfer phenomena [14, 15]. The latest research on nanofluids introduced the advanced category of fluids with superior thermal properties named hybrid nanofluids, obtained by dispersing the nanocomposite or nanoparticles of different materials into the base fluid. Hybrid nanofluids are the novel class of heat transfer fluids in the research and development phase, and extensive research must be investigated for commercial applications. The hybrid nanofluids exhibits remarkable thermophysical properties as compared to mono nanofluids [8, 11]. An individual material may not have all the rheological and thermal characteristics for a particular purpose in practical. A hybrid material unifies the physical and chemical properties of various constituent metals [16]. The feasibility of hybrid nanofluids in practical applications is straightway associated with the friction factor, pressure drop, and pumping power [17, 18]. Therefore, the viscosity of hybrid nanofluids has equal priority as thermal conductivity [19]. There is no doubt that the thermal conductivity enhanced with the addition of nanoparticles, but it also caused some serious concerns such as high pumping power and pressure drop, corrosion, stability, and increased viscosity caused by the formation of the cluster that increases the hydrodynamic diameter and decreases the specific surface area [20, 21]. The hybrid nanofluids show higher viscosity than mono nanofluids [22]. Viscosity plays a significant role in examining the pumping power supplies due to frictional effects [23, 24]. It depends on temperature, concentration, size and shape of nanoparticles, preparation method, pH value, and other parameters [25]. The increment in temperature reduces the viscosity, the increase in volume fraction intensifies the viscosity, and small-scale sphere-shaped particles decrease the viscosity [26, 27].

Rheology is a Greek term that analyzes the reaction or change of fluid form against the tension [28]. The rheological behavior of any fluid is defined as the relationship between shear stress (τ) and shear rate (γ). The shear stress is described as the tangential force applied per unit area whereas the shear rate is described as the change of shear strain per unit time. Viscosity is defined as the ratio of shear stress to shear rate. The rheological properties describe the flow properties of a fluid and the characteristics of matter defining its behavior, its reaction to deformation, and flow [29]. The fluid behavior can be grouped as Newtonian and non-Newtonian. The viscosity remains constant with the shear rate for Newtonian fluid, and the stress shows linear relation with the shear rate, and the viscosity may change with the shear rate for non-Newtonian behavior. Rheological behavior of hybrid nanofluids has an impact on the pressure drop of hybrid nanofluids [30]. It provides the purpose of nanoparticle structuring, which can help estimate thermal conductivity. The rheological behavior can be evaluated by rheometers and viscometers [31, 32]. A study by B. Finke et al. [33] on rheological behavior of nanoparticulate suspensions showed the influence of particle material and disperse properties on the viscosity; it also showed that the particle interactions caused by surface forces induce velocity differences between the particles and their surrounding fluid, which result in increased drag forces and cause the additional energy dissipation during shearing. Anoop et al. [34] studied the rheological characteristics of high-pressure Al_2O_3+Si/oil hybrid nanofluid and reported non-Newtonian and shear-thinning behavior at all shear rates. Asadi et al. [35] recommended a standard based on the rheological and thermophysical properties to analyze the heat transfer capability of MWCNT-ZnO/oil hybrid nanofluid. Due to the viscosity's importance in the flow dynamics, several analytical and experimental techniques were utilized to examine the effect of pressure on the rheological behavior of fluids [36, 37]. The equation below provides an expression for fluid's behavior where γ represents shear stress (mPa), τ represents shear rate (s^{-1}), and μ represents dynamic viscosity (mPa s).

$$\tau = \mu^{*}\gamma = \mu\frac{du}{dx} \tag{5.1}$$

5.2 Experimental and numerical studies on rheology

Several studies showed increased heat transfer capabilities with nanofluid over conventional fluids in the literature [38, 39]. In this section, a summary of experimental studies and numerical correlations on rheology will be addressed. At lower volume concentrations, Newtonian characteristics are displayed; however, nanofluids with particles and nonaqueous base fluids have represented non-Newtonian behavior [40]. Studied reported linear increment in the viscosity with increasing volume concentration. Table 5.1 shows the main findings of the literature reported.

Table 5.1 Literature review of different studies reporting viscosity.

Analysis type	Year of study	Base fluids	Particle 1	Particle 2	Findings	References
Experimental	2021	Distilled water	rGO	Co_3O_4	Increment of 70.83% with 0.2% particle loadings at 60°C was obtained for viscosity	[41]
Experimental	2021	5W40	MWCNT	TiO_2	The viscosity intensified with decreasing shear rate as well as temperature and increasing volume fraction	[42]
Experimental	2021	Water	Carbon black	Boron nitride	Non-Newtonian behavior was observed	[43]
Experimental	2021	Water	GO-TiO_2 rGO-TiO_2	Ag Ag	Graphene-based ternary hybrid nanofluids exhibit Newtonian behavior at higher volume fraction and show shear-thinning or pseudoplastic fluid characteristics at lower volume fractions	[44]
Experimental	2020	Deionized water and ethylene glycol	SiO_2	P25 TiO_2	Newtonian behavior was observed	[45]
Experimental	2020	Oil	MWCNT	MgO	Non-Newtonian fluid was observed at the temperatures of 10°C and 20°C, and it displayed Newtonian behavior from 30°C to 60°C	[46]
Experimental	2020	Water-EG	MWCNT	ZnO	Non-Newtonian behavior was observed	[47]
Experimental	2020	Water	CuO	TiO_2	The hybrid nanofluids displayed Newtonian behavior over the range of shear rate	[48]

Table 5.1 Literature review of different studies reporting viscosity—cont'd

Analysis type	Year of study	Base fluids	Particle 1	Particle 2	Findings	References
Experimental	2020	Water-EG	CuO	MWCNT	Observed Newtonian behavior for low concentration and non-Newtonian for higher concentrations	[49]
Experimental	2020	Distilled water	Graphene	Silica	Observed non-Newtonian pseudoplastic behavior	[50]
Experimental	2019	Oil	MWCNT	TiO_2	It is observed that hybrid nanofluid exhibits Newtonian fluid	[51]
Experimental	2017	Water-based emulsion	Al_2O_3	Graphene	Viscosity is increased with an increase in the nanoparticle's concentration	[52]
Experimental	2017	Water Ethylene glycol Water/EG	Co_3O_4	Graphene oxide	Water-based hybrid nanofluids has shown improvement in viscosity by 1.07 times, and for EG-based hybrid nanofluids, it was 1.42 times at 0.2% volume fraction	[53]
Experimental	2016	Ethylene glycol	MgO	MWCNT	The results showed that the hybrid nanofluid behaves as a Newtonian fluid, and dynamic viscosity increases with increasing volume concentration and decreases with the temperature increment	[54]
Experimental	2016	Engine oil	MWCNT	ZnO	The results revealed that increment in temperature reduced dynamic viscosity by 85% while it increased as the concentration increased by 45%	[55]
Experimental	2016	Ethylene glycol	Activated carbon	Graphene	4.16% maximum viscosity increment was noticed with 0.06 wt%	[56]

Continued

Table 5.1 Literature review of different studies reporting viscosity—cont'd

Analysis type	Year of study	Base fluids	Particle 1	Particle 2	Findings	References
Experimental	2016	Water	ND	Fe_3O_4	It was observed that the viscosity improvement shows nonlinear behavior with increasing volume concentrations	[57]
Experimental	2015	Water	Ag	MgO	As the volume fraction increases, the viscosity also increases	[58]
Experimental	2015	Water	MWCNT	TiO_2	It was observed that MWCNT-TiO_2 hybrid nanofluids exhibit shear thickening or dilatant behavior	[59]
Experimental	2014	Water	Silica	MWCNT	The viscosity increases with particle concentration and reduces with temperature	[60]
Experimental	2014	Water and ethylene glycol	Nanodiamond	Nickel	At 3.03 wt% of ND-Ni nanofluid, maximum viscosity increment for with 20:80% EG-water 2.4 times	[61]
Experimental	2011	Oil	Silica	Silver	These hybrid nanofluids observed lower viscosity compared to mono nanofluids	[62]
Experimental	2007	Glycerol (10 wt%) and water	Al_2O_3	MWCNT	Both Al_2O_3 nanoparticle and CNT-Al_2O_3 nanoparticle systems exhibit shear-thinning behavior. The viscosities increase monotonically with the suspension solids loading	[63]
Experimental	1998	Deionized water	SiO_2	TiO_2	With the mixing of (SiO_2/TiO_2) suspensions observed, Bingham plastic behavior and viscosity increased remarkably	[64]

Viscosity analysis models that fit well for mono nanofluids do not accurately predict the viscosity of hybrid nanofluids [65]. It is unlikely to achieve a universal correlation for viscosity that is effective for all hybrid nanofluids for different concentrations and temperature ranges [66]. Due to the behavior change with changing concentration and temperature values, hybrid nanofluids may indicate viscoelastic behavior. A study by Afrand et al. [67] reported that the rheological behavior of Fe_3O_4+Ag/EG hybrid nanofluid changes to non-Newtonian after 0.3% concentration. Sharma et al. [68] presented a typical model where viscosity depends on the temperature and volume concentration, which is expressed as

$$\frac{\mu_{hnf}}{\mu_{bf}} = 0.9653 + 77.4567\left(\frac{\varphi}{100}\right)^{1.1558}\left(\frac{T}{333}\right)^{0.6801} \tag{5.2}$$

The conventional models do not precisely predict the viscosity of hybrid nanofluids [69]. Scholars are still investigating and developing relations for better accuracy [70]. The Newtonian behavior of viscosity is linked to volume fractions and temperature values, and for the non-Newtonian case, the viscosity relations are correlated to the rate of deformation using consistency and power law indices [71]. Table 5.2 presents the different correlations to predict the viscosity of hybrid nanofluids.

Most of the correlations considered the effects of temperature, volume concentration, and shear rate. To develop more correlations for future studies, we should consider more parameters.

5.3 Effects of various parameters on the rheology of hybrid nanofluids

The rheological behavior of hybrid nanofluids is determined by the relationship between the shear rate and the shear stress of nanofluid mixture. It can be stated that viscosity is the fundamental parameter considered for determining the rheological behavior of fluids. Different parameters such as temperature, volume fraction, nanoparticle size and shape, preparation method, shear rate, shear stress, pH value, and time influence rheology (Fig. 5.1). Researchers investigate viscosity for the rheological properties of hybrid nanofluids depending on volume concentration, temperature, and shear rate (to make sure the flow is whether Newtonian or non-Newtonian).

Table 5.2 Correlation for the prediction of viscosity of hybrid nanofluids by different researchers.

Researchers	Hybrid nanofluid	Correlation	Parameters
Esfe et al. [72]	MWCNTs/ SiO_2-SAE40	$\frac{\mu_{nf}}{\mu_{bf}}=a_o+a_1\varphi+a_2\varphi^2+a_3\varphi^3$	0–2 vol.%; 25–50°C
Esfe et al. [73]	CuO/MWCNT-10w40 oil	$\mu_{nf}=633.8379+280.1511\varphi-38.4183T-6.17707\varphi T-305.838\varphi^2+0.888891T^2+0.807687\varphi^2T+0.05807\varphi T^2+166.6123\varphi^3+0.00714T^3$	0–1 vol.%; 5–55°C
Asadi et al. [74]	Al_2O_3-MWCNT/ thermal oil	$\frac{\mu_{nf}}{\mu_{bf}}=a+b\varphi$	0.125–1.5 vol. %; 25–50°C
Nabil et al. [75]	TiO_2-SiO_2/EG-water	$\frac{\mu_{nf}}{\mu_{bf}}=37\left(0.1+\frac{\varphi}{100}\right)^{1.59}\left(0.1+\frac{T}{80}\right)^{0.31}$	0.5–3 vol.%; 30–70°C
Ganji et al. [76]	TiO_2-Cu/water	$\mu_{hnf}=\frac{\mu_f}{(1-\varphi_1)^{2.5}(1-\varphi_2)^{2.5}}$	
Akbari et al. [54]	MgO-MWCNT/ water	$\frac{\mu_{hnf}}{\mu_{bf}}=\left[0.91\varphi+0.240\left(T^{-0.342}\varphi^{-0.473}\right)\right]exp\left(1.45T^{0.12}\varphi^{0.158}\right)$	0–1 vol.%; 30–60°C
Asadi et al. [55]	MWCNT/ZnO-engine oil	$\mu_{hnf}=796.8+76.26\varphi+12.88T+0.7695\varphi T+\frac{-196.9T-16.54\varphi T}{\sqrt{T}}$	0.125–1 vol. %; 5–55°C
Moldoveanu et al. [77]	Al_2O_3 + SiO_2/ water	$\mu_{nf}=0.2111\gamma^{-0.547};\ \mu_{nf}=0.2597\gamma^{-0.609}$ $Shear\ rate=\gamma=0.01-1000\ s^{-1}$	0.5–2 vol.%; 25°C
Afrand et al. [78]	MWCNTs-SiO_2/ AE40	$\frac{\mu_{nf}}{\mu_{bf}}=0.00337+exp\left(0.07731\varphi^{1.452}T^{0.3387}\right)$	0–1 vol.%; 25–60°C
Dardan et al. [79]	Al_2O_3-MWCNTs/ SAE40	$\frac{\mu_{nf}}{\mu_{bf}}=1.123+0.3251\varphi-0.08994T+0.002552T^2-0.00002386T^3+0.9695\left(\frac{T}{\varphi}\right)^{0.01719}$	0.0625–1 vol. %; 25–50°C
Asadi et al. [80]	Mg $(OH)_2$/ MWCNT-engine oil	$\frac{\mu_{nf}}{\mu_{bf}}=1604+256.8\varphi+24.73\varphi^3+1.615T^2+0.07343\varphi T^3-83.2T-7.389\varphi^2-0.01123T^3-74.19\varphi^2$	0.25–2 vol.%; 25–60°C
Sahoo et al. [81]	Al_2O_3+CuO +TiO_2/water	$\mu_{THNF}=\mu_{bf}\left(0.955-0.00271*T+1.858*\frac{\varphi}{100}+\left(705*\frac{\varphi}{100}\right)^{1.223}\right)$	0.01–0.1 vol. %; 35–50°C
Esfe et al. [82]	Ag-MgO/water	$\mu_{nf}=(1+32.795\varphi_p-7214\varphi_p^2+714600\varphi_p^3-0.1941\times10^8\varphi_p^4)\mu_f$	0–0.02 vol.%
Alarifi et al. [51]	MWCNT-TiO_2/oil	$\mu_{nf}=2.936T+\frac{2e^4}{1.68+T-(1.68\varphi)}-448.8-tan((1.68\varphi)-1.68)$	0.25–2 vol.%; 25–50°C

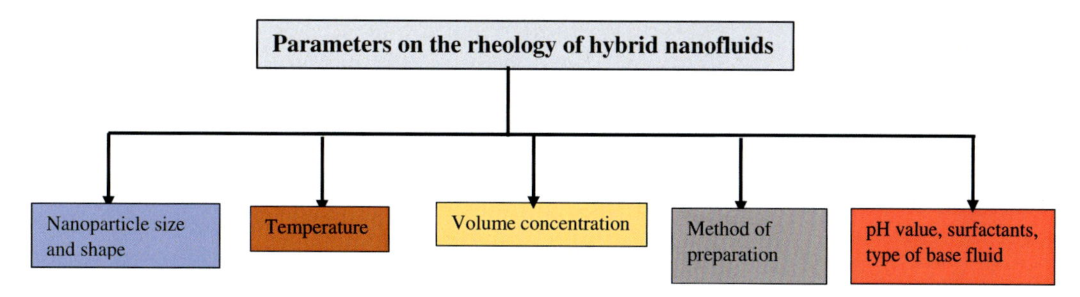

Fig. 5.1 Factors affecting the parameters on the rheology of hybrid nanofluids.

5.3.1 Temperature

Numerous researchers observed decreasing temperature trends on viscosity due to the deterioration of interparticle adhesion forces in the literature. The reason behind the reduction in viscosity with increasing temperature is when the temperature increases, the heat energy provides extra energy to separate the molecules that causes a reduction in attractive forces between molecules [83]. The rising temperature enhances the intermolecular space of nanoparticles and the base fluid because resistance against the flow and viscosity is declined. High temperature of nanofluid increases the Brownian motion of nanoparticles and lowers nanofluid viscosity [84]. Most of the literature studies inspect the viscosity of water and EG-based nanofluids at a lower temperature up to 60, while the viscosity variation of oil-based nanofluids is examined at higher temperatures. Some of the studies are mentioned below:

1. Shahsavar et al. [85] studied experimentally the viscosity of Fe_3O_4+CNT-H_2O hybrid nanofluids in the temperature range of 25–55°C. The study revealed that as the temperature increases, the viscosity reduces.
2. Said et al. [86] analyzed the viscosity of CNF+rGO-deionized water in the temperature range of 20–65°C and noticed a decrement in viscosity with increasing temperature.
3. Afrand et al. [87] investigated the dynamic viscosity of SiO_2+MWCNT-engine oil hybrid nanofluids to analyze the temperature effect in the range of 25–60°C. It was reported that the maximum improvement of viscosity was 37.4% obtained at 60°C. The results also indicated that the dynamic viscosity reduces with increasing temperature.
4. Said et al. [19] observed a decreasing trend in the viscosity with temperature for zinc+ferrite/water hybrid nanofluids.
5. Motahari et al. [88] reported a decline in viscosity of SiO_2+MWCNT-oil hybrid nanofluids with increasing temperature in the range of 40–100°C.

6. Esfe and Rostamian [89] studied the viscosity of ZnO+MWCNT (55%–45%)/engine oil hybrid nanofluids in the temperature range of 5–55°C and observed decreasing viscosity with increasing temperature.
7. Said et al. [90] experimentally studied the viscosity of nanodiamond+Fe_3O_4-H_2O+EG hybrid nanofluids and reported a reduction in viscosity with increasing temperature.
8. Esfe et al. [72] reported that the viscosity was more susceptible to lower temperature instead of higher temperature by investigating SiO_2+MWCNT-engine oil hybrid nanofluids.
9. Akilu et al. [91] studied TiO_2-CuO/C-EG nanofluids at the temperature range of 298–333 K. It was observed that a significant decrease in viscosity was noticed at higher temperatures due to weak intermolecular forces.

5.3.2 Particle size and shape

Limited literature studies are available on the effect of particle size and shape on the viscosity of hybrid nanofluids. Some studies show a decreasing trend, and some show an increasing trend with particle size. Sharma et al. [92] reported that particle size optimization could be valuable to obtain nanofluids with superior rheological properties. When nanoparticles or nanocomposites are suspended in base fluid, two key interactions are viable: solid-fluid and solid-solid. For instance, if the volume fraction is constant, decreased nanoparticle size turns into an improved overall surface area of solid-liquid and solid-solid interaction, increasing viscosity. Larger nanoparticles represented increased viscosity could cause particle agglomeration [93, 94].

However, the viscosity has a significant dependence on nanoparticle shape. The elongated nanoparticles create random clustering, which restricts both rotational and translational motion. This results in shear-thinning behavior and high viscosities of suspensions along with agglomeration. Timofeeva et al. [95] examined the effect of particle shapes on the viscosity of Al_2O_3-EG +H_2O nanofluids and noticed shear thinning behavior for cylinder-like and brick-like nanoparticles. Ferrouillat et al. [96] studied the influence of particle shape on the viscosity of ZnO-H_2O nanofluids and found that the fluid containing polygonal-shaped nanoparticle suspension exhibited greater viscosity than that of rod-shaped nanoparticles. Another reason that can change the fluid viscosity values of different nanoparticle shapes is the aerodynamics of shapes. A streamlined shape creates minimal flow resistance compared to an abnormal or sharp-edged shape that endures flow because it is challenging to dissipate energy, resulting in improved viscosity [97]. It can be concluded that spherical particles are more practical for obtaining lower viscosities—the property that is highly desirable for minimizing pumping power penalties in cooling system applications.

5.3.3 Volume concentration

Several investigations have reported a significant agreement on the effect of volume concentration that the viscosity of hybrid nanofluids increases with increasing volume concentration [98]. The viscosity increases due to the hydrodynamic collaborations between particles becoming critical as the fluid disturbance around one particle cooperate with that around other particles at higher volume fractions [99]. Normally, with increasing particle loading, the resistance against the flow is increased, a collision between nanoparticles increases as a result of random movement. The volume concentration avoids the passage of base fluid and nanoparticles on each other, and thus there is an improvement in the viscosity [32].

1. Motahari et al. [88] reported Newtonian behavior of SiO_2-MWCNT/oil hybrid nanofluids for 0.05%–1% volume concentration and observed increment in viscosity with increasing volume concentration.
2. Esfe and Rostamian [89] investigated the rheological behavior of ZnO-MWCNT/engine oil hybrid nanofluids by varying concentrations from 0.05% to 1% and exhibited non-Newtonian behavior.
3. Nabil et al. [75] experimentally investigated the viscosity of SiO_2+TiO_2-EG+H_2O hybrid nanofluids and reported that the viscosity increased with rising concentration and showed Newtonian behavior for volume concentration up to 3%.
4. Afrand et al. [67] reported Newtonian behavior for Fe_3O_4+Ag-EG hybrid nanofluids for volume concentration of less than 0.3%, whereas the concentration greater than 0.3% showed non-Newtonian behavior.
5. Soltani and Akbari [54] experimentally examined the impact of nanoparticle loading on the viscosity of MgO+MWCNT-EG hybrid nanofluids. Newtonian behavior was observed and reported that increment in viscosity was significant at higher concentrations (0.8% and 1%).
6. Yarmand et al. [100] studied the viscosity of GNP+Pt-H_2O hybrid nanofluids and observed enhancement with increasing volume concentration due to the effect on the internal shear stress of fluid.

7. Afshari et al. [101] investigated the dynamic viscosity of alumina+MWCNT-EG+H_2O hybrid nanofluids and noticed that the fluid behavior transformed from Newtonian to pseudoplastic non-Newtonian when the volume concentration increases from 0.5 vol.%.
8. Suresh et al. [102] analyzed the viscosity of alumina+Cu/water experimentally with the volume concentration ranging from 0.001 to 0.02 vol.% and observed an increment in viscosity with increasing volume concentration.
9. Said et al. [19] investigated the rheological and heat transfer properties using zinc ferrite/water hybrid nanofluids and observed an increment in viscosity with increasing particle loading.

5.3.4 Other factors

The preparation method, type of base fluid, shear rate, surfactants, and pH value are among other factors influencing the viscosity of hybrid nanofluids. Improving the thermal properties of the hybrid nanofluids needs a suitable technique for the preparation to avoid sedimentation. Nanoclusters are formed due to van der Waals forces, and a pathway with minimal thermal resistance is created for heat transfer [103]. Though nanoclustering may have an adverse impact on mono and hybrid nanofluids from two viewpoints, forming large masses causes suspension instability and lower heat transfer by forming areas without nanoparticles in fluid and improving thermal resistance. As volume concentration increases, cluster formation occurs, which leads to increased viscosity [104]. Various researchers have investigated the rheology of several hybrid nanofluids. Numerous rheological characteristics have been mentioned ranging from Newtonian to non-Newtonian shear thinning in the literature. During this process, agglomerates created in the nanofluid are cracked until they develop a well-ordered structure without the restrictions of high shear rates [105]. The additional surfactant will have a detrimental impact on rheological properties and chemical stability, so it should be regulated with supervision.

5.4 Conclusion and future outlook

1. Experimental studies show that hybrid nanofluids' viscosity increased with increased particle concentration and decreased with increasing temperature.

2. It is evident that the pressure drop and friction coefficient of the hybrid nanoparticles will augment with increasing particle loading, resulting in increased viscosity, not favorable for heat transfer performance.
3. Fabrication and application of hybrid nanofluids have been advanced in the research community. The demerits of nanofluids can be enhanced by employing hybrid nanofluids; changes in properties and performance can be observed.
4. It can be acknowledged that hybrid nanofluids are potential heat transfer fluids, and widespread experimental investigations on rheological behavior must be carried out for better performance in engineering applications.
5. Investigating the rheological characteristics of the hybrid nanofluid and viscosity impact on pressure drop, to intensity the heat transfer performance are needed to be carried out.

References

[1] Z. Said, M.H. Sajid, M.A. Alim, R. Saidur, N.A. Rahim, Experimental investigation of the thermophysical properties of Al_2O_3-nanofluid and its effect on a flat plate solar collector, Int. Commun. Heat Mass Transfer 48 (2013) 99–107.
[2] M. Sabiha, R. Saidur, S. Hassani, Z. Said, S. Mekhilef, Energy performance of an evacuated tube solar collector using single walled carbon nanotubes nanofluids, Energy Convers. Manag. 105 (2015) 1377–1388.
[3] Z. Said, A. Allagui, M.A. Abdelkareem, H. Alawadhi, K. Elsaid, Acid-functionalized carbon nanofibers for high stability, thermoelectrical and electrochemical properties of nanofluids, J. Colloid Interface Sci. 520 (2018) 50–57.
[4] Z. Said, R. Saidur, N. Rahim, Energy and exergy analysis of a flat plate solar collector using different sizes of aluminium oxide based nanofluid, J. Clean. Prod. 133 (2016) 518–530.
[5] Z. Said, R. Saidur, M. Sabiha, A. Hepbasli, N. Rahim, Energy and exergy efficiency of a flat plate solar collector using pH treated Al_2O_3 nanofluid, J. Clean. Prod. 112 (2016) 3915–3926.
[6] S.U. Choi, J.A. Eastman, Enhancing Thermal Conductivity of Fluids With Nanoparticles, Argonne National Lab, IL (United States), 1995.
[7] E. Bellos, Z. Said, C. Tzivanidis, The use of nanofluids in solar concentrating technologies: a comprehensive review, J. Clean. Prod. 196 (2018) 84–99.
[8] Z. Said, R. Saidur, M. Sabiha, N. Rahim, M. Anisur, Thermophysical properties of Single Wall Carbon Nanotubes and its effect on exergy efficiency of a flat plate solar collector, Sol. Energy 115 (2015) 757–769.
[9] Z. Said, R. Saidur, N. Rahim, Optical properties of metal oxides based nanofluids, Int. Commun. Heat Mass Transfer 59 (2014) 46–54.
[10] Z. Said, M.E.H. Assad, A.A. Hachicha, E. Bellos, M.A. Abdelkareem, D.Z. Alazaizeh, B.A. Yousef, Enhancing the performance of automotive radiators using nanofluids, Renew. Sust. Energ. Rev. 112 (2019) 183–194.
[11] Z. Said, R. Saidur, A. Hepbasli, N. Rahim, New thermophysical properties of water based TiO_2 nanofluid—the hysteresis phenomenon revisited, Int. Commun. Heat Mass Transfer 58 (2014) 85–95.

[12] R. Saidur, K.Y. Leong, H.A. Mohammed, A review on applications and challenges of nanofluids, Renew. Sust. Energ. Rev. 15 (2011) 1646–1668.
[13] Z. Said, M. Gupta, H. Hegab, N. Arora, A.M. Khan, M. Jamil, E. Bellos, A comprehensive review on minimum quantity lubrication (MQL) in machining processes using nano-cutting fluids, Int. J. Adv. Manuf. Technol. 105 (2019) 2057–2086.
[14] Z. Said, S. Arora, E. Bellos, A review on performance and environmental effects of conventional and nanofluid-based thermal photovoltaics, Renew. Sust. Energ. Rev. 94 (2018) 302–316.
[15] Z. Said, Thermophysical and optical properties of SWCNTs nanofluids, Int. Commun. Heat Mass Transfer 78 (2016) 207–213.
[16] J. Sarkar, P. Ghosh, A. Adil, A review on hybrid nanofluids: recent research, development and applications, Renew. Sust. Energ. Rev. 43 (2015) 164–177.
[17] A.A. Hachicha, B.A. Yousef, Z. Said, I. Rodríguez, A review study on the modeling of high-temperature solar thermal collector systems, Renew. Sust. Energ. Rev. 112 (2019) 280–298.
[18] M. Ehyaei, A. Ahmadi, M.E.H. Assad, A. Hachicha, Z. Said, Energy, exergy and economic analyses for the selection of working fluid and metal oxide nanofluids in a parabolic trough collector, Sol. Energy 187 (2019) 175–184.
[19] M. Gupta, V. Singh, Z. Said, Heat transfer analysis using zinc Ferrite/water (Hybrid) nanofluids in a circular tube: an experimental investigation and development of new correlations for thermophysical and heat transfer properties, Sustain. Energy Technol. Assess. 39 (2020) 100720.
[20] L.S. Sundar, Y.T. Sintie, Z. Said, M.K. Singh, V. Punnaiah, A.C. Sousa, Energy, efficiency, economic impact, and heat transfer aspects of solar flat plate collector with Al_2O_3 nanofluids and wire coil with core rod inserts, Sustain. Energy Technol. Assess. 40 (2020) 100772.
[21] N.K. Cakmak, Z. Said, L.S. Sundar, Z.M. Ali, A.K. Tiwari, Preparation, characterization, stability, and thermal conductivity of rGO-Fe_3O_4-TiO_2 hybrid nanofluid: an experimental study, Powder Technol. 372 (2020) 235–245.
[22] Z. Said, M. Ghodbane, L.S. Sundar, A.K. Tiwari, M. Sheikholeslami, B. Boumeddane, Heat transfer, entropy generation, economic and environmental analyses of linear Fresnel reflector using novel rGO-Co_3O_4 hybrid nanofluids, Renew. Energy 165 (2021) 420–437.
[23] L.S. Sundar, E.V. Ramana, Z. Said, V. Punnaiah, K.V.C. Mouli, A.C. Sousa, Properties, heat transfer, energy efficiency and environmental emissions analysis of flat plate solar collector using nanodiamond nanofluids, Diam. Relat. Mater. 110 (2020) 108115.
[24] L.S. Sundar, Z. Said, B. Saleh, M.K. Singh, A.C. Sousa, Combination of Co3O4 deposited rGO hybrid nanofluids and longitudinal strip inserts: thermal properties, heat transfer, friction factor, and thermal performance evaluations, Therm. Sci. Eng. Progr. 20 (2020) 100695.
[25] Z. Said, A.A. Hachicha, S. Aberoumand, B.A. Yousef, E.T. Sayed, E. Bellos, Recent advances on nanofluids for low to medium temperature solar collectors: energy, exergy, economic analysis and environmental impact, Prog. Energy Combust. Sci. 84 (2021) 100898.
[26] L.S. Sundar, S. Mesfin, E.V. Ramana, Z. Said, A.C. Sousa, Experimental investigation of thermo-physical properties, heat transfer, pumping power, entropy generation, and exergy efficiency of nanodiamond + Fe_3O_4/ 60:40% water-ethylene glycol hybrid nanofluid flow in a tube, Therm. Sci. Eng. Progr. 21 (2021) 100799.
[27] M. Sheikholeslami, S.A. Farshad, Z. Ebrahimpour, Z. Said, Recent progress on flat plate solar collectors and photovoltaic systems in the presence of nanofluid: a review, J. Clean. Prod. 126119 (2021).

[28] R.P. Chhabra, J.F. Richardson, Non-Newtonian Flow and Applied Rheology: Engineering Applications, Butterworth-Heinemann, 2011.
[29] Z. Said, M.A. Abdelkareem, H. Rezk, A.M. Nassef, Fuzzy modeling and optimization for experimental thermophysical properties of water and ethylene glycol mixture for Al_2O_3 and TiO_2 based nanofluids, Powder Technol. 353 (2019) 345–358.
[30] N.S. Pandya, A.N. Desai, A.K. Tiwari, Z. Said, Influence of the geometrical parameters and particle concentration levels of hybrid nanofluid on the thermal performance of axial grooved heat pipe, Therm. Sci. Eng. Progr. 21 (2021) 100762.
[31] Y. He, Y. Jin, H. Chen, Y. Ding, D. Cang, H. Lu, Heat transfer and flow behaviour of aqueous suspensions of TiO_2 nanoparticles (nanofluids) flowing upward through a vertical pipe, Int. J. Heat Mass Transf. 50 (2007) 2272–2281.
[32] A. Turgut, I. Tavman, M. Chirtoc, H.P. Schuchmann, C. Sauter, S. Tavman, Thermal conductivity and viscosity measurements of water-based TiO_2 nanofluids, Int. J. Thermophys. 30 (2009) 1213–1226.
[33] B. Finke, A. Kwade, C. Schilde, Numerical simulation of the rheological behavior of nanoparticulate suspensions, Materials 13 (2020) 4288.
[34] K. Anoop, R. Sadr, R. Yrac, M. Amani, High-pressure rheology of alumina-silicone oil nanofluids, Powder Technol. 301 (2016) 1025–1031.
[35] A. Asadi, A guideline towards easing the decision-making process in selecting an effective nanofluid as a heat transfer fluid, Energy Convers. Manag. 175 (2018) 1–10.
[36] D.M. Binding, M.A. Couch, K. Walters, The pressure dependence of the shear and elongational properties of polymer melts, J. Non-Newtonian Fluid Mech. 79 (1998) 137–155.
[37] K.R. Rajagopal, On implicit constitutive theories for fluids, J. Fluid Mech. 550 (2006) 243–249.
[38] L. Syam Sundar, E. Venkata Ramana, Z. Said, A.M. Pereira, A.C. Sousa, Heat Transfer of rGO/CO_3O_4 hybrid nanomaterial-based nanofluids and twisted tape configurations in a tube, J. Therm. Sci. Eng. Applicat. 13 (2021), 031004.
[39] A.K. Tiwari, N.S. Pandya, Z. Said, H.F. Öztop, N. Abu-Hamdeh, 4S consideration (synthesis, sonication, surfactant, stability) for the thermal conductivity of CeO_2 with MWCNT and water based hybrid nanofluid: an experimental assessment, Colloids Surf. A Physicochem. Eng. Asp. 610 (2021) 125918.
[40] A.K. Tiwari, N.S. Pandya, Z. Said, S.H. Chhatbar, Y.A. Al-Turki, A.R. Patel, 3S (Sonication, surfactant, stability) impact on the viscosity of hybrid nanofluid with different base fluids: an experimental study, J. Mol. Liq. 329 (2021) 115455.
[41] Z. Said, L.S. Sundar, H. Rezk, A.M. Nassef, H.M. Ali, M. Sheikholeslami, Optimizing density, dynamic viscosity, thermal conductivity and specific heat of a hybrid nanofluid obtained experimentally via ANFIS-based model and modern optimization, J. Mol. Liq. 321 (2021) 114287.
[42] Y.-M. Chu, M. Ibrahim, T. Saeed, A.S. Berrouk, E.A. Algehyne, R. Kalbasi, Examining rheological behavior of MWCNT-TiO_2/5W40 hybrid nanofluid based on experiments and RSM/ANN modeling, J. Mol. Liq. 333 (2021) 115969.
[43] M. Michael, A. Zagabathuni, S. Kumar Pabi, S. Ghosh, An experimental study on the rheological behavior of carbon black boron nitride hybrid nanofluids and development of a new correlation, J. Nanosci. Nanotechnol. 21 (2021) 3283–3290.
[44] M. Zayan, A.K. Rasheed, A. John, M. Khalid, A. Faris, Experimental investigation on rheological properties of water based novel ternary hybrid nanofluids, Nanoscience (2021).

[45] T. Le Ba, Z.I. Várady, I.E. Lukács, J. Molnár, I.A. Balczár, S. Wongwises, I.M. Szilágyi, Experimental investigation of rheological properties and thermal conductivity of SiO_2-P25 TiO_2 hybrid nanofluids, J. Therm. Anal. Calorim. (2020) 1–15.
[46] A. Asadi, I.M. Alarifi, H.M. Nguyen, H. Moayedi, Feasibility of least-square support vector machine in predicting the effects of shear rate on the rheological properties and pumping power of MWCNT-MgO/oil hybrid nanofluid based on experimental data, J. Therm. Anal. Calorim. 143 (2021) 1439–1454.
[47] S.-R. Yan, D. Toghraie, L.A. Abdulkareem, A.A. Alizadeh, P. Barnoon, M. Afrand, The rheological behavior of MWCNTs-ZnO/Water-Ethylene glycol hybrid non-Newtonian nanofluid by using of an experimental investigation, J. Mater. Res. Technol. 9 (2020) 8401–8406.
[48] A. Asadi, I.M. Alarifi, L.K. Foong, An experimental study on characterization, stability and dynamic viscosity of CuO-TiO_2/water hybrid nanofluid, J. Mol. Liq. 307 (2020) 112987.
[49] Z. Tian, S. Rostami, R. Taherialekouhi, A. Karimipour, A. Moradikazerouni, H. Yarmand, N.W.B.M. Zulkifli, Prediction of rheological behavior of a new hybrid nanofluid consists of copper oxide and multi wall carbon nanotubes suspended in a mixture of water and ethylene glycol using curve-fitting on experimental data, Phys. A: Stat. Mech. Applicat. 549 (2020) 124101.
[50] I. Kazemi, M. Sefid, M. Afrand, A novel comparative experimental study on rheological behavior of mono & hybrid nanofluids concerned graphene and silica nano-powders: characterization, stability and viscosity measurements, Powder Technol. 366 (2020) 216–229.
[51] I.M. Alarifi, A.B. Alkouh, V. Ali, H.M. Nguyen, A. Asadi, On the rheological properties of MWCNT-TiO_2/oil hybrid nanofluid: an experimental investigation on the effects of shear rate, temperature, and solid concentration of nanoparticles, Powder Technol. 355 (2019) 157–162.
[52] R.K. Singh, A.K. Sharma, A.R. Dixit, A.K. Tiwari, A. Pramanik, A. Mandal, Performance evaluation of alumina-graphene hybrid nano-cutting fluid in hard turning, J. Clean. Prod. 162 (2017) 830–845.
[53] L. Syam Sundar, M.K. Singh, M.C. Ferro, A.C.M. Sousa, Experimental investigation of the thermal transport properties of graphene oxide/Co_3O_4 hybrid nanofluids, Int. Commun. Heat Mass Transfer 84 (2017) 1–10.
[54] O. Soltani, M. Akbari, Effects of temperature and particles concentration on the dynamic viscosity of MgO-MWCNT/ethylene glycol hybrid nanofluid: experimental study, Phys. E: Low-dimens. Syst. Nanostruct. 84 (2016) 564–570.
[55] M. Asadi, A. Asadi, Dynamic viscosity of MWCNT/ZnO-engine oil hybrid nanofluid: an experimental investigation and new correlation in different temperatures and solid concentrations, Int. Commun. Heat Mass Transfer 76 (2016) 41–45.
[56] H. Yarmand, S. Gharehkhani, S.F.S. Shirazi, A. Amiri, E. Montazer, H.K. Arzani, R. Sadri, M. Dahari, S.N. Kazi, Nanofluid based on activated hybrid of biomass carbon/graphene oxide: synthesis, thermo-physical and electrical properties, Int. Commun. Heat Mass Transfer 72 (2016) 10–15.
[57] L.S. Sundar, E.V. Ramana, M. Graça, M.K. Singh, A.C. Sousa, Nanodiamond-Fe_3O_4 nanofluids: preparation and measurement of viscosity, electrical and thermal conductivities, Int. Commun. Heat Mass Transfer 73 (2016) 62–74.
[58] M.H. Esfe, A.A.A. Arani, M. Rezaie, W.-M. Yan, A. Karimipour, Experimental determination of thermal conductivity and dynamic viscosity of Ag-MgO/water hybrid nanofluid, Int. Commun. Heat Mass Transfer 66 (2015) 189–195.

[59] S. Abbasi, S.M. Zebarjad, S.H.N. Baghban, A. Youssefi, M.-S. Ekrami-Kakhki, Experimental investigation of the rheological behavior and viscosity of decorated multi-walled carbon nanotubes with TiO_2 nanoparticles/water nanofluids, J. Therm. Anal. Calorim. 123 (2016) 81–89.
[60] M. Baghbanzadeh, A. Rashidi, A.H. Soleimanisalim, D. Rashtchian, Investigating the rheological properties of nanofluids of water/hybrid nanostructure of spherical silica/MWCNT, Thermochim. Acta 578 (2014) 53–58.
[61] L.S. Sundar, M.K. Singh, E.V. Ramana, B. Singh, J. Grácio, A.C.M. Sousa, Enhanced thermal conductivity and viscosity of nanodiamond-nickel nanocomposite nanofluids, Sci. Rep. 4 (2014) 4039.
[62] S.S. Botha, P. Ndungu, B.J. Bladergroen, Physicochemical properties of oil-based nanofluids containing hybrid structures of silver nanoparticles supported on silica, Ind. Eng. Chem. Res. 50 (2011) 3071–3077.
[63] K. Lu, Rheological behavior of carbon nanotube-alumina nanoparticle dispersion systems, Powder Technol. 177 (2007) 154–161.
[64] W. Richmond, R. Jones, P. Fawell, The relationship between particle aggregation and rheology in mixed silica-titania suspensions, Chem. Eng. J. 71 (1998) 67–75.
[65] L.S. Sundar, K.V. Chandra Mouli, Z. Said, A.C. Sousa, Heat transfer and second law analysis of ethylene glycol-based ternary hybrid nanofluid under laminar flow, J. Therm. Sci. Eng. Applicat. 13 (2021), 051021.
[66] P.K. Kanti, K. Sharma, Z. Said, M. Gupta, Experimental investigation on thermo-hydraulic performance of water-based fly ash-Cu hybrid nanofluid flow in a pipe at various inlet fluid temperatures, Int. Commun. Heat Mass Transfer 124 (2021) 105238.
[67] M. Afrand, D. Toghraie, B. Ruhani, Effects of temperature and nanoparticles concentration on rheological behavior of Fe_3O_4-Ag/EG hybrid nanofluid: an experimental study, Exp. Thermal Fluid Sci. 77 (2016) 38–44.
[68] S. Akilu, A.T. Baheta, K.V. Sharma, Experimental measurements of thermal conductivity and viscosity of ethylene glycol-based hybrid nanofluid with TiO_2-CuO/C inclusions, J. Mol. Liq. 246 (2017) 396–405.
[69] Z. Said, L.S. Sundar, H. Rezk, A.M. Nassef, S. Chakraborty, C. Li, Thermophysical properties using ND/water nanofluids: an experimental study, ANFIS-based model and optimization, J. Mol. Liq. 330 (2021) 115659.
[70] L.S. Sundar, E.V. Ramana, Z. Said, Y.R. Sekhar, K.V.C. Mouli, A.C. Sousa, Heat transfer, energy, and exergy efficiency enhancement of nanodiamond/water nanofluids circulate in a flat plate solar collector, J. Enhanc. Heat Transfer 28 (2021).
[71] H. Eshgarf, N. Sina, M.H. Esfe, F. Izadi, M. Afrand, Prediction of rheological behavior of MWCNTs-SiO_2/EG-water non-Newtonian hybrid nanofluid by designing new correlations and optimal artificial neural networks, J. Therm. Anal. Calorim. 132 (2018) 1029–1038.
[72] M.H. Esfe, M. Afrand, W.-M. Yan, H. Yarmand, D. Toghraie, M. Dahari, Effects of temperature and concentration on rheological behavior of MWCNTs/SiO2 (20–80)-SAE40 hybrid nano-lubricant, Int. Commun. Heat Mass Transfer 76 (2016) 133–138.
[73] M.H. Esfe, F. Zabihi, H. Rostamian, S. Esfandeh, Experimental investigation and model development of the non-Newtonian behavior of CuO-MWCNT-10w40 hybrid nano-lubricant for lubrication purposes, J. Mol. Liq. 249 (2018) 677–687.
[74] A. Asadi, M. Asadi, A. Rezaniakolaei, L.A. Rosendahl, M. Afrand, S. Wongwises, Heat transfer efficiency of Al2O3-MWCNT/thermal oil hybrid nanofluid as a cooling fluid in thermal and energy management applications:

an experimental and theoretical investigation, Int. J. Heat Mass Transf. 117 (2018) 474–486.
[75] M. Nabil, W. Azmi, K.A. Hamid, R. Mamat, F.Y. Hagos, An experimental study on the thermal conductivity and dynamic viscosity of TiO_2-SiO_2 nanofluids in water: ethylene glycol mixture, Int. Commun. Heat Mass Transfer 86 (2017) 181–189.
[76] S.S. Ghadikolaei, M. Yassari, H. Sadeghi, K. Hosseinzadeh, D.D. Ganji, Investigation on thermophysical properties of Tio_2-Cu/H_2O hybrid nanofluid transport dependent on shape factor in MHD stagnation point flow, Powder Technol. 322 (2017) 428–438.
[77] G.M. Moldoveanu, C. Ibanescu, M. Danu, A.A. Minea, Viscosity estimation of Al_2O_3, SiO_2 nanofluids and their hybrid: an experimental study, J. Mol. Liq. 253 (2018) 188–196.
[78] M. Afrand, K.N. Najafabadi, N. Sina, M.R. Safaei, A.S. Kherbeet, S. Wongwises, M. Dahari, Prediction of dynamic viscosity of a hybrid nano-lubricant by an optimal artificial neural network, Int. Commun. Heat Mass Transfer 76 (2016) 209–214.
[79] E. Dardan, M. Afrand, A.M. Isfahani, Effect of suspending hybrid nano-additives on rheological behavior of engine oil and pumping power, Appl. Therm. Eng. 109 (2016) 524–534.
[80] A. Asadi, M. Asadi, A. Rezaniakolaei, L.A. Rosendahl, S. Wongwises, An experimental and theoretical investigation on heat transfer capability of Mg (OH) 2/MWCNT-engine oil hybrid nano-lubricant adopted as a coolant and lubricant fluid, Appl. Therm. Eng. 129 (2018) 577–586.
[81] R.R. Sahoo, V. Kumar, Development of a new correlation to determine the viscosity of ternary hybrid nanofluid, Int. Commun. Heat Mass Transfer 111 (2020) 104451.
[82] M. Hemmat Esfe, A.A. Abbasian Arani, M. Rezaie, W.-M. Yan, A. Karimipour, Experimental determination of thermal conductivity and dynamic viscosity of Ag-MgO/water hybrid nanofluid, Int. Commun. Heat Mass Transfer 66 (2015) 189–195.
[83] I.M. Mahbubul, T.H. Chong, S.S. Khaleduzzaman, I.M. Shahrul, R. Saidur, B.-D. Long, M.A. Amalina, Effect of ultrasonication duration on colloidal structure and viscosity of alumina-water nanofluid, Ind. Eng. Chem. Res. 53 (2014) 6677–6684.
[84] S.S. Murshed, S.-H. Tan, N.-T. Nguyen, Temperature dependence of interfacial properties and viscosity of nanofluids for droplet-based microfluidics, J. Phys. D. Appl. Phys. 41 (2008), 085502.
[85] A. Shahsavar, M. Saghafian, M. Salimpour, M. Shafii, Effect of temperature and concentration on thermal conductivity and viscosity of ferrofluid loaded with carbon nanotubes, Heat Mass Transf. 52 (2016) 2293–2301.
[86] Z. Said, M.A. Abdelkareem, H. Rezk, A.M. Nassef, H.Z. Atwany, Stability, thermophysical and electrical properties of synthesized carbon nanofiber and reduced-graphene oxide-based nanofluids and their hybrid along with fuzzy modeling approach, Powder Technol. 364 (2020) 795–809.
[87] M. Afrand, K.N. Najafabadi, M. Akbari, Effects of temperature and solid volume fraction on viscosity of SiO2-MWCNTs/SAE40 hybrid nanofluid as a coolant and lubricant in heat engines, Appl. Therm. Eng. 102 (2016) 45–54.
[88] K. Motahari, M.A. Moghaddam, M. Moradian, Experimental investigation and development of new correlation for influences of temperature and concentration on dynamic viscosity of MWCNT-SiO_2 (20–80)/20W50 hybrid nano-lubricant, Chin. J. Chem. Eng. 26 (2018) 152–158.

[89] M.H. Esfe, H. Rostamian, M.R. Sarlak, A novel study on rheological behavior of ZnO-MWCNT/10w40 nanofluid for automotive engines, J. Mol. Liq. 254 (2018) 406–413.
[90] L. Syam Sundar, S. Mesfin, E. Venkata Ramana, Z. Said, A.C.M. Sousa, Experimental investigation of thermo-physical properties, heat transfer, pumping power, entropy generation, and exergy efficiency of nanodiamond + Fe_3O_4/ 60:40% water-ethylene glycol hybrid nanofluid flow in a tube, Therm. Sci. Eng. Progr. 21 (2021) 100799.
[91] S. Akilu, A.T. Baheta, K. Sharma, Experimental measurements of thermal conductivity and viscosity of ethylene glycol-based hybrid nanofluid with TiO_2-CuO/C inclusions, J. Mol. Liq. 246 (2017) 396–405.
[92] A.K. Sharma, A.K. Tiwari, A.R. Dixit, Rheological behaviour of nanofluids: a review, Renew. Sust. Energ. Rev. 53 (2016) 779–791.
[93] S. Murshed, K. Leong, C. Yang, Determination of the effective thermal diffusivity of nanofluids by the double hot-wire technique, J. Phys. D Appl. Phys. 39 (2006) 5316.
[94] Y. He, Y. Jin, H. Chen, Y. Ding, D. Cang, H. Lu, Heat transfer and flow behaviour of aqueous suspensions of TiO_2 nanoparticles (nanofluids) flowing upward through a vertical pipe, Int. J. Heat Mass Transf. 50 (2007) 2272–2281.
[95] E.V. Timofeeva, J.L. Routbort, D. Singh, Particle shape effects on thermophysical properties of alumina nanofluids, J. Appl. Phys. 106 (2009), 014304.
[96] S. Ferrouillat, A. Bontemps, O. Poncelet, O. Soriano, J.-A. Gruss, Influence of nanoparticle shape factor on convective heat transfer and energetic performance of water-based SiO_2 and ZnO nanofluids, Appl. Therm. Eng. 51 (2013) 839–851.
[97] J.P. Meyer, S.A. Adio, M. Sharifpur, P.N. Nwosu, The viscosity of nanofluids: a review of the theoretical, empirical, and numerical models, Heat Transfer Eng. 37 (2016) 387–421.
[98] Z. Said, M. Sabiha, R. Saidur, A. Hepbasli, N.A. Rahim, S. Mekhilef, T. Ward, Performance enhancement of a flat plate solar collector using titanium dioxide nanofluid and polyethylene glycol dispersant, J. Clean. Prod. 92 (2015) 343–353.
[99] L.S. Sundar, M.K. Singh, A.C.M. Sousa, Enhanced heat transfer and friction factor of MWCNT-Fe_3O_4/water hybrid nanofluids, Int. Commun. Heat Mass Transfer 52 (2014) 73–83.
[100] H. Yarmand, S. Gharehkhani, S.F.S. Shirazi, M. Goodarzi, A. Amiri, W.S. Sarsam, M.S. Alehashem, M. Dahari, S. Kazi, Study of synthesis, stability and thermo-physical properties of graphene nanoplatelet/platinum hybrid nanofluid, Int. Commun. Heat Mass Transfer 77 (2016) 15–21.
[101] A. Afshari, M. Akbari, D. Toghraie, M.E. Yazdi, Experimental investigation of rheological behavior of the hybrid nanofluid of MWCNT-alumina/water (80%)-ethylene-glycol (20%), J. Therm. Anal. Calorim. 132 (2018) 1001–1015.
[102] S. Suresh, K.P. Venkitaraj, P. Selvakumar, M. Chandrasekar, Synthesis of Al_2O_3-Cu/water hybrid nanofluids using two step method and its thermo physical properties, Colloids Surf. A Physicochem. Eng. Asp. 388 (2011) 41–48.
[103] Z. Said, L.S. Sundar, A.K. Tiwari, H.M. Ali, M. Sheikholeslami, E. Bellos, H. Babar, Recent advances on the fundamental physical phenomena behind stability, dynamic motion, thermophysical properties, heat transport, applications, and challenges of nanofluids, Phys. Rep. (2021), https://doi.org/10.1016/j.physrep.2021.07.002. In press.
[104] A. Ghadimi, R. Saidur, H. Metselaar, A review of nanofluid stability properties and characterization in stationary conditions, Int. J. Heat Mass Transf. 54 (2011) 4051–4068.
[105] W.J. Tseng, K.-C. Lin, Rheology and colloidal structure of aqueous TiO_2 nanoparticle suspensions, Mater. Sci. Eng. A 355 (2003) 186–192.

6

Radiative transport of hybrid nanofluid

Arun Kumar Tiwari[a], Amit Kumar[a], and Zafar Said[b,c,d,*]

[a]Mechanical Engineering Department, Institute of Engineering & Technology, Dr. A.P.J. Abdul Kalam Technical University, Uttar Pradesh, Lucknow, India. [b]Department of Sustainable and Renewable Energy Engineering, University of Sharjah, Sharjah, United Arab Emirates. [c]Research Institute for Sciences and Engineering, University of Sharjah, Sharjah, United Arab Emirates [d]U.S.-Pakistan Center for Advanced Studies in Energy (USPCAS-E), National University of Sciences and Technology (NUST), Islamabad, Pakistan

**Corresponding author: zsaid@sharjah.ac.ae; zaffar.ks@gmail.com*

Chapter outline

6.1 Introduction 132
6.2 Optical properties 132
- 6.2.1 Rayleigh scattering approximation 133
- 6.2.2 Maxwell-Garnett approximation 136
- 6.2.3 Mie scattering approximation 137

6.3 Radiative transfer 140
6.4 Effect of different parameters on optical properties 143
- 6.4.1 Effect of particle size 143
- 6.4.2 Effect of volume fraction 143

6.5 Challenges and outlook 144
6.6 Summary 145
References 145

Nomenclature

X	particle size (nm)
d_p	particle diameters (nm)
λ	wavelength of light (nm)
IP	interparticle distance (nm)
σ	extinction coefficient
f_v	particle volume friction
$Q_{e\lambda}$	extinction efficiency
$Q_{a\lambda}$	absorption efficiency
$Q_{a\lambda}$	scattering efficiency
m	relative complex refractive

Hybrid Nanofluids: Preparation, Characterization and Applications. https://doi.org/10.1016/B978-0-323-85836-6.00006-5

μ	real component of the refractive index
k_{bf}	complex imaginary component of the refractive index of base fluid
A_m	solar weighted absorption factor
$I_{AM1.5}$	AM1.5 solar spectrum in light of ASTMG173
ε	dielectric function
a_n **and** b_n	Mie scattering coefficient
ψ_n **and** ξ_n	Riccati-Bessel function
S	Poynting vector
E_t	electric field
H_t	magnetic field

Subscript

a	absorption
e	extinction
s	scattering
bf	base fluid
nf	nanofluid
j **and** k	notation of two different nanoparticles
p	particle

6.1 Introduction

Hybrid nanofluids are prepared by blending two different kinds of nanoparticles in the same base fluid [1]. The hybrid nanofluids are seemed to be replacing the mono nanofluid due to having more significant advantages over the mono nanofluids, as they have a broad absorption range, lower extinction, high thermal conductivity, low-pressure drop, low frictional losses, and low pumping power compared to the mono nanofluids [2–4]. The hybrid nanofluids have been tested for various applications like solar collectors [5, 6], photovoltaic thermal applications [7], electronic component thermal management, photovoltaic thermal management [8], machine cutting, engine applications, and automotive cooling [9–11]. Different types of hybrid nanofluids have been used so far, as shown in Fig. 6.1.

6.2 Optical properties

The performance of solar collectors heavily depends upon the optical features of the working fluid [12]. These optical features of hybrid nanofluids are the functions of morphology and volume concentrations of nanoparticles. When any electromagnetic wave such as sunlight interacts with a colloidal medium containing small particles, the radiative intensity of the wave may be altered by absorption or scattering [13].

To study the optical properties of hybrid nanofluids, we will study the different parameters (i.e., absorption, transmittance,

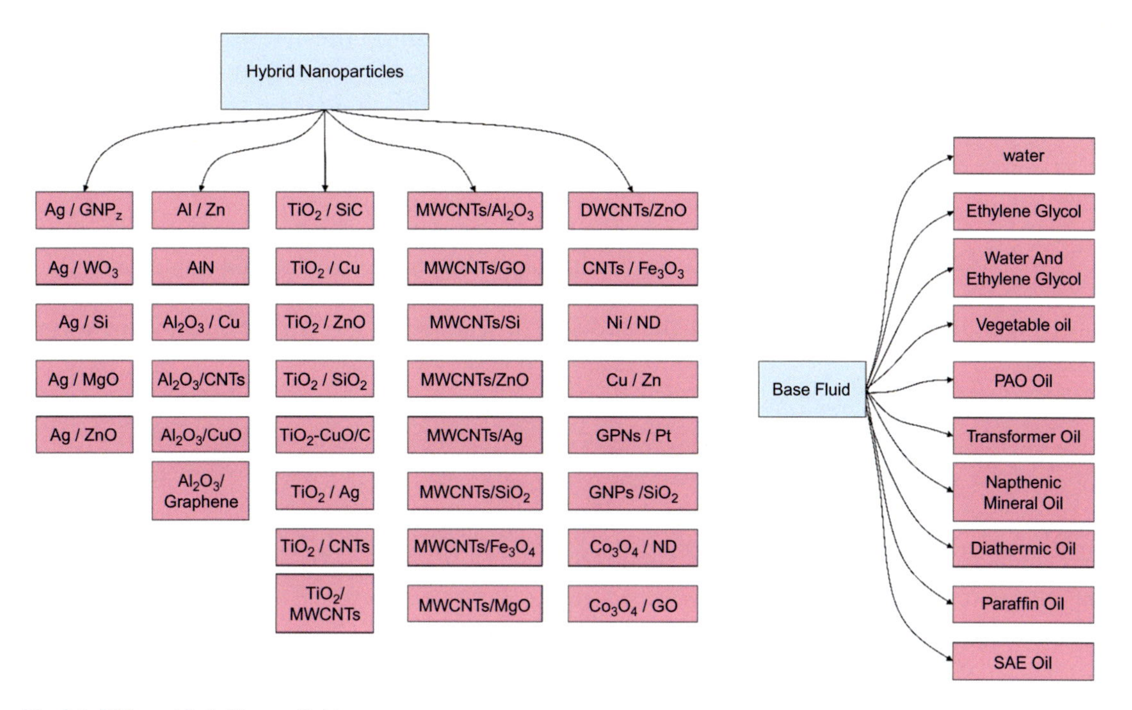

Fig. 6.1 Different hybrid nanofluids.

scattering, and extinction coefficient) of hybrid nanofluids and models related to these parameters [14]. A lot of researchers studied the optical behavior of hybrid nanofluids using different modeling approaches.

6.2.1 Rayleigh scattering approximation

The Rayleigh scattering approximation theory applies to small spherical particles having an independent scattering effect [15, 16]. However, this theory lost validity at smaller size particle $x = \pi.d_p/\lambda$, where ($x \ll 1$). Howell et al. [17] suggested that the following criterion must be satisfied for an independent scattering system

$$IP + 0.1d_p > \frac{\lambda}{2} \tag{1}$$

where IP is denoted as interparticle distance, i.e., $\frac{IP}{d_p} = 2.77 f_v^{-0.388}$ [18], d_p is denoted as particle diameter, and λ is denoted as the wavelength of light.

Small nanoparticles having $d_p < 100$ nm and low volume concentration support the independent scattering and need very simple calculation for optical properties [19]. According to Rayleigh scattering approximation, the extinction coefficient of the base fluid can be represented as [20, 21]

$$\sigma_{e\lambda,np} = \frac{3 f_v Q_{e\lambda}}{d_p} \tag{2}$$

where σ is denoted as extinction coefficient, f_v is denoted as particle volume friction, d_p is denoted as particle diameters, and $Q_{e\lambda}$ is denoted as extinction efficiency for each nanoparticle.

$Q_{e\lambda}$ is the addition of scattering efficiency ($Q_{s\lambda}$) and absorption efficiency ($Q_{a\lambda}$), which is defined as

$$Q_{e\lambda} = Q_{s\lambda} + Q_{a\lambda} \tag{3}$$

where

$$Q_{s\lambda} = \frac{8}{3} x^4 \left| \left(\frac{m^2 - 1}{m^2 + 2} \right) \right|^2 \tag{4}$$

and

$$Q_{a\lambda} = 4.x.I.m \left\{ \frac{\mathrm{m}^2 - 1}{\mathrm{m}^2 + 2} \left[1 + \frac{x^2}{15} \left(\frac{\mathrm{m}^2 - 1}{\mathrm{m}^2 + 2} \right) \frac{m^4 + 27m^2 + 38}{2m^2 + 3} \right] \right\} \tag{5}$$

where λ is denoted as incident wavelength and m is denoted as the relative complex refractive index of nanofluid $m = (\mu_{np} + i.k_{np}) / \mu_{bf}$. The scattering effect can be neglected for high absorbing transparent liquid such as water, and then the extinction coefficient of the hybrid nanofluid is equal to the absorption coefficient, which is expressed as

$$\sigma_{e\lambda,\,bf} = \frac{4\,\pi k_{bf}}{\lambda} \tag{6}$$

where k_{bf} is denoted as a complex imaginary component of the refractive index of base fluid having wavelength-dependent property, and the total extinction coefficient of the hybrid nanofluid can be written as [22]

$$\sigma_{total} = \sigma_{particles} + \sigma_{basefluid} \tag{7}$$

So far, Rayleigh scattering approximation is extensively used to evaluate the optical behavior of hybrid nanofluids. Li et al. [23] studied the optical properties of MWCNT-SiC/EG hybrid nanofluid in the direct solar absorption collector. They calculated the percentage of solar energy absorbed by hybrid nanofluid by calculating the solar weighted absorption factor (A_m), which is described as [24]

$$A_m = 1 - \frac{\int_{\lambda_{\min}}^{\lambda_{\max}} I_{AM1.5}(\lambda).\exp(-k(\lambda).l).d\lambda}{\int_{\lambda_{\min}}^{\lambda_{\max}} I_{AM1.5}(\lambda).d\lambda} \tag{8}$$

where l is denoted as penetration length and $I_{AM1.5}$ is the AM1.5 solar spectrum in light of ASTMG173 [25]. It is shown from Fig. 6.2 that for 10 cm penetration length, the addition of 0.5 wt.% MWCNT-SiC nanoparticles increase the solar weighted absorption factor (A_m) of pure EG up to 99.99% or more. Zhang et al. [26] calculated the photothermal conversion properties of CuO-MWCNT/water hybrid nanofluid for direct solar thermal energy harvest. It is depicted from Fig. 6.3 that the addition of 0.25 wt.% CuO nanoparticles in water increases A_m up to 99.99%, while further addition of MWCNT in CuO/water nanofluid shows the further increment in A_m. Taylor et al. [22] developed a physical model to predict the optical properties of graphite and high absorbing metal particle-based nanofluids using Rayleigh scattering approximation. Their experimental study accurately predicted the extinction coefficient for graphite-based nanofluid by using Rayleigh scattering approximation. The

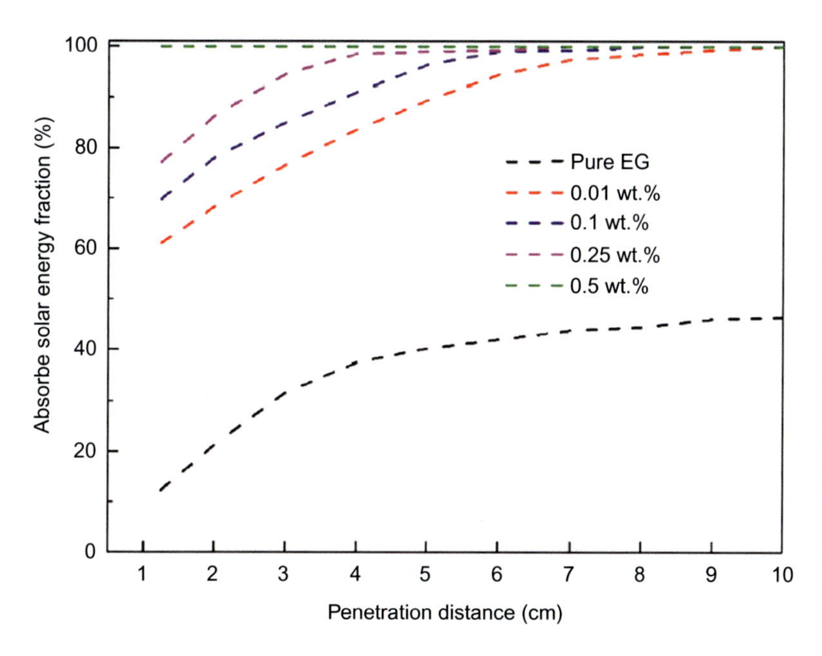

Fig. 6.2 The solar weighted absorption fraction of SiC-MWCNTs hybrid nanofluids at different mass concentrations. Copyright Elsevier (Lic # 5061780837144) (X. Li, G. Zeng, X. Lei, The stability, optical properties and solar-thermal conversion performance of SiC-MWCNTs hybrid nanofluids for the direct absorption solar collector (DASC) application, Solar Energy Mater. Solar Cells 206 (2020) 110323).

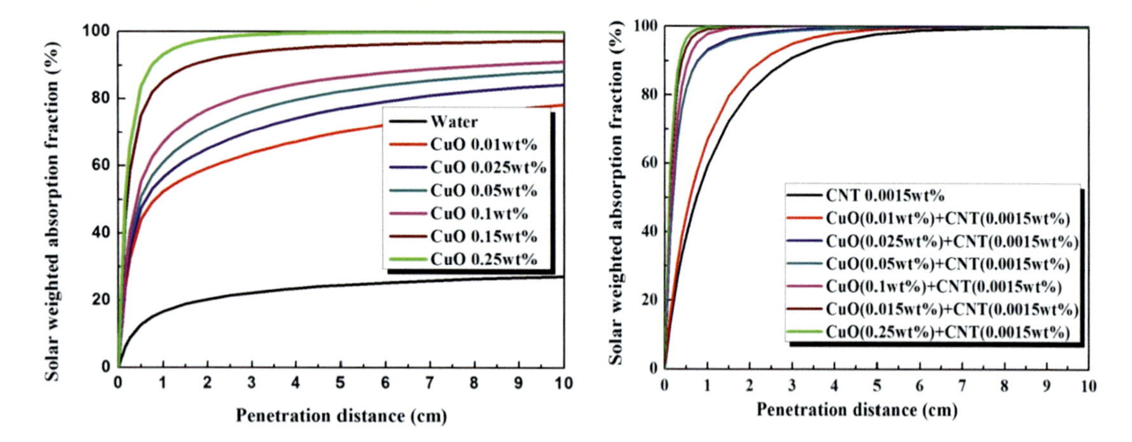

Fig. 6.3 Solar weighted absorption fraction vs the penetration distance for the CuO/H_2O nanofluids (A) and CuO-MWCNT/H_2O nanofluids (B) at different mass fractions. Copyright Elsevier (Lic # 5061781067732) (J. Qu, et al., Photo-thermal conversion properties of hybrid CuO-MWCNT/H_2O nanofluids for direct solar thermal energy harvest, Appl. Therm. Eng.147 (2019) 390–398).

extinction coefficient for metal-based nanofluid was overpredicted at the shorter wavelengths (<600–700 nm) and underpredicted at the longer wavelengths.

6.2.2 Maxwell-Garnett approximation

Maxwell-Garnett's theory is a common approach used to generate a model for finding out the optical properties of composite material. The complex refractive index can be effectively calculated by the following equation [20]:

$$\varepsilon_{nf} = \varepsilon_{bf}\left[1 + \frac{3.f_v\left(\frac{\varepsilon_{np} - \varepsilon_{bf}}{\varepsilon_{np} + 2\varepsilon_{bf}}\right)}{1 - f_v\left(\frac{\varepsilon_{np} - \varepsilon_{bf}}{\varepsilon_{np} + 2\varepsilon_{bf}}\right)}\right] = \varepsilon_1 + i.\varepsilon_2 \tag{9}$$

where f_v is denoted as the volume concentration of hybrid nanofluid while ε_{bf} and ε_{nf} are the dielectric functions of the base fluid and hybrid nanofluid, respectively. The real and imaginary components of the refractive index can be written as [20]

$$\mu_{nf} = \sqrt{\frac{\sqrt{\left(\varepsilon_1^2 + \varepsilon_2^2\right)} + \varepsilon_1}{2}} \tag{10}$$

$$k_{nf} = \sqrt{\frac{\sqrt{\left(\varepsilon_1^2 + \varepsilon_2^2\right)} - \varepsilon_1}{2}} \tag{11}$$

Using Eqs. (10) and (11), Taylor et al. [22] calculated the value of real and imaginary parts of the refractive index for water-based nanofluid at different wavelengths. The real part of the refractive index between base fluid and nanoparticles provides accurate results with minor deviations from the real refractive index at a low volume concentration. But for the imaginary part of the refractive index, this approach provides unpredicted results for extinction coefficient. Due to this large difference in extinction coefficient, the Maxwell-Garnett approximation is not the accurate approach to calculating nanofluids' extinction coefficient.

6.2.3 Mie scattering approximation

Mie scattering theory explains the interaction between electromagnetic waves and particles. This theory has no size limitation geometric optics. This theory can describe most of the spherical particle scattering system and the Rayleigh scattering system. According to Mie scattering theory, the scattering and extinction of the nanoparticles are described in terms of scattering efficiency and extinction efficiency factors, respectively [13], which are written as

$$Q_{s\lambda} = \frac{2}{x^2} \sum_{n=1}^{\infty} (2n+1)\left(|a_n|^2 + |b_n|^2\right) \tag{12a}$$

$$Q_{e\lambda} = \frac{2}{x^2} \sum_{n=1}^{\infty} (2n+1)\ \mathrm{Re}\{a_n + b_n\} \tag{12b}$$

where a_n and b_n are denoted as Mie scattering coefficients, which are the complex functions of x and $y = mx$, which is given by

$$a_n = \frac{\acute{\psi}(y)\psi_n(x) - m\psi_n(y)\acute{\psi}(x)}{\acute{\psi}(y)\xi_n(x) - m\psi_n(y)\acute{\xi}(x)} \tag{13a}$$

$$b_n = \frac{m\acute{\psi}(y)\psi_n(x) - \psi_n(y)\acute{\psi}(x)}{m\acute{\psi}(y)\xi_n(x) - \psi_n(y)\acute{\xi}(x)} \tag{13b}$$

where ψ_n and ξ_n are denoted as Riccati-Bessel functions.

As Mie scattering theory is more complex, it is rarely used to find out the optical properties of hybrid nanofluid. Rayleigh scattering theory is mainly preferred to find out the optical properties.

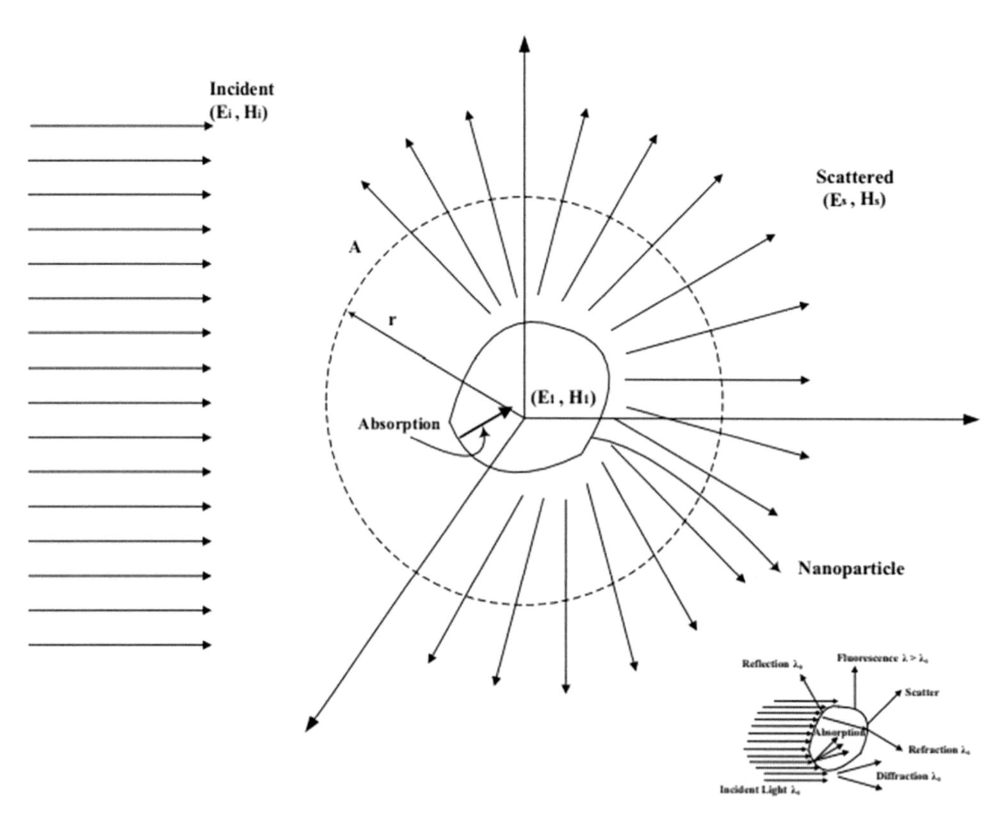

Fig. 6.4 Absorption and scattering of the light beam by a particle. Copyright Elsevier (Lic # 5061791237558) (A. Menbari, A.A. Alemrajabi, Analytical modeling and experimental investigation on optical properties of new class of nanofluids (Al2O3–CuO binary nanofluids) for direct absorption solar thermal energy, Opt. Mater. 52 (2016) 116–125).

Menbari et al. [27] developed an analytical model to calculate the optical properties of hybrid nanofluids (Al_2O_3 + CuO/water) and compared them with experimental data. As Mie scattering theory is used to calculate the extinction coefficient of similar nanoparticles dispersed in the base fluid, to calculate the extinction coefficient of hybrid nanofluids, they used the following assumption: distance between nanoparticles is greater than their diameters, as shown in Fig. 6.4.

The electromagnetic energy across the surface A is given as

$$
\begin{aligned}
W_i &= -\int (S_i.\hat{e}_r)dA \\
W_s &= -\int (S_s.\hat{e}_r)dA \\
W_e &= -\int (S_e.\hat{e}_r)dA
\end{aligned}
\tag{14}
$$

The efficiencies of absorption, scattering, and extinction are given by [13]

$$Q_{a,\lambda} = \frac{W_{a,\lambda}}{I_i.A_{particle}}$$
$$Q_{s,\lambda} = \frac{W_{s,\lambda}}{I_i.A_{particle}} \tag{15}$$
$$Q_{e,\lambda} = \frac{W_{e,\lambda}}{I_i.A_{particle}}$$

So, the Poynting vector (S) in the medium of dissimilar nanoparticles is given by

$$\begin{aligned} S = S_i + S_s + S_e &= \frac{1}{2}\mathrm{Re}(E_t \times H_t) \\ &= \frac{1}{2}\mathrm{Re}\left\{\left[\left(E_i + \sum_j E_{sj}\right) + \left(E_i + \sum_k E_{sk}\right) - E_i\right]\right. \\ &\quad \times \left.\left[\left(H_i + \sum_j H_{sj}\right) + \left(H_i + \sum_k H_{sk}\right) - H_i\right]\right\} \end{aligned} \tag{16}$$

where E_t and H_t are denoted as electric and magnetic fields, respectively, while j and k are different nanoparticles.

$$E_t = E_i + \sum_j E_{sj} + \sum_k E_{sk} \tag{17a}$$

$$H_t = H_i + \sum_j H_{sj} + \sum_k H_{sk} \tag{17b}$$

Performing some algebraic operations (addition, multiplication, and subtraction) on Eq. (16),

$$\begin{aligned} S = &\left(S_i + S_s|_j + S_e|_j + S_s|_k + S_e|_k\right) \\ &+ \frac{1}{2}\mathrm{Re}\left\{\sum_j E_{sj} \times \sum_k H_{sk} + \sum_k E_{sk} \times \sum_j H_{sj}\right\} \end{aligned} \tag{18}$$

The second term in Eq. (18) is the product of two scattering efficiency, which is negligible compared to scattering and extinction coefficient.

So, Eq. (18) can be written as

$$S = \left(S_i + S_s|_j + S_e|_j + S_s|_k + S_e|_k\right) = S_i + (S_j + S_k)_s + (S_j + S_k)_e \tag{19}$$

Using Eqs. (2), (14)–(16), and (18), we get

$$\sigma_{e,total} = \left.\frac{3f_v Q_{e\lambda}}{d_p}\right|_j + \left.\frac{3f_v Q_{e\lambda}}{d_p}\right|_k \tag{20}$$

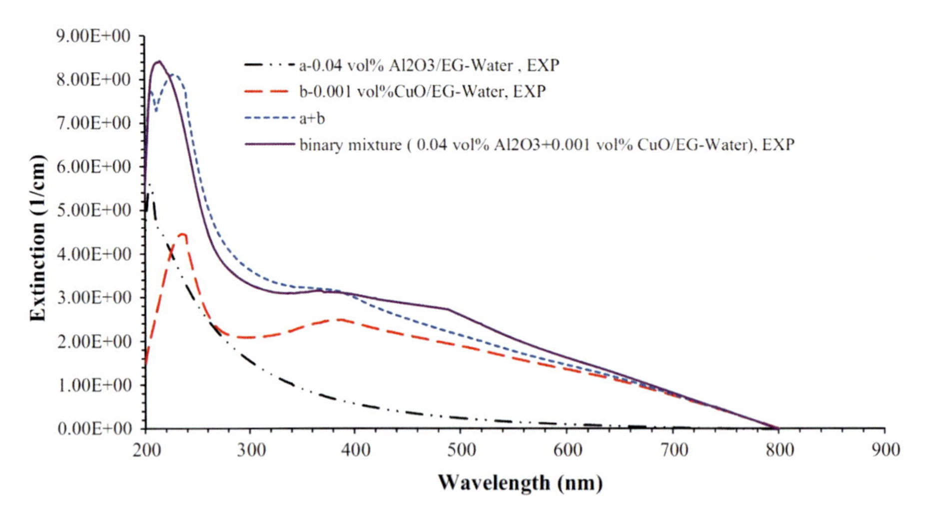

Fig. 6.5 Experimental extinction behavior of different nanofluids vs wavelength. Copyright Elsevier (Lic # 5061791237558) (A. Menbari, A.A. Alemrajabi, Analytical modeling and experimental investigation on optical properties of new class of nanofluids (Al2O3–CuO binary nanofluids) for direct absorption solar thermal energy, Opt. Mater. 52 (2016) 116–125).

$$\sigma_{s,total} = \left.\frac{3f_v Q_{s\lambda}}{d_p}\right|_j + \left.\frac{3f_v Q_{s\lambda}}{d_p}\right|_k \tag{21}$$

Based on their experimental model, they concluded that the extinction coefficient of the hybrid nanofluid is the summation of the individual extinction coefficient of the nanoparticles (as shown in Fig. 6.5); hence, they claimed that the proposed analytical model is suitable to calculate the extinction coefficient. Fig. 6.6 depicts the extinction coefficient comparison of the analytical and experimental model for different nanofluids at 0.04 vol%.

6.3 Radiative transfer

To achieve the precise value of the absorption coefficient of hybrid nanofluid, Song et al. [28] solved a simple radiation transport model. To measure the optical properties of hybrid nanofluid, they used a UV-vis spectrophotometer (U4100, Hitachi), having a lab-sphere diffuse reflector. The schematic diagram of the experimental setup is shown in Fig. 6.7.

The experiment was performed in three steps. In the first step (Fig. 6.7A), a narrow vertical slit was placed in front of the detector chamber, coinciding with the incoming radiation beam from an empty sample cell. The maximum possible distance was

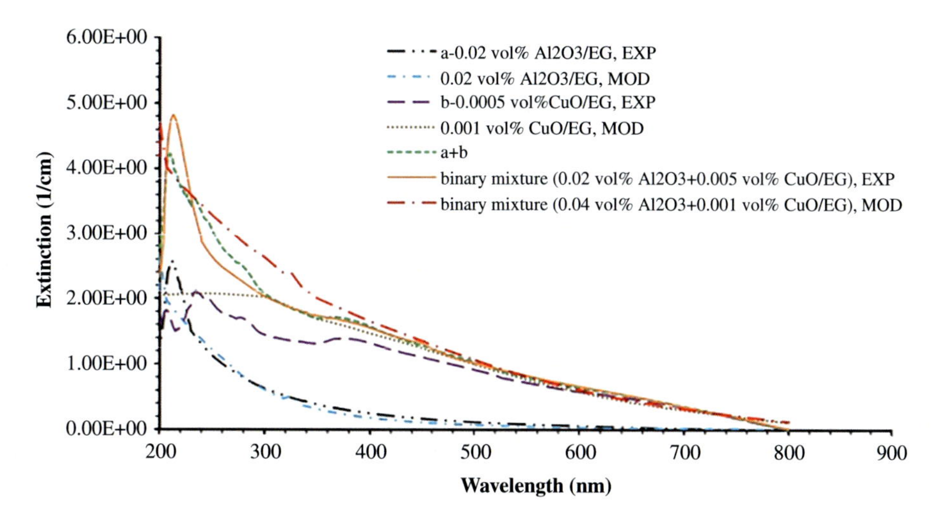

Fig. 6.6 Experimental and analytical extinction behavior for different nanoparticles dispersed in EG. Copyright Elsevier (Lic # 5061791237558) (A. Menbari, A.A. Alemrajabi, Analytical modeling and experimental investigation on optical properties of new class of nanofluids (Al2O3–CuO binary nanofluids) for direct absorption solar thermal energy, Opt. Mater. 52 (2016) 116–125).

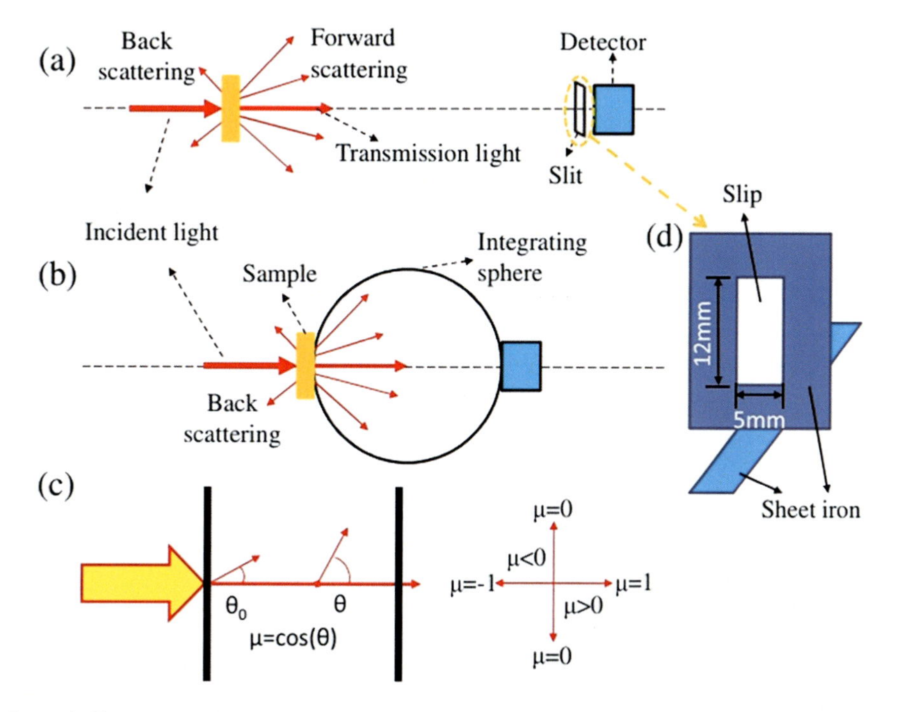

Fig. 6.7 Schematic illustration of the procedure to obtain the accurate absorption coefficient of nanofluids; the measurements of (A) extinction coefficient, (B) absorption and forward scattering coefficient; (C) working out accurate absorption coefficient; and (D) size and material of slip. Copyright Elsevier (Lic # 5061791386389) (D. Song, et al., Investigation and prediction of optical properties of alumina nanofluids with different aggregation properties, Int. J. Heat Mass Transfer 96 (2016) 430–437).

maintained between the sample and detector chamber and obtained the extinction coefficient. In the second step (Fig. 6.7B), the sample and the detector were placed very close before and after the integrating sphere, respectively, and measured all the transmitted and scattered forward radiation. All the forward radiations were collected in the integrated sphere, and backward radiation never entered the integrated sphere; thus, the absorption coefficient achieved in this step was higher than the true absorption coefficient and smaller than the extinction coefficient. In the third step (Fig. 6.7C), to obtain the pure absorption coefficient, the scattering spatial distribution function and radiative transport equation are introduced to process the experimental data.

To define the radiation fields and light distributions accurately, the Radiative transport equation (RTE) is used. Two parameters are required to solve the RTE minimum, i.e., absorption and scattering coefficient, and one scattering spatial distribution function. The radiative transport equation can be written as [29]

$$\mu\frac{\partial I_\lambda(x,\mu)}{\partial x}+\beta_\lambda I_\lambda(x,\mu)=\frac{(\beta_\lambda-k_\lambda)}{2}\int_{\acute{\mu}=-1}^{1}I_\lambda(x,\acute{\mu}).p(\mu,\acute{\mu})\,d\acute{\mu} \tag{22}$$

$$\begin{aligned} I_\lambda(0,\mu)&=I_0.\delta(\mu-\mu_0) \quad \mu>0 \\ I_\lambda(L,\mu)&=0 \qquad\qquad\quad\;\; \mu<0 \end{aligned} \tag{23}$$

where $\mu=\cos(\theta)$ and $\acute{\mu}$ is equal to the cosine of the angle between the incident and scattered rays (from -1 to 1), I is denoted as radiation intensity, and β is denoted as volumetric extinction. The phase function (p) can be written as [30]

$$p(\acute{\mu})=(8/3\pi)\left(\sqrt{1-\acute{\mu}}-\acute{\mu}\cos^{-1}\acute{\mu}\right) \tag{24}$$

This radiative transport equation is solved by the discrete ordinate method (DOM). The DOM transforms Eq. (24) into a system of algebraic equations that can be solved by computer. For the particular case of spectrophotometer cell, one single direction discretization is needed at the boundary $x=0$, for an angle close to θ_0. All other possible direction except θ_0 is neglected.

To compare the prediction achieved from the solution of Eqs. (22) and (23) with experimental values, the result can be better interpreted in terms of radiation fluxes. By definition,

$$q_\lambda=\int_{\Omega=0}^{\Omega=4\pi}I_\lambda.\Omega.d\Omega \tag{25}$$

For our model, the fluxes in the forward direction at the cell inlet ($x = 0$) and the cell outlet ($x = L$) are

$$q_\lambda^+(0) = 2\pi \int_0^1 \mu . I_\lambda(0, \mu) d\mu \tag{26}$$

$$q_\lambda^+(L) = 2\pi \int_0^1 \mu . I_\lambda(L, \mu) d\mu \tag{27}$$

where the value of I_λ is obtained from the solution of RTE, i.e., Eqs. (23) and (24).

6.4 Effect of different parameters on optical properties

The optical properties of the hybrid nanofluids have a greater impact on the thermal performance of the solar collector. Different analytical and experimental models have been discussed so far to calculate the optical properties of the nanofluid. This section will discuss the impact of different parameters on different optical properties of hybrid nanofluid.

6.4.1 Effect of particle size

It has been observed from different studies that the particle size does not have any significant effect on the optical properties of hybrid nanofluids. Chai et al. [31] studied the different parameters affecting the optical properties of nanofluids. In their study, they analyzed that the increment in the particle size of TiO_2 beyond 300 nm does not have a significant effect on the absorption coefficient; however, there was a slight increment in the scattering coefficient throughout the spectrum. It was also observed that the transmittance decreases and absorptance increases on an increment of particle size but the amount of decrement in transmittance is not equal to the increment in absorptance. Overall, it can be concluded the change in particle size is only evident on the shorter wavelength.

6.4.2 Effect of volume fraction

The volume fraction of nanoparticles has a significant effect on the extinction coefficient at a particular wavelength. In their experimental model, Ziming et al. [32] presented that the

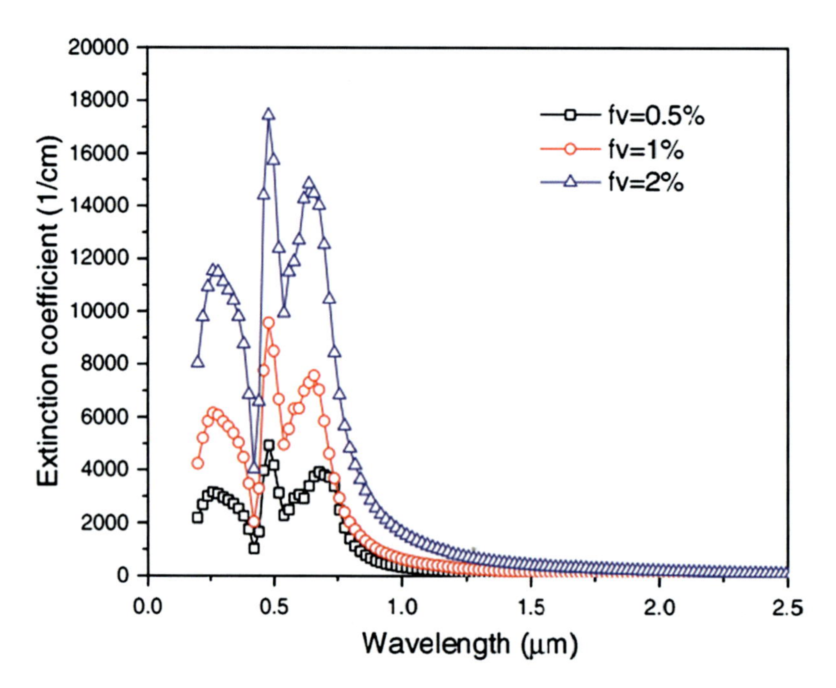

Fig. 6.8 Effect of volume fraction on extinction coefficient. Copyright Elsevier (Lic # 5061800438803) (M. Du, G.H. Tang, Optical property of nanofluids with particle agglomeration, Solar Energy 122 (2015) 864–872).

extinction coefficient increases linearly with increment in volume fraction (Eq. 25) because the number of particles per unit volume is linearly proportional to volume fraction. The relationship of volume fraction with extinction coefficient is given by

$$\sigma_{p,\lambda} = N_T \frac{\pi D^2}{4} Q_{s,\lambda} \tag{25}$$

Fig. 6.8 depicts the effect of the change in volume fraction over the extinction coefficient.

6.5 Challenges and outlook

Based on this literature study, the main challenges to using hybrid nanofluid in various applications are as follows:

- The nanoparticles and nanofluids are too costly to use in commercial applications. However, it is observed that very low volume concentrations of nanoparticles (<0.01 vol%) can enhance the optical properties of hybrid nanofluids up to a great extent. Thus, they can be used in commercial applications.

- The hybrid nanofluids have an issue related to long-term stability. The hybrid nanofluids mainly operated in a high-temperature environment. The agglomeration of nanoparticles in a hot medium has a negative impact on optical properties and thermal performance.
- In most hybrid nanofluid applications, the higher viscosity of hybrid nanofluid causes a higher pressure drop, resulting in a high pumping power requirement. Although at lower nanoparticle concentration (<0.01 vol%), it does not significantly affect the hybrid nanofluid application in the direct absorption system.
- The mechanical erosion and chemical corrosion caused by hybrid nanofluid in industrial applications are the major challenges of the hybrid nanofluid.

6.6 Summary

The hybrid nanofluid is the new class of nanofluids with better thermophysical and optical properties. The hybrid nanofluid is the dispersion of two or more two nanoparticles in conventional heat transfer fluids. The hybrid nanofluids can be used for direct absorption solar thermal system as a working fluid. So it becomes essential to study the optical properties of the hybrid nanofluid. Different nanoparticles have different properties, like Al_2O_3 nanoparticles have higher heat capacity but low absorption spectrum than CuO nanoparticles. Thus, the hybrid nanofluid consists of Al_2O_3, and CuO nanoparticles dispersed in base fluid have enhanced heat capacity and high absorption spectrum with good thermal and optical efficiencies. For direct solar absorption applications, it is essential to estimate the optical properties of the hybrid nanofluids. Different theories predict the extinction coefficient, out of which Mie scattering theory is the most suitable theory. Different theories have been discussed in the present study to calculate the extinction coefficient theoretically, and these theoretical values of extinction coefficient are compared with experimentally obtained values.

References

[1] P.K. Kanti, et al., Experimental investigation on thermo-hydraulic performance of water-based fly ash–Cu hybrid nanofluid flow in a pipe at various inlet fluid temperatures, Int. Commun. Heat Mass Transfer 124 (2021) 105238.

[2] L. Yang, et al., An updated review on the properties, fabrication and application of hybrid-nanofluids along with their environmental effects, J. Clean. Prod. 257 (2020) 120408.

[3] L.S. Sundar, et al., Heat transfer and second law analysis of ethylene glycol-based ternary hybrid nanofluid under laminar flow, J. Therm. Sci. Eng. Appl. 13 (5) (2021), 051021.
[4] L. Sundar, et al., Heat transfer of rGO/Co3O4 hybrid nanomaterial based nanofluids and twisted tape configurations in a tube, J. Therm. Sci. Eng. Appl. (2020) 1–41.
[5] Z. Said, et al., Recent advances on nanofluids for low to medium temperature solar collectors: energy, exergy, economic analysis and environmental impact, Prog. Energy Combust. Sci. 84 (2021) 100898.
[6] Z. Said, et al., Heat transfer, entropy generation, economic and environmental analyses of linear Fresnel reflector using novel rGO-Co3O4 hybrid nanofluids, Renew. Energy 165 (2021) 420–437.
[7] M. Sheikholeslami, et al., Recent progress on flat plate solar collectors and photovoltaic systems in the presence of nanofluid: a review, J. Clean. Prod. (2021) 126119.
[8] Z. Said, S. Arora, E. Bellos, A review on performance and environmental effects of conventional and nanofluid-based thermal photovoltaics, Renew. Sustain. Energy Rev. 94 (2018) 302–316.
[9] N.S. Pandya, et al., Influence of the geometrical parameters and particle concentration levels of hybrid nanofluid on the thermal performance of axial grooved heat pipe, Therm. Sci. Eng. Prog. 21 (2021) 100762.
[10] A.A. Hachicha, et al., A review study on the modeling of high-temperature solar thermal collector systems, Renew. Sustain. Energy Rev. 112 (2019) 280–298.
[11] M. Gupta, et al., Up to date review on the synthesis and thermophysical properties of hybrid nanofluids, J. Clean. Prod. 190 (2018) 169–192.
[12] Z. Said, et al., Evaluating the optical properties of TiO2 nanofluid for a direct absorption solar collector, Numer. Heat Transfer Pt A Appl. 67 (9) (2015) 1010–1027.
[13] M.F. Modest, Narrow-band and full-spectrum k-distributions for radiative heat transfer—correlated-k vs. scaling approximation, J. Quantitat. Spectrosc. Radiat. Transfer 76 (1) (2003) 69–83.
[14] M.S. Hossain, et al., Spotlight on available optical properties and models of nanofluids: a review, Renew. Sustain. Energy Rev. 43 (2015) 750–762.
[15] T.B. Gorji, A.A. Ranjbar, A review on optical properties and application of nanofluids in direct absorption solar collectors (DASCs), Renew. Sustain. Energy Rev. 72 (2017) 10–32.
[16] M. Sajid, et al., Applicability of alumina nanofluid in direct absorption solar collectors, Appl. Mech. Mater. (2015). Trans Tech Publishers.
[17] J. Meseguer, I. Grande, A. Sanz-Andres, Thermal Radiation Heat Transfer, 2012, pp. 73–86.
[18] R. Taylor, Thermal Energy Conversion in Nanofluids, 2011.
[19] Z. Said, R. Saidur, N. Rahim, Optical properties of metal oxides based nanofluids, Int. Commun. Heat Mass Transfer 59 (2014) 46–54.
[20] C.F. Bohren, D.R. Huffman, Z. Kam, Book-review—Absorption and scattering of light by small particles, Nature 306 (1983) 625.
[21] Z. Said, et al., Radiative properties of nanofluids, Int. Commun. Heat Mass Transfer 46 (2013) 74–84.
[22] R.A. Taylor, et al., Nanofluid optical property characterization: towards efficient direct absorption solar collectors, Nanoscale Res. Lett. 6 (1) (2011) 225.
[23] X. Li, G. Zeng, X. Lei, The stability, optical properties and solar-thermal conversion performance of SiC-MWCNTs hybrid nanofluids for the direct

absorption solar collector (DASC) application, Solar Energy Mater. Solar Cells 206 (2020) 110323.
[24] W. Chen, C. Zou, X. Li, Application of large-scale prepared MWCNTs nanofluids in solar energy system as volumetric solar absorber, Solar Energy Mater. Solar Cells 200 (2019) 109931.
[25] M. Victoria, et al., ASTM G173 standard tables for reference solar spectral irradiances, Jpn. J. Appl. Phys. AM15D (2012).
[26] J. Qu, et al., Photo-thermal conversion properties of hybrid CuO-MWCNT/H2O nanofluids for direct solar thermal energy harvest, Appl. Therm. Eng. 147 (2019) 390–398.
[27] A. Menbari, A.A. Alemrajabi, Analytical modeling and experimental investigation on optical properties of new class of nanofluids (Al2O3–CuO binary nanofluids) for direct absorption solar thermal energy, Opt. Mater. 52 (2016) 116–125.
[28] D. Song, et al., Investigation and prediction of optical properties of alumina nanofluids with different aggregation properties, Int. J. Heat Mass Transfer 96 (2016) 430–437.
[29] M.I. Cabrera, O.M. Alfano, A.E. Cassano, Absorption and scattering coefficients of titanium dioxide particulate suspensions in water, J. Phys. Chem. 100 (51) (1996) 20043–20050.
[30] J.R. Howell, et al., Thermal Radiation Heat Transfer, CRC Press, 2020.
[31] J.Y.H. Chai, B.T. Wong, Study of light scattering by TiO2, Ag, and SiO2 nanofluids with particle diameters of 20-60 nm, J. Nano Res. 60 (2019) 1–20.
[32] C. Ziming, et al., Investigation of optical properties and radiative transfer of sea water-based nanofluids for photocatalysis with different salt concentrations, Int. J. Hydrogen Energy 42 (43) (2017) 26626–26638.

7

Theoretical analysis and correlations for predicting properties of hybrid nanofluids

Arun Kumar Tiwari[a], Amit Kumar[a], and Zafar Said[b,c,d,*]
[a]*Mechanical Engineering Department, Institute of Engineering & Technology, Dr. A.P.J. Abdul Kalam Technical University, Uttar Pradesh, Lucknow, India.* [b]*Department of Sustainable and Renewable Energy Engineering, University of Sharjah, Sharjah, United Arab Emirates.* [c]*Research Institute for Sciences and Engineering, University of Sharjah, Sharjah, United Arab Emirates.* [d]*U.S.-Pakistan Center for Advanced Studies in Energy (USPCAS-E), National University of Sciences and Technology (NUST), Islamabad, Pakistan.*
**Corresponding author: zsaid@sharjah.ac.ae, zaffar.ks@gmail.com*

Chapter outline

7.1 Introduction 149
7.2 Different theoretical models 150
7.3 Different correlations to predict the properties of hybrid nanofluid 152
7.3.1 Thermal conductivity 152
7.3.2 Specific heat capacity 154
7.3.3 Density 160
7.3.4 Viscosity 161
7.4 Challenges and summary 164
References 166

7.1 Introduction

In recent years, the hybrid nanofluid grows into the most promising heat transfer medium, especially in solar thermal systems [1]. Hybrid nanofluids are a mixture of two or more than two nanoparticles dispersed in the liquid [2, 3]. The hybrid nanofluids seem to capture mono nanofluids' place in solar thermal

Hybrid Nanofluids: Preparation, Characterization and Applications. https://doi.org/10.1016/B978-0-323-85836-6.00007-7

application due to having some extraordinary properties [4, 5]. Hybrid nanofluids have some good properties like a wide absorption range, high thermal conductivity, low-pressure drop, low friction losses, and low pumping power [6, 7].

So far, a large number of hybrid nanofluids have been used in different heat transfer applications [8, 9]. In their review study on hybrid nanofluids, Tiwari et al. [10] mentioned the following hybrid nanofluids in their review paper: Ag-GNPs, Ag-WO_3, Ag-Si, Ag-MgO, Ag-Zn, Al_2O_3-Cu, TiO_2-SiO_2, TiO_2-CuO, TiO_2-MWCNT, MWCNT-ZnO, MWCNT-Ag, etc. This chapter will discuss the effect of different properties of hybrid nanofluid in heat transfer applications.

7.2 Different theoretical models

Different theoretical correlations have been developed so far by different researchers to predict the properties of hybrid nanofluids [11]. Yildiz et al. [12] developed a theoretical model to predict the thermal conductivity of the Al_2O_3-SiO_2/water nanofluid. They solve the flowing model as shown in Fig. 7.1. They consider a square cavity having the constant wall temperature T_h and T_c at the left and right wall of the cavity, respectively. The top and bottom surfaces of the cavity are assumed to be adiabatic. They take the following assumption to solve the problem:

- Linear temperature difference profile inside the cavity.
- Steady-state and two-dimensional flow of hybrid nanofluid.

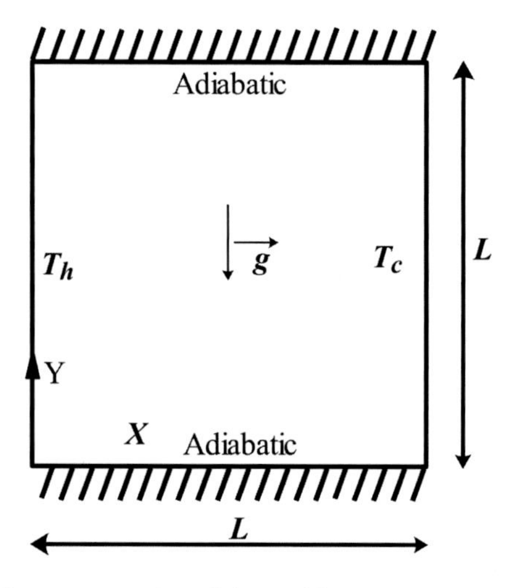

Fig. 7.1 Schematic representation of the problem.

- The density of hybrid nanofluid is treated as Boussinesq approximation while other thermophysical properties are constant.
- Incompressible and Newtonian hybrid nanofluid.
  The following governing equations were used:

$$\frac{\partial u}{\partial x}+\frac{\partial v}{\partial y}=0 \tag{7.1}$$

$$u\frac{\partial u}{\partial x}+v\frac{\partial u}{\partial y}=\frac{1}{\rho_{nf}}\left[-\frac{\partial p}{\partial x}+\mu_{nf}\left(\frac{\partial^2 u}{\partial x^2}+\frac{\partial^2 u}{\partial y^2}\right)\right] \tag{7.2}$$

$$u\frac{\partial v}{\partial x}+v\frac{\partial v}{\partial y}=\frac{1}{\rho_{nf}}\left[-\frac{\partial p}{\partial y}+\mu_{nf}\left(\frac{\partial^2 v}{\partial x^2}+\frac{\partial^2 v}{\partial y^2}\right)\right]+(\rho\beta)_{nf}.g.(T-T_c) \tag{7.3}$$

$$u\frac{\partial T}{\partial x}+v\frac{\partial T}{\partial y}=\propto_{nf}\left(\frac{\partial^2 T}{\partial x^2}+\frac{\partial^2 T}{\partial y^2}\right) \tag{7.4}$$

The density, specific heat capacity, thermal expansion coefficient, and thermal diffusivity of the nanofluid are as follows [13]:

$$\rho_{nf}=(1-\emptyset)\rho_{bf}+\emptyset.\rho_p \tag{7.5}$$

$$\rho_{hnf}=\emptyset_{p1}.\rho_{p1}+\emptyset_{p2}.\rho_{p2}+(1-\emptyset_{tot})\rho_{bf} \tag{7.6}$$

where b_f, h_{nf}, and n_f are the subscripts representing base fluid, hybrid nanofluid, nanofluid, respectively, and $\emptyset$ is denoted as volume concentration of nanofluid [8]. $p1$ and $p2$ represent two different nanoparticles of hybrid nanofluid.

$$\emptyset=\frac{\omega\rho_{bf}}{\left(1-\frac{\omega}{100}\right)\rho_{np}+\left(\frac{\omega}{100}\right)\rho_{bf}} \tag{7.7}$$

Then, the total volume concentration can be written as

$$\emptyset_{tot}=\emptyset_1+\emptyset_2 \tag{7.8}$$

Specific heat of nanofluid and hybrid nanofluid are as follows:

$$(\rho C_p)_{nf}=(1-\emptyset)\rho_{bf}+\emptyset\rho_p \tag{7.9}$$

$$(\rho C_p)_{hnf}=\emptyset_1.(\rho C_p)_{p1}+\emptyset_2.(\rho C_p)_{p2}+(1-\emptyset_{tot})\rho_{bf} \tag{7.10}$$

The thermal expansion coefficient can be defined as

$$(\rho\beta)_{nf}=(1-\emptyset)(\rho\beta)_{bf}+\emptyset(\rho\beta)_p \tag{7.11}$$

$$(\rho\beta)_{hnf}=\emptyset_{p1}.(\rho\beta)_{p1}+\emptyset_{p2}.(\rho\beta)_{p2}+(1-\emptyset_{tot})(\rho\beta)_{bf} \tag{7.12}$$

Theoretical thermal conductivity of nanofluid can be written as

$$k_{nf} = k_{bf}\frac{k_p + 2k_{bf} - 2\varnothing(k_{bf} - k_p)}{k_p + 2k_{bf} + \varnothing(k_{bf} - k_p)} \tag{7.13}$$

The variables in the governing equations can be translated into dimensionless form as follows:

$$X = \frac{x}{L},\ Y = \frac{y}{L},\ U = \frac{uL}{a_f},\ V = \frac{vL}{a_f},\ P = \frac{pL^2}{\rho_{nf}a_f^2},\ \theta = \frac{T - T_c}{T_h - T_c} \tag{7.14}$$

$$Ra = \frac{g\beta_f L^3(T_h - T_c)}{a_f.\upsilon_f},\ \Pr = \frac{\upsilon_f}{a_f} \tag{7.15}$$

By utilizing the introduced nondimensional variables, the continuity, momentum, and energy equations can be written in dimensionless form as follows:

$$\frac{\partial U}{\partial X} + \frac{\partial V}{\partial Y} = 0 \tag{7.16}$$

$$U\frac{\partial U}{\partial X} + V\frac{\partial V}{\partial Y} = -\frac{\partial P}{\partial X} + \frac{\mu_{nf}}{\rho_{nf}a_f}\left(\frac{\partial^2 U}{\partial X^2} + \frac{\partial^2 U}{\partial Y^2}\right) \tag{7.17}$$

$$U\frac{\partial V}{\partial X} + V\frac{\partial V}{\partial Y} = -\frac{\partial P}{\partial Y} + \frac{\mu_{nf}}{\rho_{nf}a_f}\left(\frac{\partial^2 V}{\partial X^2} + \frac{\partial^2 V}{\partial Y^2}\right) + \frac{(\rho\beta)_{nf}}{\rho_{nf}\beta_f}Ra.\Pr.\theta \tag{7.18}$$

$$U\frac{\partial \theta}{\partial X} + V\frac{\partial \theta}{\partial Y} = \frac{a_{nf}}{a_f}\left(\frac{\partial^2 \theta}{\partial X^2} + \frac{\partial^2 \theta}{\partial Y^2}\right) \tag{7.19}$$

7.3 Different correlations to predict the properties of hybrid nanofluid

7.3.1 Thermal conductivity

The thermal conductivity is the most significant thermophysical property of any hybrid nanofluid [14]. From a hybrid nanofluid perspective, thermal conductivity is the measure of the ability of nanofluid to conduct heat [15]. The thermal conductivity of a hybrid nanofluid depends upon multiple factors, including base fluid and its composition, types of nanoparticles, mass fraction of nanoparticles, and volume concentration of nanoparticles [16] (shown in Fig. 7.2). However, the thermal conductivity of hybrid nanofluid majorly depends on the nanoparticle concentration. In most thermal conductivity correlations, it has been observed that thermal conductivity has direct correlations with

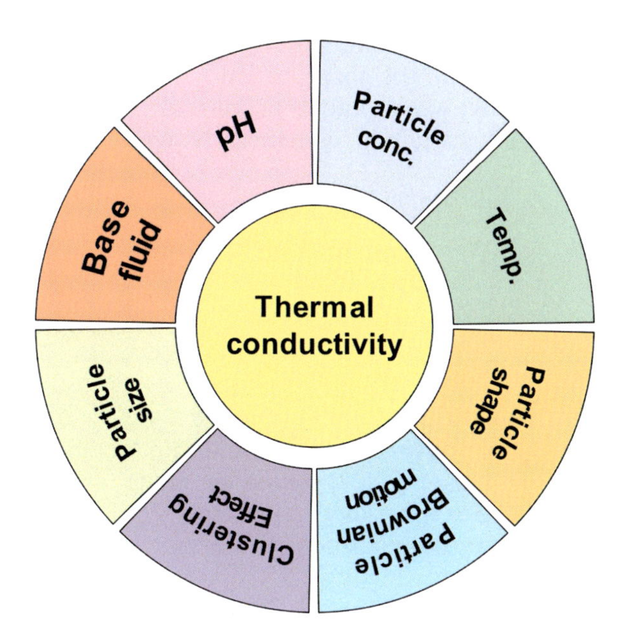

Fig. 7.2 Factors affecting the thermal conductivity [17].

concentrations of nanoparticles. The enhancement in the thermal conductivity of hybrid nanofluid results from the increment in Brownian motion and kinetic energy changes [18]. Due to the increment in Brownian motion and kinetic energy, there is an increment in molecular collision rate between the nanoparticles and base fluid, which increases the chances of chain formation between the nanoparticles and base fluid [19]. The frequency of lattice vibration increases due to increment in molecular collision rate and reduction in nanoparticles distance. Hence, the thermal conductivity of hybrid nanofluid increases with increased nanoparticle concentration.

However, this increment in nanoparticle concentration in the base fluid reduces the stability of hybrid nanofluid, reducing the thermal conductivity of hybrid nanofluid in a more extended operation period [20]. Some of these negative effects related to the stability of hybrid nanofluid can be remedied during the preparation of hybrid nanofluid. For instance, the single-step method of nanofluid preparation provides better stability to hybrid nanofluid than the two-step method. Also, optimum sonication time can improve the stability of hybrid nanofluids.

In earlier studies, the modified model of thermal conductivity calculation of nanofluid was used to predict the thermal conductivity of hybrid nanofluid. It has been seen from the various correlation that thermal conductivity is the function of particle volume concentration, the shape of particles, and temperature. The Maxwell model was modified to calculate the thermal conductivity of hybrid nanofluid, which can be written as [21]

$$k_{hnf} = k_{bf}\frac{\left(\frac{\left(\varphi_{p1}k_{p1}+\varphi_{p2}k_{p2}\right)}{\varphi_{tot}}+2k_{bf}+2\left(\varphi_{p1}k_{p1}+\varphi_{p2}k_{p2}\right)-2\varphi_{tot}k_{bf}\right)}{\frac{\left(\varphi_{p1}k_{p1}+\varphi_{p2}k_{p2}\right)}{\varphi_{tot}}+2k_{bf}-\left(\varphi_{p1}k_{p1}+\varphi_{p2}k_{p2}\right)+\varphi_{tot}k_{bf}} \tag{7.20}$$

where Ø is the solid volume fraction and $Ø = Ø_{np1} + Ø_{np2}$.

In another model to predict the thermal conductivity of hybrid nanofluid, Hamilton et al. [22] defined a shape factor $n = 3/\varphi$, where φ is denoted as sphericity defined as the ratio of the surface area of a sphere to the nanoparticles with equal volume. They gave the following formula:

$$\frac{k_{hnf}}{k_{bf}}=\frac{k_{np2}+(n-1)k_{bf}-(n-1)Ø_2\left(k_{bf}-k_{np2}\right)}{k_{np2}+(n-1)k_{bf}+Ø_2\left(k_{bf}-k_{np2}\right)} \tag{7.21}$$

$$\frac{k_{bf}}{k_f}=\frac{k_{np1}+(n-1)k_f-(n-1)Ø_1\left(k_f-k_{np1}\right)}{k_{np1}+(n-1)k_f+Ø_1\left(k_f-k_{np1}\right)} \tag{7.22}$$

Choi et al. [23] further modified the Maxwell model to predict the thermal conductivity of the hybrid nanofluid.

$$\frac{k_{hnf}}{k_{bf}}=\frac{k_{np2}+2k_{bf}-2Ø\left(k_{bf}-k_{np}\right)(1+\beta)^3}{k_{np2}+2k_{bf}-Ø\left(k_{bf}-k_{np}\right)(1+\beta)^3} \tag{7.23}$$

Other than these correlations, different researchers used different correlations to predict the thermal conductivity of hybrid nanofluids, which are shown in Table 7.1.

7.3.2 Specific heat capacity

Its thermal storage capacity defines the specific heat capacity of hybrid nanofluid [50]. It is an essential thermophysical property of hybrid nanofluid used to predict the other significant correlations like thermal conductivity and thermal diffusivity [51]. Moreover, it becomes a vital property related to convective heat transfer application to predict the heat transfer coefficient and Nusselt number. There are some other numbers in convective heat transfer that are predicted using specific heat capacity property [52]. The specific heat capacity of the hybrid nanofluid is the function of the specific heat capacity of nanoparticles and base fluid, particle volume concentration, and temperature. Under the constant

Table 7.1 Various thermal conductivity correlation of hybrid nanofluids.

Nanofluid	Thermal conductivity correlation of hybrid nanofluid	References
CNT-Al_2O_3/water	$\frac{k_{hnf}}{k_f} = 1 + A\phi + B\phi^2 + C\phi^3 + D\phi^4$ $\frac{k_{hnf}}{k_f} = \frac{-214.83 + T}{346.58 + 106.98\phi} + \frac{227.69}{T}$	[24]
ZnO-TiO_2/EG	$\frac{k_{hnf}}{k_f} = 1 + 0.004503\phi^{0.8717} T^{0.7972}$	[25]
MgO-FMWCNT/EG	$\frac{k_{hnf}}{k_f} = 0.8341 + 1.1\phi^{0.243} T^{-0.289}$	[26]
Al_2O_3-MWCNT/oil	$k_{hnf} = 0.1534 + 0.00025T + 1.1193\varnothing$	[27]
Al_2O_3-CuO/EG	$\frac{k_{hnf}}{k_f} = \frac{9.6128 + \phi}{9.3885 - 0.00010759T^2} - \frac{0.0041099}{\phi}$	[28]
SiC-TiO_2/diathermic oil	$\frac{k_{hnf} - k_f}{k_f} = 6.06\phi + 2.72$	[29]
Cu-TiO_2/EG-water	$\frac{k_{hnf}}{k_f} = 1.07 = 0.000589T - \frac{0.000814}{T.\phi} + 4.44T.\phi.$ $\cos(6.11 + 0.00673T + 4.41T\phi - 0.0414 \sin T) - 32.5\phi$	[30]
SiC-TiO_2/water	$\frac{k_{hnf}}{k_f} = 1.17(1 + R)^{-0.1151}\left(\frac{T}{80}\right)^{0.0437}$ where R = volume fraction of TiO_2 in the mixture	[31]
ZnO-Ag/water	$\frac{k_{hnf}}{k_f} = 1 + 0.0008794\phi^{0.5899} T^{1.345}$	[32]
TiO_2-SiO_2/water	$\frac{k_{hnf}}{k_f} = \left(1 + \frac{\phi}{100}\right)^{5.5}\left(0.1 + \frac{T}{80}\right)^{0.01}$	[21]
TiO_2-SiO_2/EG-water	$\frac{k_{hnf}}{k_f} = \left(1 + \frac{\phi}{100}\right)^{5.25} + \left(1 + \frac{T}{70}\right)^{0.076}$	[30]
Cu-Zn/vegetable oil	• For Cu-Zn (50:50) $\frac{k_{hnf}}{k_f} = 0.7054 + 0.009896T$ $+0.8717\phi - 6.479 \times 10^{-5}T^2$ $+0.09749T\phi - 4.714\phi^2 - 0.0002718T^2\phi$ $-0.1174T\phi^2 + 10.09\phi^3$	[33]

Continued

Table 7.1 Various thermal conductivity correlation of hybrid nanofluids—cont'd

Nanofluid	Thermal conductivity correlation of hybrid nanofluid	References
	• For Cu-Zn (75:25) $\frac{k_{hnf}}{k_f}=0.9842-0.008376T$ $-2.121\phi+2.677\times10^{-5}T^2+0.1497T\phi+2.653\phi^2$ $-0.0006927T^2\phi-0.1386T\phi^2+2.1\phi^3$ • For Cu-Zn (25:75) $\frac{k_{hnf}}{k_f}=1.321-0.01661T-4.723\phi$ $+0.000199T^2+0.2473T\phi+3.689\phi^2$ $-0.001766T^2\phi-0.1222T\phi^2-0.1045\phi^3$	
TiO_2-CuO/EG	$k_{hnf}=1+6.2299\left(\frac{\phi}{100}\right)^{0.9371}\left(\frac{T}{333}\right)^{10.2685}$	[34]
Ag-MgO/water	$\frac{k_{hnf}}{k_f}=\frac{0.1747\times10^5+\phi}{0.1747\times10^5-0.1498\times10^5\phi+0.1117\times10^7\phi^2+1.997\times10^8\phi^3}$	[35]
MWCNT-Al_2O_3/EG	$\frac{k_{hnf}}{k_f}=0.00281+5.28956\times10^{-4}T+0.17384\phi+5.31721\times10^{-3}T\phi-0.23762\phi^2$ $-5.06549\times10^{-3}T\phi^2+0.19937\phi^3$	[36]
CuO-SWCNT/water-EG	$\frac{k_{hnf}}{k_f}=1+(0.04056\phi T)-\left(0.003252(\phi T)^2\right)+\left(0.0001181(\phi T)^3\right)-\left(0.00001431(\phi T)^4\right)$	[37]
FMWCNT-Fe_3O_4/EG	$\frac{k_{hnf}}{k_f}=1+0.0162T^{0.6009}\phi^{0.7038}$	[38]
MWCNT-SiO_2/EG	$\frac{k_{hnf}}{k_f}=0.905+0.0020609\phi T+0.04375\phi^{0.09265}T^{0.3305}-0.0063\phi^3$	[39]
ZnO-MWCNT/EG-water	$\frac{k_{hnf}}{k_f}=1.024+0.5988\phi^{0.6029}\exp\left(\frac{\phi}{T}\right)-\frac{8.059\phi T^{0.2}-2.24}{6.052\phi^2+T}$	[40]

Table 7.1 Various thermal conductivity correlation of hybrid nanofluids—cont'd

Nanofluid	Thermal conductivity correlation of hybrid nanofluid	References
MWCNT-SiO_2/EG	$\frac{k_{hnf}}{k_f}=1.01+0.007685\phi T-0.5136\phi^2T^{-0.1578}+11.5\phi^3T^{-1.175}$	[36]
Al_2O_3-TiO_2/water	$\frac{k_{hnf}}{k_f}=0.995+10.097\phi_1-120.835\phi_1^2+23.227\phi_2-43.648\phi_2^2+22380.35\phi_3^2$	[41]
SWCNT-MgO/EG	$\frac{k_{hnf}}{k_f}=0.90844-0.06613\phi^{0.3}T^{0.7}+0.01266\phi^{0.31}.T$	[42]
CuO-MgO-TiO_2/water	$k_{hnf}=0.72768391+\frac{-9.6816854}{T}+\frac{283.69209}{T^2}+\frac{-4718.8713}{T^3}+\frac{43223.493}{T^4}+\frac{-175433.68}{T^5}+274.276\phi-15978.42\phi^2+41691000\phi^3-3.6405883\times10^9\phi^4$	[43]
MgO-MWCNT/water	$\frac{k_{hnf}}{k_f}=0.9787+\exp\left(0.3081\phi^{0.158}-0.002T\right)$	[44]
Al_2O_3-Fe_2O_4/10W40 engine oil	$\frac{k_{hnf}}{k_f}=0.113\times(1.011)^T\times w^{0.376}+0.921$	[45]
CeO_2-MWCNT/water	$\frac{k_{hnf}}{k_f}=1+0.580453\times\left(\frac{T}{T_0}\right)^{1.54358}\times\phi^{0.356853}$	[20]
ND-Fe_3O_4/water-EG	$\frac{k_{hnf}}{k_f}=ae^{b\phi}$	[46]
ND-Co_3O_4/EG	$\frac{k_{hnf}}{k_f}=0.01807+0.0.1593\phi-0.00107T+0.00707\phi T+1.12568\phi^2+0.00001T^2$	[47]
MgO-MWCNT/EG	$\frac{k_{hnf}}{k_f}=0.9787+\exp\left(0.3081\phi^{0.3097}-0.002T\right)$	[44]
SWCNT-ZnO/EG-water	$\frac{k_{hnf}}{k_f}=0.8707+0.179\phi^{0.179}\exp\left(0.09624\phi^2\right)+8.883\times10^{-4}.\phi T+4.435\times10^{-3}.\phi^{0.252}T$	[48]
SWCNT-Al_2O_3/EG	$\frac{k_{hnf}}{k_f}=0.008379\left[\phi^{0.4439}T^{0.9246}\right]+0.963$	[49]

temperature, the heat capacity of the hybrid nanofluid can be written using the first law of thermodynamics.

Generally, the specific heat capacity of hybrid nanofluid at a constant temperature is modeled using two simple techniques, i.e., the mixing theory and thermal equilibrium assumption. In mixing theory, the specific heat capacity of the hybrid nanofluid is based on the specific heat of the base fluid, specific heat of nanoparticles, and volume fraction. This relationship can be defined as [53]

$$C_{hnf} = \emptyset . C_{np} + (1 - \emptyset) C_{bf} \tag{7.24}$$

For the thermal equilibrium assumption, after applying the energy balance between base fluid and nanoparticles,

$$C_{hnf} = w C_{np} + (1 - w) C_{bf} \tag{7.25}$$

where w is denoted as the mass fraction of nanoparticles. Writing the mass fraction in terms of density and volume, the above relations can be written as

$$\rho_{hnf} C_{hnf} = \left(\emptyset_{np1} \rho_{np1} C_{np1} \right) + \left(\emptyset_{np2} \rho_{np2} C_{np2} \right) + \left(1 - \emptyset_{np1} - \emptyset_{np2}\right) \rho_{bf} C_{bf} \tag{7.26}$$

The specific heat capacity of hybrid nanofluid can be measured using the differential scanning calorimetry, constant room temperature, or different constant heat flux conditions or using the law of cooling. There are three parts of the standard measuring technique through the calorimetry method. A calorimetry liquid (mostly water) and a sample liquid are used during the measurement. After measuring the mass of calibrated liquid and hybrid nanofluid, the first step is to determine the heat transfer rate for the empty sample container. In the second step, the heat transfer rate of calibrated liquid is determined, and in the final step, heat transfer rate of hybrid nanofluid is determined, and the following relation is determined to calculate the specific heat capacity [54]:

$$C_{hnf} = \frac{m_{calibrationliquid}}{m_{hnf}} C_{calibrationliquid} \frac{\dot{Q}_{hnf} - \dot{Q}_{empty}}{\dot{Q}_{calibrationliquid} - \dot{Q}_{empty}} \tag{7.27}$$

Another method to calculate the specific heat capacity is the cooling method. In this method, three containers are used, i.e., container for the high-temperature sample, the adiabatic container, and the container for the low-temperature sample. The hybrid nanofluid sample is placed into the container with a known mass and specific heat and heated up until equilibrium. After that, hybrid nanofluid is moved into the adiabatic container. In the final step, the sample liquid and hybrid nanofluid are mixed, and the constant final steady temperature of the mixture is recorded. This final temperature is related to the transferred heat during cooling, which can be written as [55]

$$Q_1 = \left(C_{hnf}.m_{hnf} + C_c.m_c\right)\left(T_{hnf} - T_{mixture}\right) \tag{7.28}$$

$$Q_2 = C_{bf}.m_{bf}\left(T_{mixture} - T_{bf}\right) \tag{7.29}$$

where Q_1 is denoted as heat released from hybrid nanofluid, Q_2 is denoted as heat absorbed by the base fluid, m_{hnf} and m_{bf} are denoted as the mass of hybrid nanofluid and base fluid, respectively. According to energy conservation equations (neglecting losses), Q_1 must be equal to Q_2; then, the following relation can be drawn:

$$C_{hnf} = \frac{C_{bf}.m_{bf}\left(T_{mixture} - T_{bf}\right)}{m_{hnf}\left(T_{hnf} - T_{mixture}\right)} \tag{7.30}$$

Barbes et al. [56] studied the specific heat capacity model, and to determine it, they assume that the base fluid and nanoparticles are in thermal equilibrium and drive the specific heat capacity of the nanofluid using the first law of thermodynamics and write the following energy conservation equation:

$$C_{p,\,nf} = \omega.C_{p,np} + (1-\omega).C_{p,\,bf} \tag{7.31}$$

where $C_{p,nf}$, $C_{p,bf}$, and $C_{p,np}$ are the specific heat capacities of nanofluid, base fluid, and nanoparticles, respectively, while ω is denoted as nanoparticle mass friction. Expressing the mass in the function of

$$C_{p,\,nf} = \frac{\varnothing.\rho_{np}.C_{p,\,np} + (1-\varphi).\rho_{bf}.C_{p,\,bf}}{\varphi.\rho_{np} + (1-\varphi).\rho_{bf}} \tag{7.32}$$

Ghadikolaei et al. [57] investigated the thermophysical properties of TiO_2-Cu/water hybrid nanofluid and derive the following correlation for specific heat capacity:

$$\left(\rho C_p\right)_{hnf} = \left(\rho C_p\right)_f(1-\varnothing_2)\left[(1-\varnothing_1) + \varnothing_1\frac{\left(\rho C_p\right)_{np1}}{\left(\rho C_p\right)_f}\right] + \varnothing_2\left(\rho C_p\right)_{np2} \tag{7.33}$$

Tiwari et al. [58] investigated three metal oxides (SnO_2, CuO, and MgO) and MWCNT-based hybrid nanofluid, and based on their experimental analysis, they proposed the following correlation for the specific heat capacity:

$$Cp_{hnf} = Cp_{bf}\left(1 - 10.6364 \times \left(\frac{T}{T_0}\right)^{-0.771} \times \varnothing^{0.448} \times \left(\frac{d_{hnf}}{d_0}\right)^{-0.474}\right.$$

$$\times \left(\frac{Cp_{hnp}}{Cp_{bf}}\right)^{1.027} \times \left(\frac{\rho_{nf}}{\rho_{bf}}\right)^{-2.742} \tag{7.34}$$

where T_0 is denoted as the reference temperature (°C).

7.3.3 Density

Density is an important thermophysical property of hybrid nanofluid [59]. The density has a significant effect on the flow, Reynolds number, pressure, stability, and heat transfer performance of the hybrid nanofluid [60]. The density of hybrid nanofluid is majorly dependent upon the temperature and volume concentration of nanoparticles. The density of hybrid nanofluid increases with an increase in the volume concentration of nanoparticles and decreases with an increase in temperature [61].

Cho et al. [62] introduced the density theory of multiphase isothermal mixture of nanosized particles, i.e., nanofluid. Furthermore, the density of hybrid nanofluid is predicted by modifying the standard density model of nanofluid, which is written as [63]

$$\rho_{hnf} = (1 - \text{Ø}_1 - \text{Ø}_2)\rho_{bf} + \text{Ø}_1\rho_{np1} + \text{Ø}_2\rho_{np2} \tag{7.35}$$

where ρ is denoted as density, *hnf* is for hybrid nanofluid, *bf* is for base fluid, and *np* is denoted as a nanoparticle. Vijjha et al. [64] measured the density of Al_2O_3/EG, Sb_2O_5-SnO_2/EG-water, and ZnO/EG-water and presented the following density model for Sb_2O_5-SnO_2/EG-water hybrid nanofluid.

$$\rho_{hnf} = \frac{(0.9848\text{Ø} + 0.7382)}{100}(1 - \text{Ø})\rho_{bf} + \text{Ø}\rho_{np} \tag{7.36}$$

Ghadikolaei et al. [57] investigated the thermophysical properties of TiO_2-Cu/water hybrid nanofluid and derived the following correlation for density:

$$\rho_{hnf} = \rho_{bf}(1 - \text{Ø}_2)\left[(1 - \text{Ø}_1) + \text{Ø}_1\left(\frac{\rho_{np1}}{\rho_{bf}}\right)\right] + \text{Ø}_2\rho_{np2} \tag{7.37}$$

Said et al. [4] used rGO-Co_3O_4/water hybrid nanofluid to analyze the thermal and economic performance of a linear Fresnel reflector and predicted the following correlation to calculate combined viscosity of rGO and Co_3O_4 nanoparticles:

$$\rho(rGO - Co_3O_4)_{np} = \frac{\rho_{rGO}.W_{rGO} + \rho_{Co_3O_4}.W_{Co_3O_4}}{W_{rGO} + W_{Co_3O_4}} \tag{7.38}$$

where *np* is denoted as nanoparticles and *W* is denoted as the weight of nanoparticles in kg.

7.3.4 Viscosity

The viscosity of the hybrid nanofluid is an important thermophysical property to calculate the pumping power requirements due to the frictional effects [65]. The viscosity of the hybrid nanofluid majorly depends upon the volume concentration, temperature, size, and shapes of nanoparticles [66] (shown in Fig. 7.3). Different types of rheometers measure the viscosity of the hybrid nanofluid. During the measurement of viscosity, it is important to identify fluid behavior, i.e., Newtonian or non-Newtonian, because the Newtonian fluid exhibits a linear relationship between the shear stress and strain rate and non-Newtonian fluid exhibits a nonlinear relationship at a constant temperature. The viscosity is the function of temperature and volume concentration for a Newtonian hybrid nanofluid while it varies with strain rate and relaxation time for non-Newtonian hybrid nanofluids.

It has been seen that a viscosity model that works well for suspensions does not predict the viscosity value for hybrid nanofluid accurately. It is not possible to generate a universally accepted viscosity model for hybrid nanofluids of varying temperature ranges and volume concentrations. To solve this problem, the viscosity models are generally generated for small groups of the blend. As the classical suspension model does not predict the viscosity of hybrid nanofluid accurately, different researchers are intended to find the new viscosity model experimentally. Some of the viscosity models are displayed in Table 7.2. Other than these models,

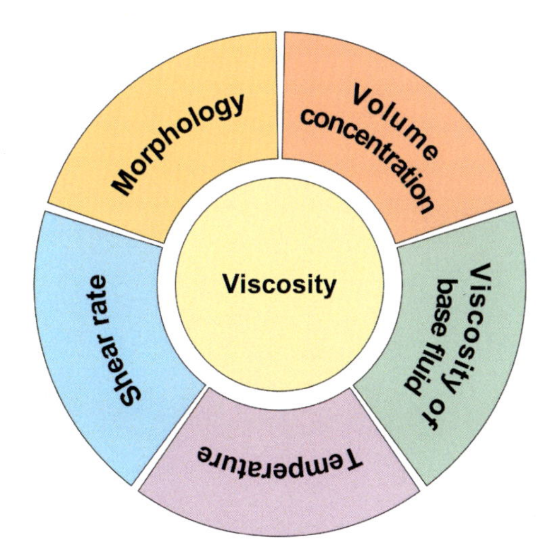

Fig. 7.3 Factors affecting the viscosity of hybrid nanofluid.

Table 7.2 Viscosity correlations for hybrid nanofluids.

Hybrid nanofluid	Viscosity correlations	Ref.
TiO_2-Cu/water	$\mu_{hnf} = \frac{\mu_f}{(1-\varnothing_1)^{2.5} + (1-\varnothing_2)^{2.5}}$	[57]
Ag-MgO/water	$\frac{\mu_{hnf}}{\mu_f} = 1 + 32.795\varnothing - 7214\varnothing^2 + 71400\varnothing^3 - 0.1941 \times 10^8\varnothing^4$	[35]
SiC-TiO_2/diathermic oil	$\frac{\mu_{hnf}}{\mu_f} = 0.312\varnothing + 1.3194$	[29]
SiO_2-TiO_2/water	$\frac{\mu_{hnf}}{\mu_f} = 1.42(1+R)^{-0.1063}\left(\frac{T}{80}\right)^{0.2321}$	[31]
TiO_2-SiO_2-/water	$\frac{\mu_{hnf}}{\mu_f} = 37\left(0.1 + \frac{\varnothing}{100}\right)^{1.59}\left(0.1 + \frac{T}{80}\right)^{0.31}$	[21]
TiO_2-CuO/EG	$\mu_{hnf} = 0.9653 + 77.4567\left[\frac{\varnothing}{100}\right]^{1.1558}\left[\frac{T}{333}\right]^{0.6881}$	[34]
MgO-MWCNT/EG	$\frac{\mu_{hnf}}{\mu_{bf}} = \left[0.91\varnothing + 0.240\left(T^{-0.342}\varnothing^{-0.473}\right)\right]\exp\left(1.45T^{0.12}\varnothing^{0.158}\right)$	[67]
MWCNT-ZnO/engine oil	$\mu_{hnf} = 796.8 + 76.26\varnothing + 12.88T + 0.7695\varnothing T + \frac{-196.9T - 16.54\varnothing T}{\sqrt{T}}$	[68]
Al_2O_3-CuO-TiO_2/water	$\mu_{hnf} = \mu_{bf}\left(0.955 - 0.00271T + 1.858\frac{\varnothing}{100} + \left(705\frac{\varnothing}{100}\right)^{1.223}\right)$	[69]
SiO_2-oleic acid/liquid paraffin	$\frac{\mu_{hnf}}{\mu_{bf}} = 6.8376 + 15.2522w + 0.038779 - 2.63029w^2$, where w is the mass fraction (wt.%)	[70]
MWCNT-SiO_2/water	• Model 1 $$\mu_{hnf} = 15.88\varnothing^{0.8514}T^{-1.189}\dot{\gamma}^{-0.5693}$$ • Model 1 $$\mu_{hnf} = m\dot{\gamma}^{n-1}$$ where $m = 0.02048 + 2.189\exp\left(-\frac{1.083}{\varnothing} - 0.03327T\right)$ and $n = 0.686\varnothing^{(-0.0906-0.001474T)} - 0.006267T$	[69]
ND-Co_3O_4/EG	$\frac{\mu_{hnf}}{\mu_{bf}} = 0.50437 + 4.3886\varnothing - 0.04183T - 0.26697\varnothing.T + 22.66087\varnothing^2 - 0.00121T^2$ $+ 0.003325\varnothing^2.T + 0.00332T^2.\varnothing - 0.00001T^3$	[49]

Table 7.2 Viscosity correlations for hybrid nanofluids—cont'd

Hybrid nanofluid	Viscosity correlations	Ref.
MWCNT-ZnO/SAE40	$T = 25\ °C,\ \frac{\mu_{hnf}}{\mu_{bf}} = 1.0087 + 0.1553\varnothing - 0.0334\varnothing^2 + 0.0631\varnothing^3$	[71]
	$T = 30\ °C,\ \frac{\mu_{hnf}}{\mu_{bf}} = 1.0085 + 0.2499\varnothing - 0.2865\varnothing^2 + 0.2043\varnothing^3$	
	$T = 35\ °C,\ \frac{\mu_{hnf}}{\mu_{bf}} = 1.0223 + 0.5341\varnothing - 0.6313\varnothing^2 + 0.366\varnothing^3$	
	$T = 40\ °C,\ \frac{\mu_{hnf}}{\mu_{bf}} = 1.0382 + 0.5376\varnothing - 0.5013\varnothing^2 + 0.261\varnothing^3$	
	$T = 45\ °C,\ \frac{\mu_{hnf}}{\mu_{bf}} = 1.013 + 0.6448\varnothing - 0.9427\varnothing^2 + 0.5225\varnothing^3$	
SiO_2-MWCNT/SAE40	$\frac{\mu_{hnf}}{\mu_{bf}} = a_0 + a_1\varnothing + a_2\varnothing^2 + a_3\varnothing^3 + a_4\varnothing^4$	[72]

	a_0	a_1	a_2	a_3	a_4
$T = 25\ °C$	0.9556	0.9841	−4.4687	7.8779	−4.0731
$T = 30\ °C$	1.023	−0.1613	1.6674	−2.6513	1.3953
$T = 35\ °C$	0.9956	0.1194	0.7286	−1.5119	0.9488
$T = 40\ °C$	1.0133	−0.1316	1.5444	−2.326	1.2019
$T = 45\ °C$	1.007	−0.1187	1.6253	−2.5437	1.3528
$T = 50\ °C$	1.0182	−0.2347	1.593	−1.7341	0.7139

Akilu et al. [34] presented a viscosity model for Newtonian hybrid nanofluids, assuming viscosity is the function of volume concentration only:

$$\mu_{hnf} = \mu_{bf}\left(0.9653 + 77.4567\left(\frac{Ø}{100}\right)^{1.1558}\left(\frac{T}{333}\right)^{0,6801}\right. \tag{7.39}$$

The Newtonian viscosity model of hybrid nanofluid is the function of temperature and volume concentration, while it is the function of the rate of deformation using consistency and power indices for non-Newtonian hybrid nanofluid. Eshgarf et al. [73] predicted the following new correlations of viscosity for MWCNT-SiO_2/water-EG of non-Newtonian hybrid nanofluid, which is given as

$$\mu_{hnf} = m\dot{\gamma}^{n-1} \tag{7.40}$$

where consistency index

$$m = 0.01125 + \left(\frac{38.19 - 0.3T}{7.655 + 0.6953T}\right)(0.01138Ø + 0.5592Ø^2 - 0.3613Ø^3 + 0.07Ø^4 \tag{7.41}$$

and power index

$$n = 0.8543 + \left(\frac{-3.03 + 1.418T}{15.8 + 0.391T}\right)9 - 0.7366Ø + 0.8519Ø^2 - 0.455Ø^3 + 0.08871Ø^4 \tag{7.42}$$

Also, there are some other factors, i.e., shape of nanoparticles, base fluid, and friction, which affect the viscosity of hybrid nanofluid. Sahoo et al. [74] investigated these effects using Al_2O_3-CuO/EG-PG hybrid nanofluid.

7.4 Challenges and summary

The hybrid nanofluid is the new class of nanofluids with better thermophysical and optical properties. The hybrid nanofluid is the dispersion of two or more two nanoparticles in conventional heat transfer fluids. Different experimental studies on hybrid nanofluid indicate that it will replace the mono nanofluid due to having enhanced heat transfer capability, especially in automobile, electro-mechanical, manufacturing, and solar thermal collectors.

Hybrid nanofluids have a superior thermal conductivity than mono nanofluid that make it the better heat transfer medium. The thermal conductivity of hybrid nanofluid increases with an increase in volume concentration of nanoparticles in the base fluid, leading to an increase in viscosity. The viscosity of hybrid nanofluids is majorly dependent on two factors, i.e., nanoparticle volume concentration and temperature. The increment in nanoparticle concentrations leads to an increment in viscosity, whereas the increment in temperature causes a reduction in viscosity. The main drawback of the rise in viscosity is a higher pressure drop, which requires higher pumping power. As a result of this, the overall thermal performance of the system decreased and operational cost increased. The shape and size of the nanoparticles also have a significant influence on viscosity.

The stability of the hybrid nanofluid is an important parameter that is to be considered. The increment in nanoparticle concentration can reduce the stability of the nanofluid. The optimization of different parameters of hybrid nanofluid is a bigger challenge for researchers. The thermophysical properties of hybrid nanofluid depend upon the nanoparticles' volume concentration, temperature, and size and shape of the nanoparticles, but the optimization of these values to obtain better thermophysical properties is a big challenge. There are also some other challenges to the hybrid nanofluid preparation, which are listed below:

- The classical models and empirical correlations of thermal conductivity and viscosity prediction failed to accurately predict the thermal conductivity and viscosity of hybrid nanofluid. These models and correlations are valid only for the specific hybrid nanofluid and for a narrow range of nanoparticles' volume concentration and temperature. However, these models provide some excellent results for the specific conditions, but they increase the number of correlations in the literature and create difficulties in numerical model prediction.
- Further experimental and theoretical studies are needed to have a more generalized model for the thermophysical property prediction of hybrid nanofluids, which can be used for a wide range of operating conditions and hybrid nanofluids.
- Most of the models represented by research for hybrid nanofluid are focused on thermal conductivity and viscosity prediction. Very few have studied the density and specific heat capacity of the hybrid nanofluids. Therefore, there should be more focused studies on density and specific heat capacity to understand hybrid nanofluids better.

References

[1] Z. Said, et al., Recent advances on nanofluids for low to medium temperature solar collectors: energy, exergy, economic analysis and environmental impact, Progr. Energy Combust. Sci. 84 (2021), 100898.
[2] L.S. Sundar, et al., Heat transfer and second law analysis of ethylene glycol-based ternary hybrid nanofluid under laminar flow, J. Therm. Sci. Eng. Appl. 13 (5) (2021), 051021.
[3] L. Sundar, et al., Heat transfer of rGO/Co_3O_4 hybrid nanomaterial based nanofluids and twisted tape configurations in a tube, J. Therm. Sci. Eng. Appl. (2020) 1–41.
[4] Z. Said, et al., Heat transfer, entropy generation, economic and environmental analyses of linear Fresnel reflector using novel rGO-Co3O4 hybrid nanofluids, Renew. Energy 165 (2021) 420–437.
[5] M. Sheikholeslami, et al., Recent progress on flat plate solar collectors and photovoltaic systems in the presence of nanofluid: a review, J. Clean. Prod. (2021) 126119.
[6] P.K. Kanti, et al., Experimental investigation on thermo-hydraulic performance of water-based fly ash–Cu hybrid nanofluid flow in a pipe at various inlet fluid temperatures, Int. Commun. Heat Mass Transfer 124 (2021) 105238.
[7] Z. Said, et al., Optimizing density, dynamic viscosity, thermal conductivity and specific heat of a hybrid nanofluid obtained experimentally via ANFIS-based model and modern optimization, J. Mol. Liq. 321 (2021) 114287.
[8] N.S. Pandya, et al., Influence of the geometrical parameters and particle concentration levels of hybrid nanofluid on the thermal performance of axial grooved heat pipe, Therm. Sci. Eng. Prog. 21 (2021) 100762.
[9] A.A. Hachicha, et al., A review study on the modeling of high-temperature solar thermal collector systems, Renew. Sustain. Energy Rev. 112 (2019) 280–298.
[10] A.K. Tiwar, et al., A review on the application of hybrid nanofluids for parabolic trough collector: recent progress and outlook, J. Clean. Prod. (2021) 126031.
[11] L.S. Sundar, et al., Combination of Co3O4 deposited rGO hybrid nanofluids and longitudinal strip inserts: thermal properties, heat transfer, friction factor, and thermal performance evaluations, Therm. Sci. Eng. Prog. 20 (2020) 100695.
[12] Ç. Yıldız, M. Arıcı, H. Karabay, Comparison of a theoretical and experimental thermal conductivity model on the heat transfer performance of Al2O3-SiO2/water hybrid-nanofluid, Int. J. Heat Mass Transfer 140 (2019) 598–605.
[13] M. Mahmoodi, Numerical simulation of free convection of a nanofluid in L-shaped cavities, Int. J. Therm. Sci. 50 (9) (2011) 1731–1740.
[14] L.S. Sundar, et al., Energy, efficiency, economic impact, and heat transfer aspects of solar flat plate collector with Al2O3 nanofluids and wire coil with core rod inserts, Sustain. Energy Technol. Assess. 40 (2020) 100772.
[15] N.K. Cakmak, et al., Preparation, characterization, stability, and thermal conductivity of rGO-Fe3O4-TiO2 hybrid nanofluid: an experimental study, Powder Technol. 372 (2020) 235–245.
[16] Z. Said, et al., Stability, thermophysical and electrical properties of synthesized carbon nanofiber and reduced-graphene oxide-based nanofluids and their hybrid along with fuzzy modeling approach, Powder Technol. 364 (2020) 795–809.
[17] H. Babar, H.M. Ali, Towards hybrid nanofluids: preparation, thermophysical properties, applications, and challenges, J. Mol. Liq. 281 (2019) 598–633.

[18] Z. Said, et al., Fuzzy modeling and optimization for experimental thermophysical properties of water and ethylene glycol mixture for Al2O3 and TiO2 based nanofluids, Powder Technol. 353 (2019) 345–358.
[19] Z. Said, S. Arora, E. Bellos, A review on performance and environmental effects of conventional and nanofluid-based thermal photovoltaics, Renew. Sustain. Energy Rev. 94 (2018) 302–316.
[20] A.K. Tiwari, et al., 4S consideration (synthesis, sonication, surfactant, stability) for the thermal conductivity of CeO2 with MWCNT and water based hybrid nanofluid: an experimental assessment, Colloids Surf. A Physicochem. Eng. Asp. 610 (2021) 125918.
[21] M. Nabil, et al., An experimental study on the thermal conductivity and dynamic viscosity of TiO2-SiO2 nanofluids in water: ethylene glycol mixture, Int. Commun. Heat Mass Transfer 86 (2017) 181–189.
[22] R.L. Hamilton, O. Crosser, Thermal conductivity of heterogeneous two-component systems, Ind. Eng. Chem. Fund. 1 (3) (1962) 187–191.
[23] W. Yu, S. Choi, The role of interfacial layers in the enhanced thermal conductivity of nanofluids: a renovated Maxwell model, J. Nanopart. Res. 5 (1) (2003) 167–171.
[24] M.H. Esfe, et al., Study on thermal conductivity of water-based nanofluids with hybrid suspensions of CNTs/Al 2 O 3 nanoparticles, J. Therm. Anal. Calorim. 124 (1) (2016) 455–460.
[25] D. Toghraie, V.A. Chaharsoghi, M. Afrand, Measurement of thermal conductivity of ZnO–TiO 2/EG hybrid nanofluid, J. Therm. Anal. Calorim. 125 (1) (2016) 527–535.
[26] M. Afrand, Experimental study on thermal conductivity of ethylene glycol containing hybrid nano-additives and development of a new correlation, Appl. Therm. Eng. 110 (2017) 1111–1119.
[27] A. Asadi, et al., Heat transfer efficiency of Al2O3-MWCNT/thermal oil hybrid nanofluid as a cooling fluid in thermal and energy management applications: an experimental and theoretical investigation, Int. J. Heat Mass Transfer 117 (2018) 474–486.
[28] A. Parsian, M. Akbari, New experimental correlation for the thermal conductivity of ethylene glycol containing Al2O3–Cu hybrid nanoparticles, J. Therm. Anal. Calorim. 131 (2) (2018) 1605–1613.
[29] B. Wei, et al., Thermo-physical property evaluation of diathermic oil based hybrid nanofluids for heat transfer applications, Int. J. Heat Mass Transfer 107 (2017) 281–287.
[30] K. Hamid, et al., Improved thermal conductivity of TiO2–SiO2 hybrid nanofluid in ethylene glycol and water mixture, in: IOP Conference Series: Materials Science and Engineering, IOP Publishing, 2017.
[31] K.A. Hamid, et al., Experimental investigation of thermal conductivity and dynamic viscosity on nanoparticle mixture ratios of TiO2-SiO2 nanofluids, Int. J. Heat Mass Transfer 116 (2018) 1143–1152.
[32] N.N. Esfahani, D. Toghraie, M. Afrand, A new correlation for predicting the thermal conductivity of ZnO–Ag (50%–50%)/water hybrid nanofluid: an experimental study, Powder Technol. 323 (2018) 367–373.
[33] S. Mechiri, V. Vasu, A. Venu Gopal, Investigation of thermal conductivity and rheological properties of vegetable oil based hybrid nanofluids containing Cu–Zn hybrid nanoparticles, Exp. Heat Transfer 30 (3) (2017) 205–217.
[34] S. Akilu, A.T. Baheta, K. Sharma, Experimental measurements of thermal conductivity and viscosity of ethylene glycol-based hybrid nanofluid with TiO2-CuO/C inclusions, J. Mol. Liq. 246 (2017) 396–405.

[35] M. Benzema, et al., Second law analysis of MHD mixed convection heat transfer in a vented irregular cavity filled with Ag–MgO/water hybrid nanofluid, J. Therm. Anal. Calorim. 137 (3) (2019) 1113–1132.
[36] M.H. Esfe, S. Esfandeh, M. Rejvani, Modeling of thermal conductivity of MWCNT-SiO2 (30: 70%)/EG hybrid nanofluid, sensitivity analyzing and cost performance for industrial applications, J. Therm. Anal. Calorim. 131 (2) (2018) 1437–1447.
[37] S.H. Rostamian, et al., An inspection of thermal conductivity of CuO-SWCNTs hybrid nanofluid versus temperature and concentration using experimental data, ANN modeling and new correlation, J. Mol. Liq. 231 (2017) 364–369.
[38] S.S. Harandi, et al., An experimental study on thermal conductivity of F-MWCNTs–Fe3O4/EG hybrid nanofluid: effects of temperature and concentration, Int. Commun. Heat Mass Transfer 76 (2016) 171–177.
[39] M.H. Esfe, et al., Thermal conductivity enhancement of SiO2–MWCNT (85: 15%)–EG hybrid nanofluids, J. Therm. Anal. Calorim. 128 (1) (2017) 249–258.
[40] M.H. Esfe, et al., Experimental evaluation, sensitivity analyzation and ANN modeling of thermal conductivity of ZnO-MWCNT/EG-water hybrid nanofluid for engineering applications, Appl. Therm. Eng. 125 (2017) 673–685.
[41] G.M. Moldoveanu, et al., Al2O3/TiO2 hybrid nanofluids thermal conductivity, J. Therm. Anal. Calorim. 137 (2) (2019) 583–592.
[42] M.H. Esfe, A. Alirezaie, M. Rejvani, An applicable study on the thermal conductivity of SWCNT-MgO hybrid nanofluid and price-performance analysis for energy management, Appl. Therm. Eng. 111 (2017) 1202–1210.
[43] S. Mousavi, F. Esmaeilzadeh, X. Wang, Effects of temperature and particles volume concentration on the thermophysical properties and the rheological behavior of CuO/MgO/TiO 2 aqueous ternary hybrid nanofluid, J. Therm. Anal. Calorim. 137 (3) (2019) 879–901.
[44] M. Vafaei, et al., Evaluation of thermal conductivity of MgO-MWCNTs/EG hybrid nanofluids based on experimental data by selecting optimal artificial neural networks, Physica E 85 (2017) 90–96.
[45] M.T. Sulgani, A. Karimipour, Improve the thermal conductivity of 10w40-engine oil at various temperature by addition of Al2O3/Fe2O3 nanoparticles, J. Mol. Liq. 283 (2019) 660–666.
[46] L.S. Sundar, et al., Nanodiamond-Fe3O4 nanofluids: preparation and measurement of viscosity, electrical and thermal conductivities, Int. Commun. Heat Mass Transfer 73 (2016) 62–74.
[47] M.H. Esfe, M.H. Hajmohammad, Thermal conductivity and viscosity optimization of nanodiamond-Co3O4/EG (40: 60) aqueous nanofluid using NSGA-II coupled with RSM, J. Mol. Liq. 238 (2017) 545–552.
[48] M.H. Esfe, A.A.A. Arani, M. Firouzi, Empirical study and model development of thermal conductivity improvement and assessment of cost and sensitivity of EG-water based SWCNT-ZnO (30%: 70%) hybrid nanofluid, J. Mol. Liq. 244 (2017) 252–261.
[49] M.H. Esfe, et al., Estimation of thermal conductivity of ethylene glycol-based nanofluid with hybrid suspensions of SWCNT–Al2O3 nanoparticles by correlation and ANN methods using experimental data, J. Therm. Anal. Calorim. 128 (3) (2017) 1359–1371.
[50] E. Bellos, Z. Said, C. Tzivanidis, The use of nanofluids in solar concentrating technologies: a comprehensive review, J. Clean. Prod. 196 (2018) 84–99.
[51] M. Gupta, et al., Up to date review on the synthesis and thermophysical properties of hybrid nanofluids, J. Clean. Prod. 190 (2018) 169–192.
[52] M. Jamei, et al., On the specific heat capacity estimation of metal oxide-based nanofluid for energy perspective—a comprehensive assessment of data analysis techniques, Int. Commun. Heat Mass Transfer 123 (2021) 105217.

[53] H. O'Hanley, et al., Measurement and model validation of nanofluid specific heat capacity with differential scanning calorimetry, Adv. Mech. Eng. 4 (2012) 181079.
[54] G.J. Tertsinidou, et al., New measurements of the apparent thermal conductivity of nanofluids and investigation of their heat transfer capabilities, J. Chem. Eng. Data 62 (1) (2017) 491–507.
[55] Y. Gao, et al., Experimental investigation of specific heat of aqueous graphene oxide Al2O3 hybrid nanofluid, Therm. Sci. 00 (2019) 381.
[56] B. Barbés, et al., Thermal conductivity and specific heat capacity measurements of Al2O3 nanofluids, J. Therm. Anal. Calorim. 111 (2) (2013) 1615–1625.
[57] S. Ghadikolaei, et al., Investigation on thermophysical properties of Tio2–Cu/H2O hybrid nanofluid transport dependent on shape factor in MHD stagnation point flow, Powder Technol. 322 (2017) 428–438.
[58] A.K. Tiwari, et al., Experimental comparison of specific heat capacity of three different metal oxides with MWCNT/water-based hybrid nanofluids: proposing a new correlation, Appl. Nanosci. (2020) 1–11.
[59] Z. Said, et al., Acid-functionalized carbon nanofibers for high stability, thermoelectrical and electrochemical properties of nanofluids, J. Colloid Interface Sci. 520 (2018) 50–57.
[60] L.S. Sundar, et al., Experimental investigation of thermo-physical properties, heat transfer, pumping power, entropy generation, and exergy efficiency of nanodiamond+ Fe3O4/60: 40% water-ethylene glycol hybrid nanofluid flow in a tube, Therm. Sci. Eng. Prog. 21 (2021) 100799.
[61] M. Gupta, et al., A review on thermophysical properties of nanofluids and heat transfer applications, Renew. Sustain. Energy Rev. 74 (2017) 638–670.
[62] B.C. Pak, Y.I. Cho, Hydrodynamic and heat transfer study of dispersed fluids with submicron metallic oxide particles, Exp. Heat Transfer Int. J. 11 (2) (1998) 151–170.
[63] B. Takabi, S. Salehi, Augmentation of the heat transfer performance of a sinusoidal corrugated enclosure by employing hybrid nanofluid, Adv. Mech. Eng. 6 (2014) 147059.
[64] R. Vajjha, D. Das, B. Mahagaonkar, Density measurement of different nanofluids and their comparison with theory, Petro. Sci. Technol. 27 (6) (2009) 612–624.
[65] Z. Said, R. Saidur, Thermophysical properties of metal oxides nanofluids, in: Nanofluid heat and mass transfer in engineering problems, 2017.
[66] A.K. Tiwari, et al., 3S (Sonication, surfactant, stability) impact on the viscosity of hybrid nanofluid with different base fluids: an experimental study, J. Mol. Liq. 329 (2021) 115455.
[67] O. Soltani, M. Akbari, Effects of temperature and particles concentration on the dynamic viscosity of MgO-MWCNT/ethylene glycol hybrid nanofluid: experimental study, Physica E 84 (2016) 564–570.
[68] M. Asadi, A. Asadi, Dynamic viscosity of MWCNT/ZnO–engine oil hybrid nanofluid: an experimental investigation and new correlation in different temperatures and solid concentrations, Int. Commun. Heat Mass Transfer 76 (2016) 41–45.
[69] R.R. Sahoo, V. Kumar, Development of a new correlation to determine the viscosity of ternary hybrid nanofluid, Int. Commun. Heat Mass Transfer 111 (2020) 104451.
[70] Z. Li, et al., Experimental study of temperature and mass fraction effects on thermal conductivity and dynamic viscosity of SiO2-oleic acid/liquid paraffin nanofluid, Int. Commun. Heat Mass Transfer 110 (2020) 104436.
[71] M. Hemmat Esfe, et al., Examination of rheological behavior of MWCNTs/ZnO-SAE40 hybrid nano-lubricants under various temperatures and solid volume fractions, Exp. Therm. Fluid Sci. 80 (2017) 384–390.

[72] M. Afrand, K. Nazari Najafabadi, M. Akbari, Effects of temperature and solid volume fraction on viscosity of SiO2-MWCNTs/SAE40 hybrid nanofluid as a coolant and lubricant in heat engines, Appl. Therm. Eng. 102 (2016) 45–54.
[73] H. Eshgarf, et al., Prediction of rheological behavior of MWCNTs–SiO 2/EG–water non-Newtonian hybrid nanofluid by designing new correlations and optimal artificial neural networks, J. Therm. Anal. Calorim. 132 (2) (2018) 1029–1038.
[74] V. Kumar, R.R. Sahoo, Viscosity and thermal conductivity comparative study for hybrid nanofluid in binary base fluids, Heat Transfer Asian Res. 48 (7) (2019) 3144–3161.

8

Brief overview of the applications of hybrid nanofluids

M. Sheikholeslami[a,b], Elham Abohamzeh[c], Z. Ebrahimpour[a,b], and Zafar Said[d,e,f,*]

[a]Department of Mechanical Engineering, Babol Noshirvani University of Technology, Babol, Iran. [b]Renewable Energy Systems and Nanofluid Applications in Heat Transfer Laboratory, Babol Noshirvani University of Technology, Babol, Iran. [c]Department of Energy, Materials, and Energy Research Center (MERC), Karaj, Iran. [d]Department of Sustainable and Renewable Energy Engineering, University of Sharjah, Sharjah, United Arab Emirates. [e]Research Institute for Sciences and Engineering, University of Sharjah, Sharjah, United Arab Emirates. [f]U.S.-Pakistan Center for Advanced Studies in Energy (USPCAS-E), National University of Sciences and Technology (NUST), Islamabad, Pakistan

Corresponding author: zsaid@sharjah.ac.ae, zaffar.ks@gmail.com

Chapter outline

8.1 Introduction 172
8.2 Electronics cooling 172
8.3 Solar collectors 177
8.4 Heat exchangers 181
8.5 Engine cooling 185
8.6 Refrigeration 188
8.7 Machining 190
8.8 Desalination 192
8.9 Challenges and outlook 194
8.10 Summary 195
References 196

Hybrid Nanofluids: Preparation, Characterization and Applications. https://doi.org/10.1016/B978-0-323-85836-6.00008-9

8.1 Introduction

Nanofluids, invented by Choi [1], are colloids composed of nanoparticles and a pure fluid. Compared to the base fluids, the conductivities of nanoparticles are commonly higher while their size is considerably lower than 100 nm [2]. The thermal properties of base fluids improve substantially with the application of nanoparticles [3]. Different fluids such as polymeric solutions, lubricants and oils, organic liquids, water, bio-fluids, and other common liquids can be considered as the base fluids [4, 5]. The added nanoparticles may be metal oxides, metal carbides, metal nitrides (AIN, SiN, etc.), chemically stable metals (e.g., copper, gold), and carbon in different forms (e.g., fullerene, CNT, graphite, diamond) [6–12]. A very new type of nanofluids is hybrid nanomaterial [13, 14]. To produce these materials, different kinds of nanoparticles (more than one) or hybrid (composite) nanoparticles are suspended in the pure fluid. The chemical and physical features of different materials are combined in the hybrid material simultaneously, providing the characteristics in a homogeneous phase [15–18]. The effective viscosity [19] and density [20] of hybrid nanofluids may be in the same order for mono-nanofluids, while their thermal conductivity might be substantially higher than mono nanofluids considering synergistic effects [21]. The outstanding enhancement in thermal transfer properties of nanofluids led researchers to use them in various engineering applications, including nuclear cooling [22, 23], desalination [24], machining [25], refrigeration [26], engine cooling [27], heat exchangers (HEX) [28], solar collectors [29, 30], and electronics cooling [31], as illustrated in Fig. 8.1.

A great deal of studies on nanofluids has been performed for the investigation of their heat transfer and thermophysical properties [32, 33]. However, there have been relatively a few studies on their commercial applications [34–37]. United States electric power industry initiated utilizing nanofluids as cooling agents in 2008, leading to the annual saving of 10–30 trillion Btu energy. This amount of energy-saving results in reducing No_x, CO_2, and 21,000 metric tons of SiO_2 [38]. Suresh et al. [39] scrutinized the thermal properties of the Al_2O_3-Cu nanocomposite. They observed a 14% improvement in heat transfer. Nine et al. [40] used hybrid nanofluids containing Al_2O_3-MWNTs and investigated the thermal characteristics at 1–6% volume concentrations. Jia et al. [41] applied a hydrothermal approach for preparing Fe_3O_4-CNT.

8.2 Electronics cooling

Devices like electronic processors are extremely sensitive to temperatures. Since temperature values have adverse effects on

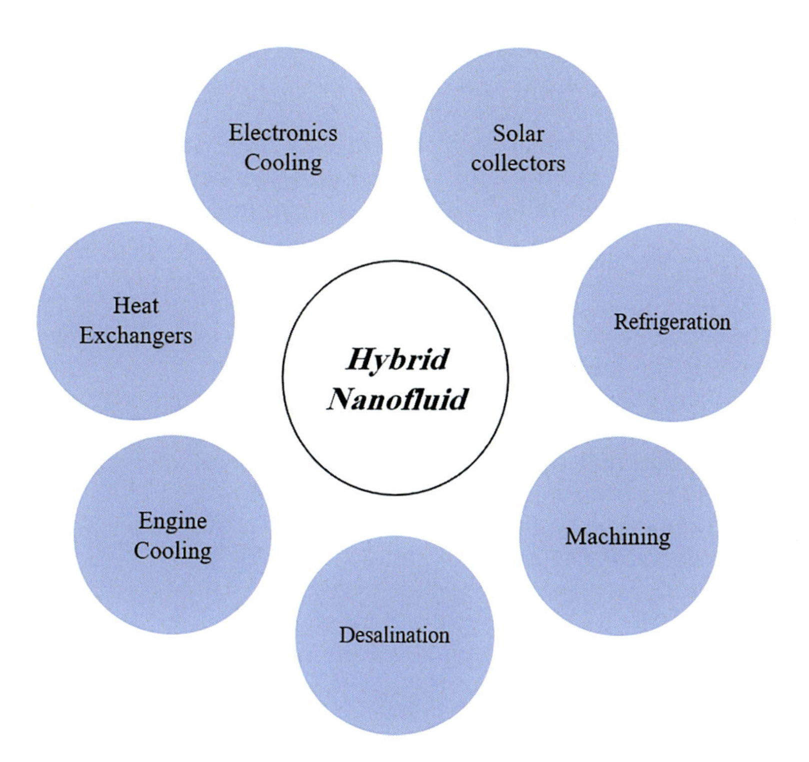

Fig. 8.1 Different usages of hybrid nanofluids.

their energy efficiency and lifetime, thermal management and cooling of these devices are very important. Until now, one important challenge in microelectronics technology and computer systems is the cooling of central processor units (CPUs). The hybrid nanofluids as new kinds of nanofluids are prepared with the dispersion of different nanoparticles in mixture form or composite state. The improvement of thermal conductivities, enhancing heat transfer rates, and benefiting several particular attributes are the main objectives for preparing hybrid nanomaterials. In fact, using two or more dissimilar materials leads to combining their features, resulting in promising applications. Using hybrid nanofluids for cooling electronic chips has been investigated in some studies. Selvakumar and Suresh [42] employed the copper-alumina hybrid nanofluids in a channel. Based on their results, with the application of hybrid nanomaterial, a significant improvement in conduction was reported compared with the case of using the base liquid. In addition, the increase of required pumping power was lower than the increase of convection when hybrid nanomaterial is used instead of water. The enhancement in thermal conductivities as 30% was examined by Sarbolookzadeh Harandi et al. [43] experimentally. They suggested a particular correlation for the prediction of thermal conductivities in the case of employing hybrid nanofluids. Based on their outputs, the increase of concentration of solid leads to enhancing the thermal conductivity ratios.

Moreover, more changes were noticed in thermal conductivity ratios with volume fractions of solid at higher temperatures. In addition, thermal conductivity ratios were more strongly affected by the temperature at higher volume fractions of solid. Nimmagadda and Venkatasubbaiah [44] investigated the improvement in thermal transport using hybrid nanofluids containing Ag and hybrid nanoparticles statistically for achieving excellent heat transfer properties and becoming an acceptable alternative with lower cost. Bahiraei and Heshmatian [45] considered three heat sinks containing hybrid nanofluid with Ag nanoparticles on graphene nanoplatelets and studied the efficacy and second law attributes of hybrid nanofluids for CPU cooling. They compared the performance of a new distributor heatsink with that of two conventional liquid blocks. Based on their outputs, the efficiency of the new distributor heat sink was higher from both thermal efficiency and irreversibility characteristics. Moreover, compared with H_2O, the application of the nanofluid was more advantageous. Fig. 8.2 illustrates average temperatures on the surface of CPU versus *Re*

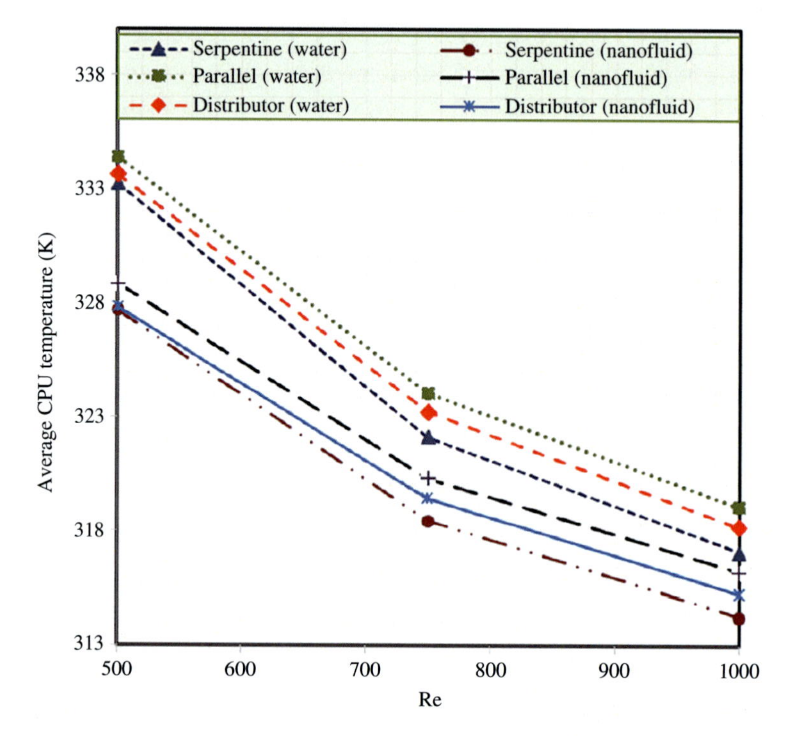

Fig. 8.2 The average temperature on the CPU surface versus the *Re* number for both water and nanofluid in different liquid blocks. From M. Bahiraei, S. Heshmatian, Efficacy of a novel liquid block working with a nanofluid containing graphene nanoplatelets decorated with silver nanoparticles compared with conventional CPU coolers, Appl. Therm. Eng. 127 (2017) 1233–1245. Copyright Elsevier (Lic# 5064631177341).

in the case of using H_2O and nanofluid for different liquid blocks. According to this diagram, the average temperature reduced considerably on the surface of the CPU by adding nanoparticles. For lower Reynolds numbers, the reduction in average temperature was more significant. At $Re = 500$, the greatest decrease was observed. Additionally, the worst cooling performance was noticed for the parallel flow liquid block, while the best cooling was observed for the serpentine liquid block for both nanofluid and water and under a constant Reynolds number. Furthermore, it can be understood from the figure that the influence of the liquid block type was more significant when Re numbers were higher.

An entropy generation study was performed by Ahammed et al. [46] on the cooling of the electronic components using hybrid nanofluids. They dispersed the graphene, alumina, and their hybrid in the base fluid separately and conducted the test. They observed that applying graphene/water nanofluids led to an 88.62% augmentation in the value of k_{nf}, and this increment was 63.13% and 31.89% for hybrid nanofluids and alumina/water, respectively. According to Fig. 8.3, the viscosity declined by

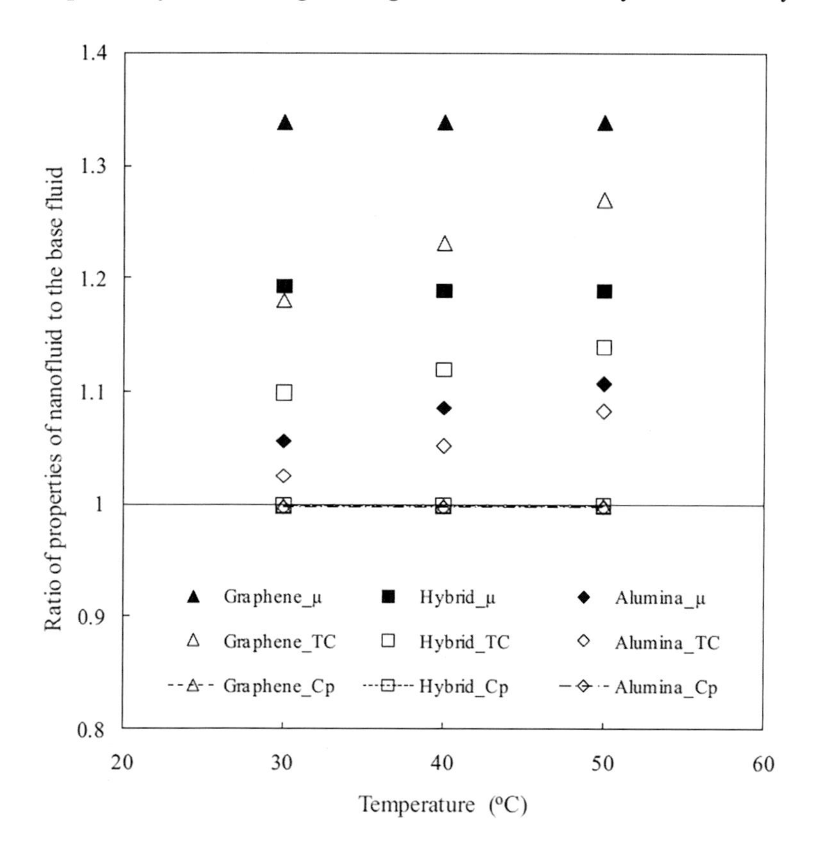

Fig. 8.3 Altering in the ratio of thermophysical characteristics of nanofluids. From N. Ahammed, L.G. Asirvatham, S. Wongwises, Entropy generation analysis of graphene–alumina hybrid nanofluid in multiport minichannel heat exchanger coupled with thermoelectric cooler, Int. J. Heat Mass Transf. 103 (2016) 1084–1097. Copyright Elsevier (Lic# 5064631402931).

augmenting the temperature, and its value was greater for hybrid nanofluids in comparison with that of the pure fluid.

Numerical simulations were conducted by Kumar and Sarkar [47] for measuring thermal transport and pressure drop features of mini channel heat sink containing Al_2O_3-MWNTs/water hybrid and Al_2O_3/water nanofluids. The Reynold number, channel aspect ratio, and hydraulic diameter were important factors for this study. When nanofluids at a concentration of 0.01 vol% were used, the highest increase (15.6%) in heat transfer coefficient was noticed in the mini channel with 0.5 mm depth. For three heat sinks with the dimension of $3 \times 1 \times 0.5 mm^3$, the profile of temperature for Al_2O_3-MWNT hybrid nanofluid with 0.01 vol% and water along the channel length can be observed in Fig. 8.4.

Bahiraei and Heshmatian [48] scrutinized the efficiency of two heat sink operating with hybrid nanomaterial comprising silver and graphene nanoparticles numerically. They observed that the influence of heat transfer on irreversibility is much more compared to the friction factor. Moreover, they concluded that nanofluid could be considered as the best option for electronics cooling applications according to the first and second law of thermodynamics. Khoshvaght-Aliabadi and Nozan [49] investigated the influence of the geometry of mini channels on Nusselt numbers utilizing water as coolant. They considered sinusoidal, trapezoidal, and triangular shapes and observed the maximum value for trapezoidal shape. With the application of water as the coolant, a 13%–14% increase was noticed in the Nusselt number; however,

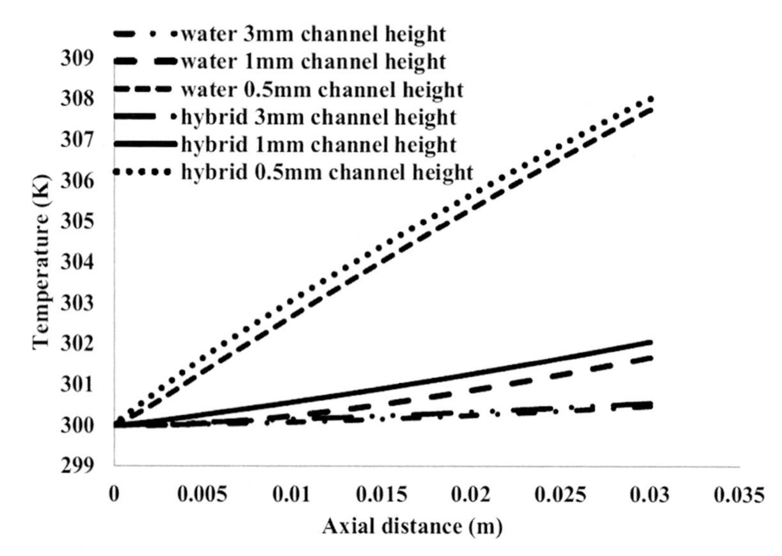

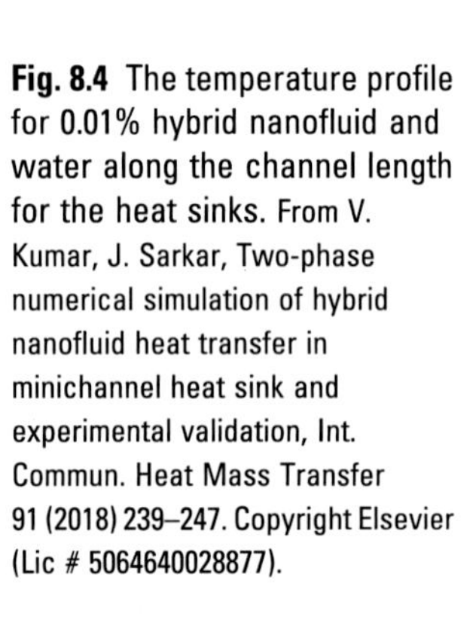

Fig. 8.4 The temperature profile for 0.01% hybrid nanofluid and water along the channel length for the heat sinks. From V. Kumar, J. Sarkar, Two-phase numerical simulation of hybrid nanofluid heat transfer in minichannel heat sink and experimental validation, Int. Commun. Heat Mass Transfer 91 (2018) 239–247. Copyright Elsevier (Lic # 5064640028877).

better results can be obtained for thermal properties in the case of applying hybrid nanofluids. Krishna et al. [50] performed simulations to evaluate the pressure drop and thermal transport features for hybrid nanofluid laminar flow inside a circular microchannel heat sink (MCHS). The heat sink performance was studied utilizing hybrid nanofluid containing CuO and MWNT nanoparticles, considering volume fractions of 1%–3% and various flow rates. To define the features of hybrid nanomaterial, the law of correlations and mixtures from previous literature was applied. Based on the outputs, heat transfer enhanced with utilizing hybrid nanomaterial instead of mono nanofluid and also increasing the volume fraction provided better convection. The greathest value for Nu has been obtained to utilize hybrid nanofluid with a 3% concentration. Heidarshenas et al. [51] scrutinized the stability and efficiency of a new hybrid nanofluid containing ionic liquid-alumina. With coating Al_2O_3 nanoparticles with various dimensions by ionic liquid [BMIM]PF6, the hybrid system was prepared. Compared to water, using the hybrid nanofluid can provide up to 26% improvement in Nusselt number. Comparing the results with a previous experimental study conducted in the same system but applying alumina nanofluid revealed higher stability of hybrid nanofluids. According to the outputs of the conducted study, it can be proved that hybrid nanofluids containing ionic liquid-alumina are an appropriate alternative as an operating fluid, particularly heat transfer devices with small scale considering their facile preparation method durability and high stability.

8.3 Solar collectors

Over recent years, the great potential of solar concentrators and collector devices for converting solar energy into thermal energy has been revealed by several studies [52]. Indeed, solar radiation can be converted into heat that can be transferred to media like air, solar fluid, or water [53]. Consequently, solar heat can be utilized for backup heating systems and also heating water. One promising approach for enhancing heat transfer features is the application of nanofluids as novel heat transfer media [54, 55]. The thermal properties can be considerably enhanced in the case of employing nanoparticles for solar energy applications [56]. Considering the superior heat transfer of hybrid nanomaterial, they have been regarded as a great substitute for mono nanofluids in a variety of fields, including solar energy [57, 58]. Farajzadeh et al. [59] scrutinized the impact of dispersing $TiO_2$and alumina nanopowders and also a combination of these nanoparticles into the base fluid on the solar collector performance numerically and experimentally. Based on the outputs, 19%, 21%, and 26% increments in efficiency were observed with usingTiO_2, Al_2O_3, and their combinations, respectively. Moreover, the performance of the collector has been enhanced by about 5%. Karimi [60] used hybrid nanofluids containing Fe_3O_4/SiO_2 nanoparticles and performed entropy and energy analyses of direct absorption solar collector. They fabricated hybrid nanofluids with various concentrations and volume ratios and selected the suitable samples to be utilized as working fluids in the collector. A similar behavior was seen for changes in exergy and energy performance of the collector with volume fractions and flow rates. Moreover, a decrease was noticed in the number of dimensionless entropy generation by concentration. The lowest minimum number of entropy generation was reported in the volume fraction of 2000 ppm and a flow rate of 0.0225 kg/s. Chen et al. [61] investigated the application of hybrid nanofluid containing antimony-doped tin oxide (ATO) and copper (II) oxide nanoparticles for enhancing solar thermal conversion property. Based on the results, the solar thermal utilization efficiencies of 80.7%, 81.3%, and 92.5% were defined for using ATO, CuO, and two-component nanofluids, respectively. Bellos and Tzivanidis [62] considered hybrid nanofluids and mono nanofluids and studied their application in parabolic trough collectors (PTCs). They tested the operation of LS-2 PTC with Syltherm 800, whereas the explored nanofluids were as follows: 1.5% TiO_2/oil, 1.5% Al_2O_3, 3% TiO_2/oil, and 3% Al_2O_3/oil. For having a better comparison, the same volume concentrations have been considered for nanoparticles in all the cases. Based on the outputs, the improvement in the thermal efficiency for mono nanofluids was up to 0.7%, while this improvement reached up to 1.8% for the hybrid nanofluid. For various working fluids, the thermal efficiency at various inlet temperatures in the range of 300–650 K is depicted in Fig. 8.5. The values of thermal efficiency were higher when nanofluids were used compared to the case with oil, proving the improvement. The highest value for thermal efficiency was obtained for the case with hybrid nanofluid.

A review of the performance of solar units that operate with hybrid nanofluids was performed by Shah and Muhammad Ali [63]. They also reviewed the efficiency of mono nanofluids-based solar energy devices. In this study, the preparation techniques and properties of hybrid nanomaterial and their impacts on the operating factors of solar devices were investigated. Based on the reviewed studies, significant improvement in the efficiency and output power was noticed for these systems. They also mentioned

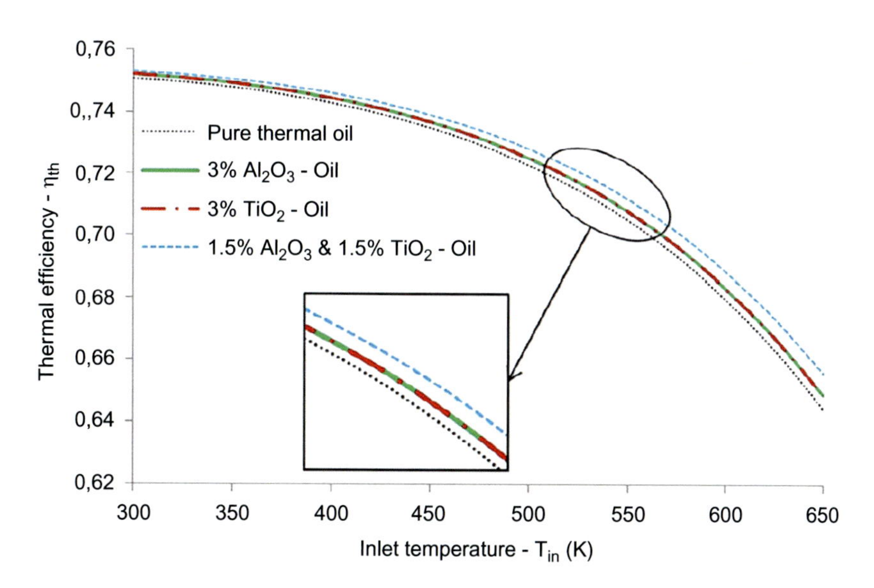

Fig. 8.5 Thermal efficiency for all the investigated heat transfer fluids at various inlet temperatures. From E. Bellos, C. Tzivanidis, Thermal analysis of parabolic trough collector operating with mono and hybrid nanofluids. Sustainable Energy Technol. Assess. 26 (2018) 105–115. Copyright Elsevier (Lic # 5064640209442).

the challenges that should be addressed to further develop and commercialize hybrid nanofluids-based devices. These issues include increased pumping power, rheological issues, increased friction factor, and instability. Subsequently, the ecologic and economic benefits of utilizing binary nanofluids in solar devices are discussed. Verma et al. [64] scrutinized collector efficiency for the case of using nanofluids instead of conventional heat transfer fluid, water. H_2O-based hybrid nanomaterial containing MWNTs, MgO, and CuO was selected. They conducted an experimental study under specified ambient conditions for various flow rates (0.5–2.0 LPM) and different concentrations in the range of 0.25%–2.0%. The entropy generation resulting from heat transfer because of changes in temperature was enhanced, and entropy generation, due to irreversibility, was suppressed. The energetic efficiency of 70.55% and exergetic efficiency of 71.54% were obtained to employ MgO hybrid nonmaterial in the solar collector. At the same conditions, the energetic and exergetic efficiency of 69.11% and 70.63% were defined for the CuO hybrid nanofluid. Compared to CuO hybrid, better performance was observed for MgO hybrid nanofluid that was also closer to MWNTs/water fluid. The variations of the collector efficiency with changes in solar intensity are exhibited in Fig. 8.6.

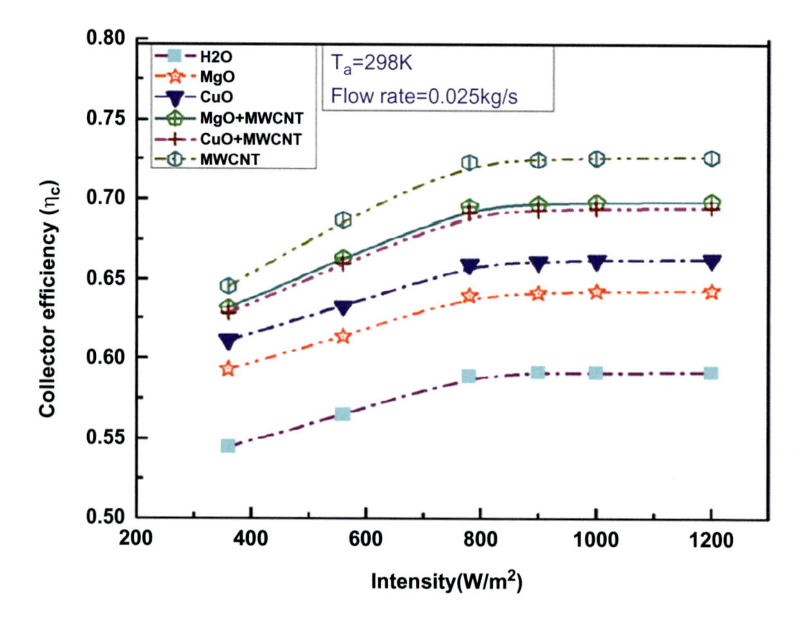

Fig. 8.6 The efficiency of collector for different solar intensity. From S.K. Verma, et al., Performance analysis of hybrid nanofluids in flat plate solar collector as an advanced working fluid, Sol. Energy 167 (2018) 231–241. Copyright Elsevier (Lic # 5064640376401).

Both elevated solar-thermal conversion efficiency and improved stability were achieved by Li et al. [65] to apply hybrid nanomaterial in the direct absorption solar collector (DASC). The stability of hybrid nanofluids was confirmed. In both near-infrared and visible regions, great solar irradiation absorption capacity was displayed by hybrid nanofluids. So, the effectiveness of hybrid nanofluids for DASC applications was confirmed, where 99.9% solar energy can be absorbed by MWNT-SiC nanofluid with 0.5 wt% at just 1 cm path length. Moreover, with increasing mass concentration, the conversion efficiency of hybrid nanofluid is enhanced. For performance, the highest value was defined to be 97.3% for MWNT-SiC nanofluid with 1 wt% at 10 min, which was 48.6% greater compared to pure ethylene glycol. The possible applications of MWNT-SiC nanofluid for low-temperature DASC systems were discussed. Fig. 8.7A shows the variations of bulk temperature for MWNT-SiC nanofluids versus mass concentration. In addition, Fig. 8.7B exhibits the solar-thermal conversion efficiency versus mass concentration.

Hybrid nanofluids' flow parameters and three-dimensional heat transfer under turbulence conditions in PTC were studied by Ekiciler et al. [66]. As working fluids, Ag-MgO/Syltherm 800, Ag-TiO_2/Syltherm 800, and Ag-ZnO/Syltherm 800 with volume fractions of 1.0%, 2.0%, 3.0%, and 4.0% were utilized. Reynolds numbers in the range of 10,000–80,000 were considered. With increasing volume fractions and Reynolds number, heat transfer

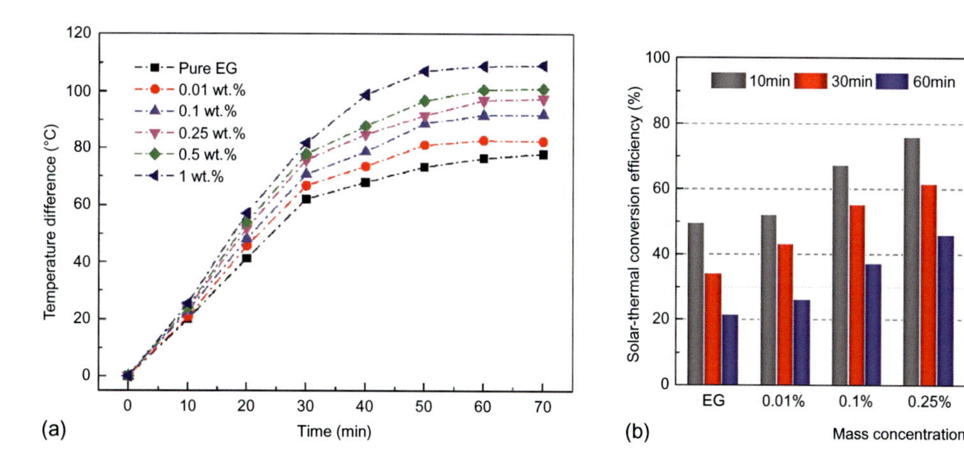

Fig. 8.7 (A) Temperature variations versus irradiation time for MWNT-SiC nanofluids; (B) the solar-thermal conversion efficiency versus irradiation time. From X. Li, G. Zeng, X. Lei, The stability, optical properties and solar-thermal conversion performance of SiC-MWCNTs hybrid nanofluids for the direct absorption solar collector (DASC) application, Sol. Energy Mater. Sol. Cells, 2020. 206: p. 110323 Copyright Elsevier (Lic # 5064640524957).

enhances. One important parameter for PTC is thermal efficiency that reduces with augmenting Reynolds numbers and increments with rising volume fractions. Ag-MgO/Syltherm 800 (4.0 wt%) was determined as the hybrid nanofluid with the maximum effectiveness. The convection factor for various hybrid nanomaterial and Syltherm 800 versus volume fractions and Reynolds numbers is depicted in Fig. 8.8. According to the outcomes, the convection heat transfer coefficient augments with increasing nanoparticle volume fractions in all hybrid nanofluids.

Utilizing nanofluid and hybrid nanofluids in absorber tubes was reviewed in the study investigated by Tiwari et al. [33]. They observed that the optical and thermal features could be improved using nanofluids. In addition, intensifying of thermal properties was noticed by replacing mono nanofluid with hybrid nanofluids. Their study made progress in understanding the optimization of alteration factors to obtain the greatest possible improvement in results and can offer considerably sufficient understanding to select particular methods for enhancing the total efficiency of PTC.

8.4 Heat exchangers

There have been several attempts to enhance the performance of heat exchangers with changing plate surface textures, for instance, by creating corrugations and waviness on the surface

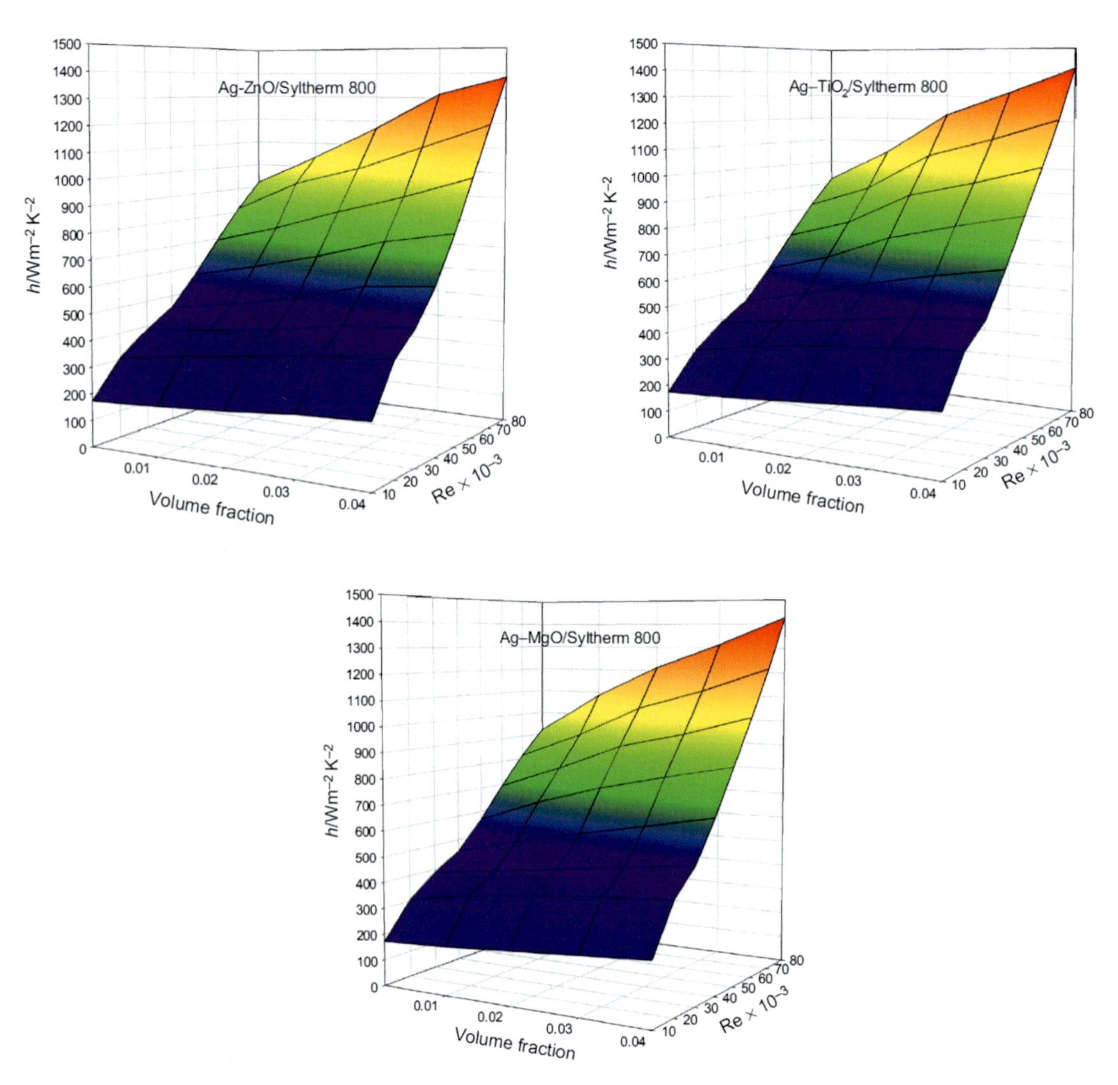

Fig. 8.8 Convection heat transfer coefficient distribution in terms of *Re* and concentration. From R. Ekiciler, et al., Effect of hybrid nanofluid on heat transfer performance of parabolic trough solar collector receiver, J. Therm. Anal. Calorim. 143(2) (2021) 1637–1654. Copyright Springer Nature (Lic # 5064640739817).

of plates, leading to increased heat transfer coefficients and, therefore, thermal performance [67]. In addition, it is possible to enhance heat transfer properties by changing the needed thermophysical characteristics of heat transfer fluids. Achieving this improvement is possible by dispersing one or more nanopowder into base fluids [42, 43]. Based on the literature, engineered hybrid nanomaterials prepared with adding two or more nanopowders in base fluids possess excellent thermal conductivities compared with base fluids and adaptable characteristics that make them appropriate for several applications [68]. They have the potential of decreasing thermal resistance and increasing thermal conductivity [69, 70]. Allahyar et al. [71] studied mono and hybrid nanomaterial efficiency inside a coiled HEX at laminar flow operating conditions and constant wall temperature. Different concentrations of 0.1–0.4 vol% were considered for nanoparticles. With the application of 0.4 wt% hybrid nanofluid, the highest rate of 31.58% was achieved for heat transfer that was greater than that of the distilled water. For hybrid nanofluid, the highest efficiency factor was around 2.55, indicating the outstanding efficiency of the suggested approach for heat transfer improvement in heat exchangers. According to Fig. 8.9, all obtained values for the performance factor were higher than one, which proves the substantial improvement in the system performance by utilizing hybrid nanofluids in the helical.

Irreversibilities due to friction and heat transfer for the flow of non-Newtonian hybrid nanomaterial within a double-pipe HEX were evaluated by Bahiraei et al. [72]. Two types of nanoparticles, namely gum arabic-coated CNTs and ammonium hydroxide-coated nanoparticles, were considered, and for preparing nanofluids, these nanoparticles were added into the base fluid. Based on the outputs, as the temperature of the water on the annulus

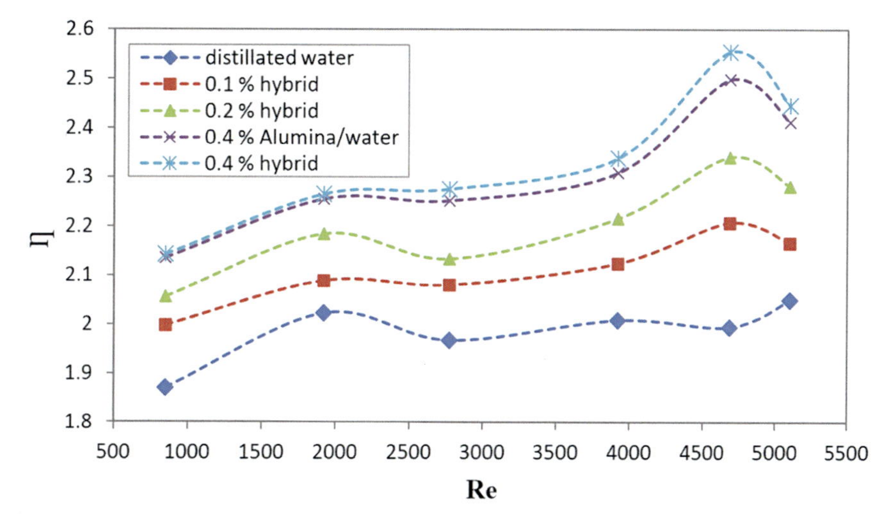

Fig. 8.9 Performance factor in different Reynolds numbers using mono and hybrid nanofluids. From H.R. Allahyar, F. Hormozi, B. ZareNezhad, Experimental investigation on the thermal performance of a coiled heat exchanger using a new hybrid nanofluid, Exp. Thermal Fluid Sci. 76 (2016) 324–329. Copyright Elsevier (Lic # 5064640888414).

side increased, frictional entropy production decreased, while thermal entropy production increased. The highest amount of thermal entropy generation was observed for high magnetite concentration at maximum CNT concentration, while for low magnetite concentration, it occurred at minimum CNT concentration. Huang et al. [73] considered a chevron corrugated-plate HEX operating with a hybrid nanomaterial containing MWNTs and alumina nanopowder and conducted an experimental study to investigate pressure drop as well as thermal transport properties of the hybrid nanofluids. They compared the outputs obtained for the case with hybrid nanofluid with those of the water and water-based Al_2O_3 nanofluid. Based on the similar flow velocity comparisons, a slightly higher heat transfer coefficient was noticed for hybrid nanofluid. Moreover, at a specified pumping power, the maximum heat transfer coefficient was achieved for hybrid nanofluids. Hybrid nanofluid of water-based Al_2O_3nanomaterial and water-based MWNT nanofluid (sample 1: water/alumina (1.89 vol% 25 mL) and water/MWNT (0.278 vol% 10 mL) and sample 2: water/alumina (5.0 vol% 25 mL) and water/MWNT (0.278 vol% 10 mL)) were prepared by Wu et al. [74]. The viscosity and k_{nf} of these mixtures were studied experimentally. They observed a 7.2% improvement in thermal conductivity for sample 1 and 7.9% enhancement for sample 2, indicating that a noticeable rise in the volume fraction of Al_2O_3did not provide a noticeable improvement in the thermal conductivity of the mixture. Ahammed et al. [46] conducted research on a two-pass multiport mini channel HEX coupled with a thermoelectric cooler to evaluate entropy generation of hybrid nanofluid as the working fluid. Nanofluids were prepared with dispersing graphene (5 nm), alumina (50 nm), and also a mixture of them in similar fractions with volume concentrations of 0.1% in the base fluid. For the channel, the aspect ratio and hydraulic diameter were 0.689 and 1.184 mm, respectively. According to the outputs, when the water-based graphene nanofluid was used, 17.32% improvement in the coefficient of performance (COP) and cooling capacity was observed compared with other tested nanofluid combinations. In the same way, the convective heat transfer coefficient improved 88.62%, and the temperature of the device decreased about 4.7°C in the case of using pure water-based graphene nanofluid as the coolant. Between the investigated nanofluids, water-based graphene nanofluid exhibited superior performance concerning thermodynamic, heat transfer, and exergy analysis. Labib et al. [75] considered a combination of alumina nanoparticles with water-based CNT nanofluid and performed a numerical study to evaluate this combined/hybrid nanofluid convective heat transfer. They observed an important augmentation in heat transfer performance applying the nanofluid mixture. They proposed that the higher shear shinning behavior of CNT might be the cause of this heat transfer improvement. Several experiments were conducted by Bhattad et al. [76] to augment the exergetic and energetic efficacy of a counter-flow plate HEX applying various hybrid nanofluids as working fluid. They studied several combinations, namely MWNT-Al_2O_3, CuO-Al_2O_3, MgO-Al_2O_3, AlN-Al_2O_3, and SiC-Al_2O_3 in 1:4 volume ratios and also Al_2O_3 particles with 0.1 vol% concentration dispersed in DI water. The highest increase in the irreversibility was observed for CuO-Al_2O_3 (1,4) hybrid nanofluid and about 1.6%. The greatest performance was noticed for MWNT-Al_2O_3 (1:4), suggesting that this hybrid nanofluid can be an excellent alternative for enhancing thermal performance. The hydrothermal performance of the double tube HEX was investigated experimentally by Singh and Sarkar [77], with the application of various wire coil inserts and twisted tapes and using MWNT/Al_2O_3 hybrid nanofluid as working fluid. They conducted the experiments considering volume concentration of 0.01%, nanofluid flow rates of 5–25 lpm, and Reynolds number in the range of 8000–40,000. They observed that with the decrease of width ratio and rise of depth ratio, both pressure drop and heat transfer coefficient increased. Ponangi et al. [78] conducted research to regulate how the addition of graphene oxide (GO), carboxyl graphene (CG), and new hybrid nanoparticles into a conventional hot fluid can affect the performance of compact HEX. For 0.04 vol% of GO and 0.02 vol% of CG, Nusselt number enhances up to 4.92 and 3.59 times, respectively. This enhancement was noticed to be 11 times for 0.0075 vol% of CG and 0.005 vol% of GO hybrid nanomaterial. The outputs of this study indicated a substantial improvement in the compact heat exchanger performance concerning pressure drop and effectiveness. At a fraction of 0.005 vol% of CG and 0.005 vol% of GO, 66% decrease in pressure drop and 232% enhancement in efficiency were observed. Functionalized graphene and hybrid nanoparticles can provide the opportunity to augment the efficiency of compact heat-exchanger and/or reducing the dimension.

8.5 Engine cooling

The thermal management of automobile engines is regarded extremely important since it has substantial effects on automobile performance concerning emission maintenance, material selection, and economic parameters [79]. With the thermal management in automobile engines, energy consumption can be reduced, leading to minimizing losses and enhancing thermal efficiency. For automotive cooling, a promising approach is employing the mixture of H_2O and ethylene glycol instead of H_2O. Improving the engine cooling and automotive performance is possible by adding nanoparticles into the engine oil, resulting in the decrease of heat losses and reduction of the radiator dimension and weight. Engine oil is used in a variety of engineering fields, including power cars, engine generators, bearings, and equipment with moving parts. The efficiency is decreased by the friction existing between the moving parts in mechanical systems since the kinetic energy is converted to heat. Oils' primary function is to decrease friction between pieces that move in opposite directions. Furthermore, the sections heated because of friction can be cooled by engine oil [80]. The influences of nanoparticle concentrations and temperature on the μ of SAE40 oil-based hybrid nanofluids containing of MWNT and SiO_2 nanopowder were studied experimentally by Afrand et al. [81]. According to viscosity measurements, SAE40-based hybrid nanomaterial containing MWNTs and nanopowders acts as a Newtonian fluid at all given temperatures and volume fractions. The μ decreases with rising temperature and augments with augmenting volume fractions, according to the findings. Furthermore, the outputs showed that 37.4% was the maximum improvement of the hybrid nanofluid viscosity. Esfe et al. [82] accomplished research to explore the thermal conductivity of double-wall carbon nanotubes (DWNT)-ZnO/EG hybrid nanofluid. The highest relative thermal conductivity of 24.9% was achieved at a concentration of 1.9% and temperature of 50°C. The superior performance of hybrid nanofluids in terms of cost and thermal conductivity improvement was proved considering qualitative performance and economic evaluation. Dardan et al. [83] prepared fuel oil with different amounts of hybrid nanoadditives and evaluated their rheological behavior. The tests were carried out at volume fractions of 0%–1.0% and temperatures ranging from 25 to 50°C. At all giventemperatures and volume fractions, SAE40-based hybrid nanofluid containing MWNTs and Al_2O_3 nanoparticles acts as a Newtonian fluid, according to viscosity measurements. The hybrid nanofluid viscosity decreased with rising temperature and augmented with increasing concentrations of nanoadditives, according to the experiments. Furthermore, the outputs showed that 46% was the maximum improvement of the hybrid nanofluid viscosity. According to the viscosity sensitivity analysis, the sensitivity of viscosity to temperature fluctuations was negligible, but its sensitivity to volume fraction changes is higher. Karimi and Afrand [84] used ethylene glycol/MWNTs-MgO hybrid nanofluid to study the performance of horizontal and vertical tube radiators. A vertical tube radiator exhibited superior performance with 10% higher efficiency based on the outputs. The thermal conductivity of hybrid nanofluids containing engine oil and WO_3-MWNTs was evaluated by Soltani et al. [85]. Nanofluids were prepared by dispersing MCNT and nanoparticles in the engine oil, applying a two-step approach. Diameters of MNWTs range from 20 to 30 nm. The outputs showed that with increasing nanopowders' volume fractions and temperature, the thermal conductivity of hybrid nanofluid increased. It should be mentioned that thermal conductivity was more strongly affected by volume fraction than temperature. The highest value for thermal conductivity, with 19.85% improvement compared to the base fluid, was obtained at a volume fraction of 0.6% and $T = 60°C$ and was 19.85% compared with the base fluid. They also introduced two mathematical models for the thermal conductivity of the hybrid nanofluid. Ali et al. [86] used nanolubricants containing TiO_2 and alumina hybrid nanoparticles and investigated wear rate and frictional power losses in contact with cylinder and piston ring. In comparison with the base fluid (oil), 17% and 40%–51% reduction in wear rate and frictional losses were reported, respectively. They also added oleic acid as a surfactant and observed an 18% reduction in friction coefficient. Sahoo and Sarkar [87] investigated the improvement in the convective heat transfer of ethylene glycol brine-based different nanofluids such as TiO_2CuO, SiC, Cu, and Ag in Al_2O_3 nanofluids, as the coolant in a louver-finned automobile radiator. The maximum heat transfer rate and effectiveness and also pumping power were observed for 1% Ag hybrid nanofluid (0.5% Al_2O_3 and 0.5% Ag). However, the highest performance index was noticed for SiC+ Al_2O_3 suspended, and therefore, it was selected as the best coolant. With the application of Ag hybrid nanofluids in a radiator with the same dimension and rate of heat transfer, pumping power augments, resulting in increased thermal efficiency of the engine, and decreasing engine fuel consumption. For a similar heat transfer rate and coolant flow, employing Ag hybrid nanofluids led to reducing the radiator dimension and increasing pumping power, resulting in the decrease of cost, weight, and dimension of the radiator. Fig. 8.10. illustrates the performance index variations versus volume fractions at a coolant flow rate of 120 L/min.

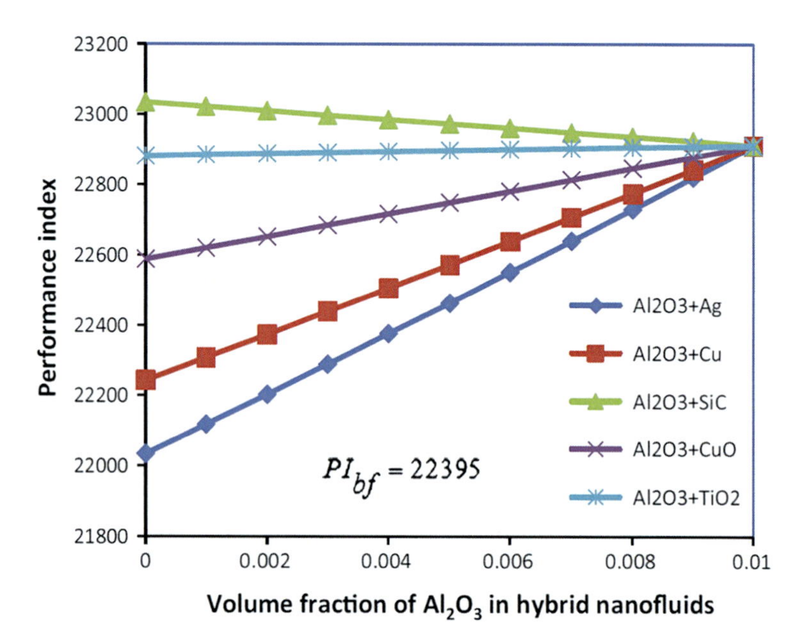

Fig. 8.10 Performance index variation versus volume fractions. From R.R. Sahoo, J. Sarkar, Heat transfer performance characteristics of hybrid nanofluids as coolant in louvered fin automotive radiator, Heat Mass Transf. 53(6) (2017) 1923–1931. Copyright Springer Nature (Lic # 5064641052835).

8.6 Refrigeration

For refrigeration systems, it is possible to enhance heat transfer and achieve high values of heat transfer coefficients by adding nanoparticles into refrigerants [88]. Two significant parameters in defining the efficiency of nanorefrigerants are viscosity and thermal conductivity. The application of different types of nanorefrigerants as a substitute for conventional fluids for measuring the refrigeration system performance has been investigated in several previous research studies. Some studies have focused on modifying the setup design, but the investigation of different fluids containing nanoparticles has also gained much attention. The performance of the HVAC system increases because of adding nanoparticles. Bhattad and Sarkar [89] studied the influence of the combination, size, and shape of nanoparticles on exergetic efficiency, thermal performance factor, performance index, coefficient of performance, pump work, and heat transfer area. For exergetic efficiency, the best performance was observed for silica-copper hybrid nanofluid, while hybrid nanofluid of copper oxide-copper had the best performance. The highest amount of reduction (5.9%) in effective heat transfer area was noticed for propylene glycol brine-based hybrid nanofluid containing copper oxide nanoparticles. They observed that with increasing particle size, percentage variation in performance index and heat transfer

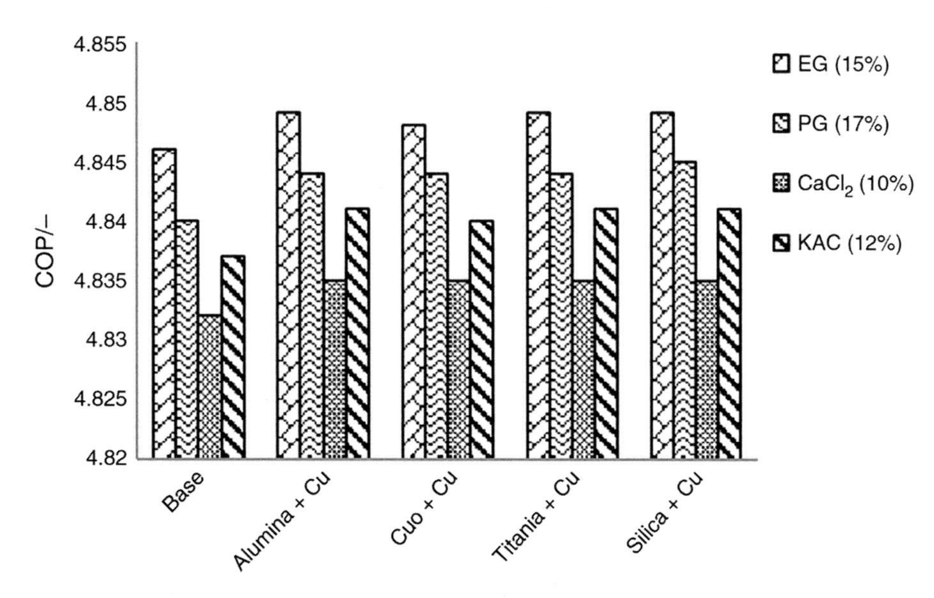

Fig. 8.11 Coefficient of performance for different cases. From A. Bhattad, J. Sarkar, Effects of nanoparticle shape and size on the thermohydraulic performance of plate evaporator using hybrid nanofluids, J. Therm. Anal. Calorim. (2019) 1–13. Copyright Springer Nature (Lic # 5064641211485).

area decreased, while thermal performance factor increased. Particles with brick shape exhibited maximum variations in performance index and heat transfer area, while particles with platelet shape showed worse performance. Fig. 8.11. indicates the performance index variations.

Ahmed et al. [90] prepared ternary hybrid composite nanofluids based on TiO_2/DW+Al_2O_3+ZnO with varying concentrations of 0.1, 0.075, 0.05, and 0.025, applying the sonochemical approach and studied heat transfer development in a square HEX. The highest amount of heat transfer, up to 900–5700 W/m^{2}K, was reported for ternary hybrid composite nanofluids containing TiO_2/DW+Al_2O_3+ZnO with concentrations of 0.1 wt%. This value was 89% higher in comparison with heat transfer in the case of using distilled H_2O. For concentrations of 0.025, 0.05, and 0.075 wt%, the reported values for heat transfer improvement were 900–2750 W/m^{2}K, 900–3350 W/m^{2}K, and 900–3870 W/m^{2}K, respectively. Based on the results of their work, ternary hybrid composite nanofluids containing metal oxides can be regarded as appropriate candidates for possible applications as nanocoolants considering their superior thermophysical properties. The application of brine-based hybrid nanofluids in counterflow corrugated plate evaporator as a secondary refrigerant was proposed by Bhattad et al. [91]. For specific cooling load, they assessed various sorts of Al_2O_3 hybrid nanofluids comprising various nanoparticles (MCNT, silver, and copper) at the same volume and the total volume concentration of 0.8% suspended in different brines. With decreasing application temperature, a reduction in exergetic performance was noticed. For the propylene glycol brine-based hybrid nanofluid containing MWNT/alumina, the highest increase in exergetic efficiency and irreversibility distribution ratio (IDR) and the lowest decrease in nondimensional exergy destruction and irreversibility were reported. The results of their investigation revealed the great potential of secondary refrigerants. Safaei et al. [92] investigated the process of developing, stabilizing, and measuring of thermal conductivity of antifreeze-$SiO_2/CoFe_2O_4$ hybrid nanofluids. The outputs obtained from the photography capturing method revealed producing the best conditions with long-term stability by utilizing CMC surfactant with 0.1 nanoparticles mass ratio. A KD2-Pro thermal analyzer was used for measuring nanofluid thermal conductivity at mass fractions of 0.1%–1.5% and temperature ranges from 25°C to 50°C. It was observed that along with the increase of nanoparticles' mass fraction up to 37.7%, thermal conductivity increased. An empirical correlation with high accuracy was proposed for calculating the thermal conductivity of the nanoantifreeze with the curve fitting technique of laboratory data. It was concluded that this nanoantifreeze could be considered as a promising candidate with great heat transfer potential in refrigeration and thermal devices, considering its thermal properties. Ahmed and Mimi [93] used nanofluids (TiO_2, Al_2O_3, and a water-based hybrid nanofluid containingTiO_2 and Al_2O_3) and studied the performance characteristics of a vapor compression chiller experimentally. For TiO_2/H_2O, concentration ratios of 1, 0.6, 0.4, 0.3, and 0.1, and forAl_2O_3/H_2O, concentration ratios of 1, 0.6, 0.4, 0.3, 0.2, 0.1, and 0.05 were taken into account. Five different concentrations of 0.1/0.1, 0.2/0.1, 0.3/0.1, 0.3/0.2, and 0.3/0.3 wt% were considered for Al_2O_3/TiO_2 hybrid nanoparticles. Based on the results of experiments, lower elapsed time for cooling of the chiller system fluid and higher coefficient of performance were observed for Al_2O_3/H_2Onanofluid. Greater values of the refrigeration effect and lower values of compression ratio were provided for Al_2O_3/H_2O than for TiO_2/H_2O.

8.7 Machining

A huge amount of energy is produced per unit volume during the grinding process, and converting a large portion of this energy into heat leads to producing a considerable amount of heat that can damage workpieces. It is probable for the workpiece to undergo different distortion, cracks, and residual stresses. A machine of small size cannot work unless the working fluid can properly circulate as lubrication. Many researchers utilized minimum quantity lubrication (MQL) in their studies [94, 95]. Consequently, considering the cooling and lubrication challenges, hybrid nanofluids were used since small lubrication results in improper performance of the machine. Owing to the enhanced thermal properties of hybrid nanofluids, their applications in the manufacturing sector as nanolubricants have been considered by researchers. Jamil et al. [96] studied oil-based hybrid nanofluids containing MWNTs and alumina nanoparticles utilizing the minimum quantity lubrication (MQL) approach for turning Ti-6Al-4V. The main aim of their study was to compare the effect of methods of MQL based on hybrid nanofluid and cryogenic CO_2 for turning Ti-6Al-4V. The Taguchi L9 orthogonal array was applied for designing experiments, while the examined variables were cooling technique, feed rate, and cutting speed. They utilized vegetable oil-based hybrid nanofluid containing MWNTs and alumina (Al_2O_3) nanoparticles. According to the results, an increase of 23% was observed in the tool life, while a reduction of 11.8% in cutting force and 8.72% in the average surface roughness was noticed with applying hybrid nanoadditives compared to the cryogenic cooling. However, at low and high levels of feed rate and cutting speed, an 11.2% reduction in cutting temperature was observed for the cryogenic technique compared to the MQL-hybrid nanofluids. A hybrid nanocutting fluid by hybridization of Al_2O_3 and MWNT nanoparticles was developed by Sharma et al. [97] with concentrations of 1.25, 0.75, and 0.25 vol%. The tests were conducted on the base nanofluid (Al_2O_3nanofluid) and hybrid nanofluid to assess their thermophysical characteristics. Based on the outputs, the wear was decreased by augmenting the fraction of nanopowders in cutting fluid, and the lowest wear was noticed for hybrid nanofluid.

Consequently, the minimum quantity lubrication (MQL) technique was used for evaluating their performance as a cutting fluid during the turning of AISI 304 steel concerning surface roughness and machining forces. Based on the outputs, hybrid nanofluids showed superior performance in comparison with alumina nanoparticles mixed cutting fluid. Sahid et al. [98] scrutinized the impact of hybrid nanomaterial on coated carbide cutting tool performance. They considered various factors, including cutting speed and feed rate. The results revealed that better machineability could be provided, and machining costs can be reduced using ethylene glycol-based hybrid nanofluids containing ZnO andTiO_2 nanoparticles. Gajrani et al. [99] mixed hybrid nanoparticles of calcium fluoride and molybdenum disulfide in vegetable oil to study machining performance. Different parameters of surface roughness, surface adhesion, tool wear, and nanofluids concentrations were considered for performing the experiments. They observed better performance at a hybrid nanofluid concentration of 0.3%. According to outputs, 37% improvement was observed in surface finish, while friction coefficient, feed force, and tool cutting decreased 11%, 28%, and 17%, respectively. Eltaggaz et al. [100] studied hybrid nanoparticles of gamma and alumina utilizing MQL during austempered ductile iron machining. They examined the influence of hybrid nanofluids on tool wear behavior. The best tool life was achieved by combining MQL-nanofluids at a feed rate of 0.2 mm/rev and a cutting speed of 120 m/min. Esfe et al. [101] performed a review study on the employment of hybrid and conventional nanocutting fluids in different machining processes, including grinding, drilling, milling, and turning. Based on their study, when nanofluids are used as cutting fluids, an improvement can be observed in cooling and lubrication as compared to conventional cutting liquids, and some parameters can define the levels of this improvement. The optimum performance for each nanofluid in the turning process can be achieved in a particular nanoparticle volume fraction, flow rate, and pressure to reach. In most of the machining processes, optimal cases were repeated when concentrations of nanoparticles were between 0.25% and 0.5%. Moreover, in comparison with base fluids and conventional nanofluids, more positive effects were exhibited by hybrid nanofluids. Based on the reports, 10%–40% and sometimes 50%–70% positive changes were observed in important parameters, including surface roughness, tool wear, cutting force, and cutting temperature, with the application of nanoparticles in turning processes.

8.8 Desalination

Liveliness, sustainability, and state-of-the-art technologies are generically evoked by the phrases "Nano Energy" and "freshwater." The application of nanofluids to produce freshwater from seawater has been reported to be highly beneficial to the environment. According to the reports, large quantities of carbon dioxide can be decreased by using water-based copper oxide nanofluid [102]. Rabbi and Sahin [103] used two hybrid nanofluids to analyze the improvement of the solar still performance. The application of two conventional nanofluids and hot water was also examined for investigating the enhancement of the solar still performance. They linked conventional nanofluids, hot water, and hybrid nanofluids to the solar system with HEX positioned below the solar still basin. Through mathematical modeling, they investigated thermal and exergy efficiencies and the yield of pure distilled water. Although the performance of solar still was improved by using hot water in HEX, a better performance was observed through the application of conventional nanofluids, and the best performance was achieved by employing hybrid nanofluids. When H_2O-based hybrid nanofluid containing SiO_2 and Al_2O_3 nanopowders was utilized, the water yield was 4.99 kg/m^2 per day, and exergy efficiency and thermal efficiency were 0.82% and 37.76%, respectively. For the performance improvements, some factors were observed to be extremely important. These factors are ambient temperature, solar radiation, nanoparticle concentrations, the inlet temperature of the nanomaterial, nanofluid flow rate, and basin water depth. An experimental study was conducted by Sadeghi and Nazari [104] to examine the thermoelectric cooling channel solar performance still integrated with a concentrating evacuated tube collector utilizing deionized water-based hybrid nanofluid containing $Ag@Fe_3O_4$nanoparticles. Based on the outputs, both energy efficiency and productivity were enhanced by integrating the concentrating solar collector into the thermoelectric cooling channel solar still and the application of $Ag@Fe_3O_4$ nanoparticles. Compared with the traditional solar still, 117% improvement in energy efficiency and 218% increase in daily productivity were observed with the application of hybrid nanofluids with a 0.08% volume concentration, and the modified solar still fitted with the solar collector. Moreover, 0.019 ($\$/L/m^2$) was estimated for the cost of freshwater production, and a payback period of 369 days was calculated for the application of the presented modified solar still. A novel fractional model for the simulation of the conventional solar still performance was presented by El-Gazar et al. [105] on the basis of the Riemann Liouville fractional derivative, which exhibited how the application of hybrid nanofluids can affect the desalination system. Hybrid nanofluid containing copper oxide and alumina with a 0.025% concentration for each nanoparticle was used to implement this model. According to the outputs, utilizing hybrid nanofluids led to raising the still daily productivity by 21.7% in winter and 27.2% in summer compared to the case without nanoparticles. Moreover, 23.212% and 49.54% improvements in the still average energy efficiency were observed in winter and summer. Additionally, utilizing hybrid nanofluids led to raising the average exergy efficiency by 13.4% and 22.5% in winter and summer, respectively.

8.9 Challenges and outlook

1. Stability problem is regarded as one important challenge that is probable in the case of hybrid nanofluids [106]. Compared to mono nanofluids, dispersing two types of nanoparticles can be more challenging regarding the stability problem [107]. It is obvious that the modeling of thermal conductivity is more complicated for hybrid nanofluids. It is necessary to properly prepare hybrid nanofluids with the selection of appropriate nanomaterial pair for getting acceptable aspect ratios, enhanced synergistic effect, and the thermal path between pair, resulting in improved heat transfer. The process of producing hybrid nanofluids is more complicated, leading to higher costs for nanofluid production, which can be regarded as an obstacle in front of applying hybrid nanofluids in the industry.
2. One significant challenge for nanoparticle production is controlling the form and dimension of nanoparticles, particularly for practical applications, including the utilization of nanofluids in liquid blocks, which significantly influences thermal performance.
3. Further research is required for developing models, which consider several parameters needed for predicting hybrid nanofluid thermal conductivity with higher precision [108].
4. A suitable solution as hybrid nanofluids to be applied in liquid blocks can be employing magnetic nanomaterials with carbon-based nanopowders in a pure fluid. Using magnetic nanoparticles for the functionalization of CNTs can benefit the attributes of CNTs as well as magnetic nanoparticles, leading to the production of advanced materials with new thermal, physical, and chemical properties. It is possible to synthesize a new type of hybrid nanofluids with the controllability of magnetic nanomaterials and high thermal conductivity of CNTs through the combination of CNTs and magnetic nanoparticles. In this case, the hybrid nanofluid is controllable when there is an external magnetic field and a great thermal conductivity. So, the usage of hybrid nanomaterial can be suggested in various heat sinks in the future [109, 110].
5. For improving the performance and developing possible applications, both experimental and theoretical investigations should be considered in future studies [111].

6. Some important challenges in further developing commercializing nanofluids are the proper selection of hybrid nanofluids, pumping power, stability, and production cost [112]. Nanofluids are anticipated to have an important influence on applications by understanding these difficulties [113]. Despite the challenges mentioned above, the importance of surface modification for great suspension, thermal stability, and effective thermal transport of hybrid nanofluids has been revealed through the rapid progress of nanotechnology and nanoscience. However, transforming nanofluid technology from experiments on small scales to industrial commercialization and production can be facilitated and systematized by addressing all the above-mentioned challenges.
7. Considering the random movement of nanofluids, a parallel direction for the flow and hybrid nanoparticle is probable, which can degrade the heat transfer. Testing more mixtures of hybrid nanofluids is required in the future [114]. The further augmentation of heat transfer might be accomplished by employing hybrid nanofluid mixtures.

8.10 Summary

Some important parameters that affect the thermal conductivity of hybrid nanofluids are the type of base fluid, stability, sonication time, size and shape of nanopowders, temperature, and volume fractions. According to the previous studies, increasing volume fraction and temperature lead to the improvement of thermal conductivity. Clustering of nanoadditives has been introduced as the cause of thermal conductivity enhancement at high volume fractions. The increase of thermal conductivity at high temperatures can be attributed to increasing molecules' collision rates because of high kinetic energy. Researchers who are working in the area of electronic systems cooling have found several advantages. Control of the variable utilized for improving the nanofluids' thermal conductivity is complicated. It is essential to revise the cooling system thermal design for an electronic system. Utilizing nanofluids in the microchannel heat sink (MCHS) is impeded by the two important parameters: production cost and stability. Ongoing research on microchannel heat sinks for real cooling structures is under huge development. Based on studies conducted recently, hybrid nanofluids have been found the best option among available working fluids for optimizing the performance of the solar system because of the individual nanoparticles' synergistic effect [115]. Considering thermal and optical properties, excellent performance expedition (above 200%) has been exhibited by hybrid nanofluids-based solar devices. Utilizing hybrid nanofluids has led to evolving the factors that determine solar devices' performance, including photothermal conversion efficiency, solar weighted absorption, and extinction coefficient. Moreover, increasing the nanoparticle concentration up to a specific point gives rise to the optical performance of hybrid nanofluids. Although the pressure drop rate and heat transfer rate are strongly affected by nanofluid concentrations, the effect on the inlet temperature of the cooling fluid on them is small. The mass fraction and temperature of nanofluids, as two important factors, improved the thermal conductivity (k_{nf}), and the k_{nf} at higher temperatures was inherently higher. When the k_{nf} enhances at higher temperatures, the variations are more intense, and the angle of the changes is steeper. The collision speed increases with increasing temperature of the nanofluid, and the collision number in nanofluids increases with rising concentrations. As an output, the thermal conductivity improves by these mechanisms. At higher temperatures, more intense improvement is observed in k_{nf}. The increase of temperature with enhancing the mechanisms for the base fluid and nanoparticles' thermal conductivity and Brownian motions of particles improves the thermal conductivity of nanofluids.

References

[1] S.U. Choi, J.A. Eastman, Enhancing Thermal Conductivity of Fluids with Nanoparticles, Argonne National Lab, IL (United States), 1995.
[2] E. Bellos, Z. Said, C. Tzivanidis, The use of nanofluids in solar concentrating technologies: a comprehensive review, J. Clean. Prod. 196 (2018) 84–99.
[3] Z. Said, et al., Thermophysical properties of single wall carbon nanotubes and its effect on exergy efficiency of a flat plate solar collector, Sol. Energy 115 (2015) 757–769.
[4] T. Gao, et al., Mechanics analysis and predictive force models for the single-diamond grain grinding of carbon fiber reinforced polymers using CNT nano-lubricant, J. Mater. Process. Technol. 290 (2021) 116976.
[5] X. Wang, et al., Vegetable oil-based nanofluid minimum quantity lubrication turning: academic review and perspectives, J. Manuf. Process. 59 (2020) 76–97.
[6] J. Sarkar, A critical review on convective heat transfer correlations of nanofluids, Renew. Sust. Energ. Rev. 15 (6) (2011) 3271–3277.
[7] L. Wu, et al., Improved breakdown strength of poly (vinylidene fluoride)-based composites by using all ball-milled hexagonal boron nitride sheets without centrifugation, Compos. Sci. Technol. 190 (2020) 108046.
[8] Y. Yang, et al., New pore space characterization method of shale matrix formation by considering organic and inorganic pores, J Nat Gas Sci Eng 27 (2015) 496–503.
[9] Q. Wang, B. Liu, Z. Wang, Investigation of heat transfer mechanisms among particles in horizontal rotary retorts, Powder Technol. 367 (2020) 82–96.

[10] P. Wang, K. Vafai, Modeling and analysis of an efficient porous media for a solar porous absorber with a variable pore structure, J. Sol. Energy Eng. 139 (5) (2017).
[11] X. Chen, et al., Enhanced photoresponsivity of a GaAs nanowire metal-semiconductor-metal photodetector by adjusting the fermi level, ACS Appl. Mater. Interfaces 11 (36) (2019) 33188–33193.
[12] H. Li, et al., Optical properties of quasi-type-II structure in GaAs/GaAsSb/GaAs coaxial single quantum-well nanowires, Appl. Phys. Lett. 113 (23) (2018) 233104.
[13] M. Gupta, et al., Up to date review on the synthesis and thermophysical properties of hybrid nanofluids, J. Clean. Prod. 190 (2018) 169–192.
[14] Z. Said, S. Arora, E. Bellos, A review on performance and environmental effects of conventional and nanofluid-based thermal photovol.Taics, Renew. Sust. Energ. Rev. 94 (2018) 302–316.
[15] T. Gao, et al., Dispersing mechanism and tribological performance of vegetable oil-based CNT nanofluids with different surfactants, Tribol. Int. 131 (2019) 51–63.
[16] Y. Zhang, et al., Experimental evaluation of the lubrication performance of MoS2/CNT nanofluid for minimal quantity lubrication in Ni-based alloy grinding, Int. J. Mach. Tools Manuf. 99 (2015) 19–33.
[17] M. Sui, et al., Temperature of grinding carbide with castor oil-based MoS2 nanofluid minimum quantity lubrication, J. Thermal. Sci. Eng. Appl. 13 (5) (2021), 051001.
[18] Z. Duan, et al., Milling force and surface morphology of 45 steel under different Al 2 O 3 nanofluid concentrations, Int. J. Adv. Manuf. Technol. 107 (3) (2020) 1277–1296.
[19] Z. Said, et al., New thermophysical properties of water based TiO2 nanofluid—the hysteresis phenomenon revisited, Int. Commun. Heat Mass Transfer 58 (2014) 85–95.
[20] Z. Said, et al., Acid-functionalized carbon nanofibers for high stability, thermoelectrical and electrochemical properties of nanofluids, J. Colloid Interface Sci. 520 (2018) 50–57.
[21] Z. Said, et al., Stability, thermophysical and electrical properties of synthesized carbon nanofiber and reduced-graphene oxide-based nanofluids and their hybrid along with fuzzy modeling approach, Powder Technol. 364 (2020) 795–809.
[22] B.A. Khuwaileh, et al., On the performance of nanofluids in APR 1400 PLUS7 assembly: Neutronics, Ann. Nucl. Energy 144 (2020) 107508.
[23] F.I. Alhamadi, et al., Performance of Nano-fluids as coolants/moderator in APR1400–Neutronics case study, Transactions 120 (1) (2019) 524–527.
[24] A. Rafiei, et al., Solar desalination system with a focal point concentrator using different nanofluids, Appl. Therm. Eng. 174 (2020) 115058.
[25] Z. Said, et al., A comprehensive review on minimum quantity lubrication (MQL) in machining processes using nano-cutting fluids, Int. J. Adv. Manuf. Technol. 105 (5) (2019) 2057–2086.
[26] S. Rahman, et al., Performance enhancement of a solar powered air conditioning system using passive techniques and SWCNT/R-407c nano refrigerant, Case Stud. Therm. Eng. 16 (2019) 100565.
[27] Z. Said, et al., Enhancing the performance of automotive radiators using nanofluids, Renew. Sust. Energ. Rev. 112 (2019) 183–194.
[28] Z. Said, et al., Heat transfer enhancement and life cycle analysis of a Shell-and-tube heat exchanger using stable CuO/water nanofluid, Sustainable Energy Technol. Assess. 31 (2019) 306–317.

[29] A.A. Hachicha, et al., A review study on the modeling of high-temperature solar thermal collector systems, Renew. Sust. Energ. Rev. 112 (2019) 280–298.

[30] Z. Said, et al., Evaluating the optical properties of TiO2 nanofluid for a direct absorption solar collector, Numer. Heat Transf.; A: Appl. 67 (9) (2015) 1010–1027.

[31] M. Gupta, V. Singh, Z. Said, Heat transfer analysis using zinc ferrite/water (hybrid) nanofluids in a circular tube: an experimental investigation and development of new correlations for thermophysical and heat transfer properties, Sustainable Energy Technol. Assess. 39 (2020) 100720.

[32] M. Ghodbane, et al., Performance assessment of linear Fresnel solar reflector using MWCNTs/DW nanofluids, Renew. Energy 151 (2020) 43–56.

[33] A.K. Tiwar, et al., A review on the application of hybrid nanofluids for parabolic trough collector: Recent progress and outlook, J. Clean. Prod. (2021) 126031.

[34] M.A. Carpenter, M.A. Geletkanycz, W.G. Sanders, Upper echelons research revisited: antecedents, elements, and consequences of top management team composition, J. Manag. 30 (6) (2004) 749–778.

[35] D. Zhenjing, et al., Milling surface roughness for 7050 aluminum alloy cavity influenced by nozzle position of nanofluid minimum quantity lubrication, Chin. J. Aeronaut. (2020).

[36] X.-K. Li, et al., Ammonia decomposition over Ru and Ni catalysts supported on fumed SiO2, MCM-41, and SBA-15, J. Catal. 236 (2) (2005) 181–189.

[37] J. Zhang, et al., Convective heat transfer coefficient model under nanofluid minimum quantity lubrication coupled with cryogenic air grinding Ti–6Al–4V, Int. J. Precis. Eng. Manuf.—Green Technol. (2020) 1–23.

[38] J. Roubort, Nanofluids Industrial Cooling, North America, ANL Michellin, 2009.

[39] S. Suresh, et al., Effect of Al2O3–cu/water hybrid nanofluid in heat transfer, Exp. Thermal Fluid Sci. 38 (2012) 54–60.

[40] M.J. Nine, et al., Investigation of Al2O3-MWCNTs hybrid dispersion in water and their thermal characterization, J. Nanosci. Nanotechnol. 12 (6) (2012) 4553–4559.

[41] B. Jia, L. Gao, J. Sun, Self-assembly of magnetite beads along multiwalled carbon nanotubes via a simple hydrothermal process, Carbon 45 (7) (2007) 1476–1481.

[42] P. Selvakumar, S. Suresh, Use of ${\rm Al}_{2}{\rm O}_{3}\hbox{--}{\rm cu}$/water hybrid Nanofluid in an electronic heat sink, IEEE Trans. Compon. Packag. Manuf. Technol. 2 (10) (2012) 1600–1607.

[43] S.S. Harandi, et al., An experimental study on thermal conductivity of F-MWCNTs–Fe3O4/EG hybrid nanofluid: effects of temperature and concentration, Int. Commun. Heat Mass Transfer 76 (2016) 171–177.

[44] R. Nimmagadda, K. Venkatasubbaiah, Conjugate heat transfer analysis of micro-channel using novel hybrid nanofluids (Al2O3+ ag/water), Eur. J. Mech. B. Fluids 52 (2015) 19–27.

[45] M. Bahiraei, S. Heshmatian, Efficacy of a novel liquid block working with a nanofluid containing graphene nanoplatelets decorated with silver nanoparticles compared with conventional CPU coolers, Appl. Therm. Eng. 127 (2017) 1233–1245.

[46] N. Ahammed, L.G. Asirvatham, S. Wongwises, Entropy generation analysis of graphene–alumina hybrid nanofluid in multiport minichannel heat exchanger coupled with thermoelectric cooler, Int. J. Heat Mass Transf. 103 (2016) 1084–1097.

[47] V. Kumar, J. Sarkar, Two-phase numerical simulation of hybrid nanofluid heat transfer in minichannel heat sink and experimental validation, Int. Commun. Heat Mass Transfer 91 (2018) 239–247.

[48] M. Bahiraei, S. Heshmatian, Thermal performance and second law characteristics of two new microchannel heat sinks operated with hybrid nanofluid containing graphene–silver nanoparticles, Energy Convers. Manag. 168 (2018) 357–370.
[49] M. Khoshvaght-Aliabadi, F. Nozan, Water cooled corrugated minichannel heat sink for electronic devices: effect of corrugation shape, Int. Commun. Heat Mass Transfer 76 (2016) 188–196.
[50] V.M. Krishna, et al., Numerical investigation of heat transfer and pressure drop for cooling of microchannel heat sink using MWCNT-CuO-water hybrid nanofluid with different mixture ratio, Mater. Today: Proc. 42 (2021) 969–974.
[51] A. Heidarshenas, et al., Experimental investigation of heat transfer enhancement using ionic liquid-Al2O3 hybrid nanofluid in a cylindrical microchannel heat sink, Appl. Therm. Eng. 191 (2021) 116879.
[52] A. Ejaz, et al., Concentrated photovol.taics as light harvesters: Outlook, recent progress, and challenges, Sustainable Energy Technol. Assess. 46 (2021) 101199.
[53] E. Bellos, C. Tzivanidis, Z. Said, Investigation and optimization of a solar-assisted pumped thermal energy storage system with flat plate collectors, Energy Convers. Manag. 237 (2021) 114137.
[54] M. Sheikholeslami, et al., Recent progress on flat plate solar collectors and photovol.taic systems in the presence of nanofluid: a review, J. Clean. Prod. (2021) 126119.
[55] M. Sheikholeslami, et al., Modification for helical turbulator to augment heat transfer behavior of nanomaterial via numerical approach, Appl. Therm. Eng. 182 (2021) 115935.
[56] R. Prasher, P. Bhattacharya, P.E. Phelan, Thermal conductivity of nanoscale colloidal solutions (nanofluids), Phys. Rev. Lett. 94 (2) (2005), 025901.
[57] N.A.C. Sidik, et al., A review on preparation methods, stability and applications of hybrid nanofluids, Renew. Sust. Energ. Rev. 80 (2017) 1112–1122.
[58] Z. Said, et al., Heat transfer, entropy generation, economic and environmental analyses of linear Fresnel reflector using novel rGO-Co3O4 hybrid nanofluids, Renew. Energy 165 (2021) 420–437.
[59] E. Farajzadeh, S. Movahed, R. Hosseini, Experimental and numerical investigations on the effect of Al2O3/TiO2H2O nanofluids on thermal efficiency of the flat plate solar collector, Renew. Energy 118 (2018) 122–130.
[60] M. Karami, Experimental investigation of first and second laws in a direct absorption solar collector using hybrid Fe 3 O 4/SiO 2 nanofluid, J. Therm. Anal. Calorim. 136 (2) (2019) 661–671.
[61] N. Chen, et al., Complementary optical absorption and enhanced solar thermal conversion of CuO-ATO nanofluids, Sol. Energy Mater. Sol. Cells 162 (2017) 83–92.
[62] E. Bellos, C. Tzivanidis, Thermal analysis of parabolic trough collector operating with mono and hybrid nanofluids, Sustainable Energy Technol. Assess. 26 (2018) 105–115.
[63] T.R. Shah, H.M. Ali, Applications of hybrid nanofluids in solar energy, practical limitations and challenges: a critical review, Sol. Energy 183 (2019) 173–203.
[64] S.K. Verma, et al., Performance analysis of hybrid nanofluids in flat plate solar collector as an advanced working fluid, Sol. Energy 167 (2018) 231–241.
[65] X. Li, G. Zeng, X. Lei, The stability, optical properties and solar-thermal conversion performance of SiC-MWCNTs hybrid nanofluids for the direct absorption solar collector (DASC) application, Sol. Energy Mater. Sol. Cells 206 (2020) 110323.
[66] R. Ekiciler, et al., Effect of hybrid nanofluid on heat transfer performance of parabolic trough solar collector receiver, J. Therm. Anal. Calorim. 143 (2) (2021) 1637–1654.

[67] P.K. Kanti, et al., Experimental investigation on thermo-hydraulic performance of water-based fly ash–cu hybrid nanofluid flow in a pipe at various inlet fluid temperatures, Int. Commun. Heat Mass Transfer 124 (2021) 105238.
[68] J. Sarkar, P. Ghosh, A. Adil, A review on hybrid nanofluids: recent research, development and applications, Renew. Sust. Energ. Rev. 43 (2015) 164–177.
[69] H. Babar, H.M. Ali, Towards hybrid nanofluids: preparation, thermophysical properties, applications, and challenges, J. Mol. Liq. 281 (2019) 598–633.
[70] G. Huminic, A. Huminic, Hybrid nanofluids for heat transfer applications–a state-of-the-art review, Int. J. Heat Mass Transf. 125 (2018) 82–103.
[71] H.R. Allahyar, F. Hormozi, B. ZareNezhad, Experimental investigation on the thermal performance of a coiled heat exchanger using a new hybrid nanofluid, Exp. Thermal Fluid Sci. 76 (2016) 324–329.
[72] M. Bahiraei, M. Berahmand, A. Shahsavar, Irreversibility analysis for flow of a non-Newtonian hybrid nanofluid containing coated CNT/Fe3O4 nanoparticles in a minichannel heat exchanger, Appl. Therm. Eng. 125 (2017) 1083–1093.
[73] D. Huang, Z. Wu, B. Sunden, Effects of hybrid nanofluid mixture in plate heat exchangers, Exp. Thermal Fluid Sci. 72 (2016) 190–196.
[74] Z. Wu, et al., A comparative study on thermal conductivity and rheology properties of alumina and multi-walled carbon nanotube nanofluids, Front. Heat Mass Transf. 5 (18) (2014) 1–10.
[75] M.N. Labib, et al., Numerical investigation on effect of base fluids and hybrid nanofluid in forced convective heat transfer, Int. J. Therm. Sci. 71 (2013) 163–171.
[76] A. Bhattad, J. Sarkar, P. Ghosh, Hydrothermal performance of different alumina hybrid nanofluid types in plate heat exchanger, J. Therm. Anal. Calorim. 139 (6) (2020) 3777–3787.
[77] S.K. Singh, J. Sarkar, Hydrothermal performance comparison of modified twisted tapes and wire coils in tubular heat exchanger using hybrid nanofluid, Int. J. Therm. Sci. 166 (2021) 106990.
[78] B.R. Ponangi, V. Krishna, K.N. Seetharamu, Performance of compact heat exchanger in the presence of novel hybrid graphene nanofluids, Int. J. Therm. Sci. 165 (2021) 106925.
[79] M.W. Wambsganss, Thermal management concepts for higher-efficiency heavy vehicles, SAE Trans. (1999) 41–47.
[80] M.Z. Saidi, et al., Enhanced tribological properties of wind turbine engine oil formulated with flower-shaped MoS2 nano-additives, Colloids Surf. A Physicochem. Eng. Asp. 620 (2021) 126509.
[81] M. Afrand, K.N. Najafabadi, and M. Akbari, effects of temperature and solid vol.ume fraction on viscosity of SiO2-MWCNTs/SAE40 hybrid nanofluid as a coolant and lubricant in heat engines, Appl. Therm. Eng. 102 (2016) 45–54.
[82] M.H. Esfe, et al., Experimental evaluation, new correlation proposing and ANN modeling of thermal properties of EG based hybrid nanofluid containing ZnO-DWCNT nanoparticles for internal combustion engines applications, Appl. Therm. Eng. 133 (2018) 452–463.
[83] E. Dardan, M. Afrand, A.H.M. Isfahani, Effect of suspending hybrid nano-additives on rheological behavior of engine oil and pumping power, Appl. Therm. Eng. 109 (2016) 524–534.
[84] A. Karimi, M. Afrand, Numerical study on thermal performance of an air-cooled heat exchanger: effects of hybrid nanofluid, pipe arrangement and cross section, Energy Convers. Manag. 164 (2018) 615–628.
[85] F. Soltani, D. Toghraie, A. Karimipour, Experimental measurements of thermal conductivity of engine oil-based hybrid and mono nanofluids with tungsten oxide (WO3) and MWCNTs inclusions, Powder Technol. 371 (2020) 37–44.

[86] M.K.A. Ali, et al., Reducing frictional power losses and improving the scuffing resistance in automotive engines using hybrid nanomaterials as nano-lubricant additives, Wear 364 (2016) 270–281.
[87] R.R. Sahoo, J. Sarkar, Heat transfer performance characteristics of hybrid nanofluids as coolant in louvered fin automotive radiator, Heat Mass Transf. 53 (6) (2017) 1923–1931.
[88] O.A. Alawi, N.A.C. Sidik, H.A. Mohammed, A comprehensive review of fundamentals, preparation and applications of nanorefrigerants, Int. Commun. Heat Mass Transfer 54 (2014) 81–95.
[89] A. Bhattad, J. Sarkar, Effects of nanoparticle shape and size on the thermo-hydraulic performance of plate evaporator using hybrid nanofluids, J. Therm. Anal. Calorim. (2019) 1–13.
[90] W. Ahmed, et al., Heat transfer growth of sonochemically synthesized novel mixed metal oxide ZnO+ Al2O3+ TiO2/DW based ternary hybrid nanofluids in a square flow conduit, Renew. Sust. Energ. Rev. 145 (2021) 111025.
[91] A. Bhattad, J. Sarkar, P. Ghosh, Exergetic analysis of plate evaporator using hybrid nanofluids as secondary refrigerant for low-temperature applications, Int. J. Exergy 24 (1) (2017) 1–20.
[92] M.R. Safaei, et al., Effects of cobalt ferrite coated with silica nanocomposite on the thermal conductivity of an antifreeze: new nanofluid for refrigeration condensers, Int. J. Refrig. 102 (2019) 86–95.
[93] M.S. Ahmed, A.M. Elsaid, Effect of hybrid and single nanofluids on the performance characteristics of chilled water air conditioning system, Appl. Therm. Eng. 163 (2019) 114398.
[94] J.S. Nam, et al., Optimization of environmentally benign micro-drilling process with nanofluid minimum quantity lubrication using response surface methodology and genetic algorithm, J. Clean. Prod. 102 (2015) 428–436.
[95] M.S. Najiha, M.M. Rahman, K. Kadirgama, Performance of water-based TiO2 nanofluid during the minimum quantity lubrication machining of aluminium alloy, AA6061-T6, J. Clean. Prod. 135 (2016) 1623–1636.
[96] M. Jamil, et al., Effects of hybrid Al 2 O 3-CNT nanofluids and cryogenic cooling on machining of Ti–6Al–4V, Int. J. Adv. Manuf. Technol. 102 (9) (2019) 3895–3909.
[97] A.K. Sharma, et al., Measurement of machining forces and surface roughness in turning of AISI 304 steel using alumina-MWCNT hybrid nanoparticles enriched cutting fluid, Measurement 150 (2020) 107078.
[98] N.S.M. Sahid, et al., Experimental investigation on the performance of the TiO 2 and ZnO hybrid nanocoolant in ethylene glycol mixture towards AA6061-T6 machining, Int. J. Automot. Mech. Eng. 14 (1) (2017).
[99] K.K. Gajrani, et al., Thermal, rheological, wettability and hard machining performance of MoS2 and CaF2 based minimum quantity hybrid nano-green cutting fluids, J. Mater. Process. Technol. 266 (2019) 125–139.
[100] A. Eltaggaz, et al., Hybrid nano-fluid-minimum quantity lubrication strategy for machining austempered ductile iron (ADI), Int. J. Interact. Des. Manuf. 12 (4) (2018) 1273–1281.
[101] M.H. Esfe, M. Bahiraei, A. Mir, Application of conventional and hybrid nanofluids in different machining processes: A critical review, Adv. Colloid Interf. Sci. (2020) 102199.
[102] L. Sahota, G.N. Tiwari, Exergoeconomic and enviroeconomic analyses of hybrid double slope solar still loaded with nanofluids, Energy Convers. Manag. 148 (2017) 413–430.
[103] H.M.F. Rabbi, A.Z. Sahin, Performance improvement of solar still by using hybrid nanofluids, J. Therm. Anal. Calorim. 143 (2) (2021) 1345–1360.

[104] G. Sadeghi, S. Nazari, Retrofitting a thermoelectric-based solar still integrated with an evacuated tube collector utilizing an antibacterial-magnetic hybrid nanofluid, Desalination 500 (2021) 114871.
[105] E.F. El-Gazar, et al., Fractional modeling for enhancing the thermal performance of conventional solar still using hybrid nanofluid: energy and exergy analysis, Desalination 503 (2021) 114847.
[106] A.K. Tiwari, et al., 3S (sonication, surfactant, stability) impact on the viscosity of hybrid nanofluid with different base fluids: an experimental study, J. Mol. Liq. 329 (2021) 115455.
[107] N.S. Pandya, et al., Influence of the geometrical parameters and particle concentration levels of hybrid nanofluid on the thermal performance of axial grooved heat pipe, Therm. Sci. Eng. Prog. 21 (2021) 100762.
[108] Z. Said, et al., Optimizing density, dynamic viscosity, thermal conductivity and specific heat of a hybrid nanofluid obtained experimentally via ANFIS-based model and modern optimization, J. Mol. Liq. 321 (2021) 114287.
[109] L.S. Sundar, et al., Effect of core-rod diameter on wire coil inserts for heat transfer and friction factor of high-prandtl number magnetic fe 3 o 4 nanofluids ina fully developed laminar flow, Heat Transf. Res. (2021) **52**(3).
[110] L.S. Sundar, et al., Combination of Co3O4 deposited rGO hybrid nanofluids and longitudinal strip inserts: thermal properties, heat transfer, friction factor, and thermal performance evaluations, Therm. Sci. Eng. Prog. 20 (2020) 100695.
[111] M. Jamei, et al., On the specific heat capacity estimation of metal oxide-based nanofluid for energy perspective–a comprehensive assessment of data analysis techniques, Int. Commun. Heat Mass Transfer 123 (2021) 105217.
[112] A.A. Hachicha, et al., On the thermal and thermodynamic analysis of parabolic trough collector technology using industrial-grade MWCNT based nanofluid, Renew. Energy 161 (2020) 1303–1317.
[113] A.K. Tiwari, et al., Experimental comparison of specific heat capacity of three different metal oxides with MWCNT/water-based hybrid nanofluids: proposing a new correlation, Appl. Nanosci. (2020) 1–11.
[114] L. Sundar, et al., Heat transfer of rGO/Co3O4 hybrid nanomaterial based nanofluids and twisted tape configurations in a tube, J. Therm. Sci. Eng. Appl. (2020) 1–41.
[115] A.A. Minea, W.M. El-Maghlany, Influence of hybrid nanofluids on the performance of parabolic trough collectors in solar thermal systems: recent findings and numerical comparison, Renew. Energy 120 (2018) 350–364.

9

Recent advances in the prediction of thermophysical properties of nanofluids using artificial intelligence

Mehdi Jamei[a] and Zafar Said[b,c,d,*]

[a]*Faculty of Engineering, Shohadaye Hoveizeh Campus of Technology, Shahid Chamran University of Ahvaz, Dashte Azadegan, Iran.* [b]*Department of Sustainable and Renewable Energy Engineering, University of Sharjah, Sharjah, United Arab Emirates.* [c]*Research Institute for Sciences and Engineering, University of Sharjah, Sharjah, United Arab Emirates.* [d]*U.S.-Pakistan Center for Advanced Studies in Energy (USPCAS-E), National University of Sciences and Technology (NUST), Islamabad, Pakistan*

**Corresponding author: zsaid@sharjah.ac.ae, zaffar.ks@gmail.com*

Chapter outline

9.1 Introduction 203
9.2 Modeling structure using AI methods 208
9.2.1 Data preprocessing 209
9.2.2 Introduction to artificial intelligence methods 210
9.3 Sensitivity analysis 225
9.4 Summary 226
References 226

9.1 Introduction

Many researchers have considered the data-driven model (DDM) methods in recent decades to solve complex and non-linear engineering problems: finance, marketing, management, and environmental sciences. These methods can establish a connection between the independent and dependent parameters of a system with significant accuracy without the need for the

Hybrid Nanofluids: Preparation, Characterization and Applications. https://doi.org/10.1016/B978-0-323-85836-6.00009-0

physical behavior of a phenomenon. Recently, artificial intelligence (AI) methods have been widely welcomed due to the weakness of traditional regression-based methods and their low accuracy in nonlinear problems related to the study of thermophysical properties of nanofluids [1, 2].

In recent years, various investigations have been devoted to applying AI in estimating the thermophysical properties of nanofluids and energy applications [3–6], most of which have focused on single nanofluids. Among the available thermophysical properties, the coefficient of estimation of thermal conductivity using AI methods has been considered by researchers, which has recently been widely developed. The most prominent writings of recent years are summarized in Table 9.1. Also, in recent years, various AI methods have been used to estimate the viscosity, as one of the most important thermophysical properties of nanofluids, which can estimate a wide range of nanofluid types.

Table 9.2 describes recent literature in this field. Regarding estimating the specific heat coefficient of nanofluids using AI methods, both for a specific type of nanofluids and for a wide range of them, valuable efforts have been made, which are summarized in Table 9.3. However, in the field of hybrid nanofluids, AI methods have been more limited or have been used mainly for a specific type of nanofluid with a limited number of input parameters such as volume fraction and temperature [30].

The application of AI on the assessment of the thermophysical properties of hybrid nanofluids is more limited than single nanofluids, and most of the literature is related to a specific type (types) of hybrid nanofluids. In more recent years, specialized studies focusing on AI on a wide range of nanofluids have been performed to estimate their thermophysical properties. Table 9.4 reports some of the recent applications of AI-based models for predicting thermophysical properties of hybrid nanofluids.

Jamei et al. [43] provided the robust models including genetic programming (GP), M5Tree, and multilayer linear regression (MLR) models to predict the EG-based hybrid nanofluids' thermal conductivity. In this research, various nanoparticles, including metal oxides and nano carbons dispersed in the EG, were considered to model the thermal conductivity of hybrid nanofluids. The input variables were volume fraction, temperature, mean size, density, and thermal conductivity of each nanoparticle. Also, Pourrajab et al. [44] developed three robust models, namely,

Table 9.1 The highlight of literature on AI-based studies on thermal conductivity of nanofluids using AI approaches.

Nanofluids	Model inputs	ML methods	Reference
Various nanofluids	Temperature, volume fraction, thermal conductivity of nanoparticles	ANN-based method	Hojjat et al. [7]
Various nanofluids containing silica particles	Nanoparticle's volume fraction, temperature, average diameter of nanoparticles, thermal conductivity of base fluids	MLP-ANN and GMDH	Maleki et al. [8]
Various nanofluids containing CuO nanoparticles	volume fraction, temperature, average diameter of nanoparticles, thermal conductivity of base fluids	ANN and GMDH	Komeilibirjandi [9]
$Mg(OH)_2$-EG	Temperature, volume fraction	MLP-ANN	Esfe et al. [10]
Various nanofluids	Temperature, nanoparticle's volume fraction, average diameter of nanoparticles, thermal conductivity of nanoparticles, and base fluids	RBF-NN, MLP-ANN, LSSVM, ANFIS	Zhang et al. [11]
TiO_2-deionized water	Temperature, mass fraction, and diameter of nanoparticles	MLP-ANN, ANFIS, LSSVM, and RBF-NN	Ahmadi et al. [12]
Various nanofluids	Temperature, volume fraction, average diameter of nanoparticles, thermal conductivity of nanoparticles, and base fluids	ANN	Ahmadloo et al. [13]
Al_2O_3-water	Temperature, nanoparticle's volume fraction	MLP-ANN	Esfe et al. [14]
Various nanofluids	Temperature, volume fraction, average diameter of nanoparticles, thermal conductivity of nanoparticles, and base fluids	LLSVM-ISA, RBF, and KNN	Naseri et al. [15]
Ferromagnetic nanofluid	Temperature, volume fraction, and mean diameter of nanoparticles	ANN	Esfe et al. [16]
Various nanofluids containing ZnO particles	Temperature, volume fraction, average diameter of nanoparticles, thermal conductivity of base fluids	ANN, GMDH, and MARS	Maleki et al. [17]

LSSVM, least square support vector machine; *GMDH*, group method of data handling; *RBF-NN*, radial based function neural network; *MLP*, multilayer perceptron neural network; *MARS*, multivariate adaptive regression splines; *ANFIS*, adaptive network-based fuzzy inference system; *ISA*, improved simulated annealing.

locally weighted linear regression (LWLR), linear genetic programming (LGP), and GEP, and accurate estimation of the thermal conductivity ratio of several kinds of hybrid nanofluids; fluids were water, EG, and various volume percentages of EG/water. They used volume fraction, temperature, mean size, density, and

Table 9.2 The published literature on AI-based studies for prediction of the viscosity of nanofluids.

Nanofluids	Model inputs	ML methods	Reference
Various nanofluids	Temperature, mean size of nanoparticles, density, volume fraction, and base fluid viscosity	LSSVM, ANFIS, and GEP	Barati-Harooni. et al. [18]
Various nanofluids	Temperature, mean size of nanoparticles, density, volume fraction, and base fluid viscosity	RBF-NN	Barati-Harooni. et al. [19]
Various nanofluids	Temperature, mean size of nanoparticles, density, volume fraction, and base fluid viscosity	MLP	Heidari et al. [20]
Various nanofluids	Temperature, mean size of nanoparticles, volume fraction, density, and base fluid viscosity	Deep neural network	Changdar et al. [21]
Various nanofluids	Temperature, mean size of nanoparticles, volume fraction, density, and base fluid viscosity	Committee machine intelligent system (CMIS)	Sarapardeh et al. [22]
Various nanofluids	Temperature, mean size of nanoparticles, volume fraction, density, and base fluid viscosity	RF, MLP, and SVR	Gholizadeh et al. [1]
Various nanofluids	Temperature, volume concentration, and mean size of nanoparticles	FCM-ANFIS	Mehrabi et al. [23]

GEP, gene expression programming; *RF*, random forest; *SVR*, support vector regression.

Table 9.3 The published literature on AI-based studies for prediction of the specific heat capacity of nanofluids.

Nanofluids	Model inputs	ML methods	Reference
CuO/water	Nanoparticle's volume fraction, fluid temperature	ANN and SVR	Alade et al. [24]
Al_2O_3/EG, CuO/EG	SHC of base fluid, SHC of nanoparticles, the temperature of base fluid, nanoparticle's volume fraction	Bayesian-SVR	Alade et al. [25]
TiO_2/molten salt, Al_2O_3/molten salt, SiO_2/molten salt	Temperature, nanoparticle's weight fraction	ANN	Hassan et al. [26]
Al_2O_3/water	Nanoparticle's volume fraction, SHC of nanoparticles	Hybrid genetic algorithm-SVR	Alade et al. [27]
Various nanofluids	SHC of base fluid, SHC of nanoparticles, temperature, Nanoparticle's volume fraction, mean size of nanoparticles	KELM, MARS, M5Tree, and GEP	Jamei et al. [28]
Various nanofluids	SHC of base fluid, temperature, nanoparticle's volume fraction, mean size of nanoparticles	GPR, GRNN, and RF	Jamei et al. [29]

KELM, kernel extreme learning machine; *MARS*, multivariate adaptive regression spline; *M5Tree*, M5 tree model; *GPR*, Gaussian process regression; *GRNN*, generalized regression neural network.

Table 9.4 Summary of the application of AI in estimating the thermal conductivity of individual hybrid nanofluids.

Nanoparticles	Base fluid	Model inputs	Methodology	Reference
Cu-Zn	Water	Temperature, volume fraction	ANFIS	Balla et al. [31]
Cu-TiO_2	Water-EG	Temperature, volume fraction	MLP	Esfe et al. [32]
CuO-SWCNT	EG-water	Temperature, volume fraction	MLP	Rostamian et al. [33]
MgO-MWCNT	EG	Temperature, volume fraction	ANN	Vafaei et al. [34]
SWCNT-ZnO	EG-water	Temperature, volume fraction	ANN	Esfe et al. [35]
ZnO-MWCNT	EG-water	Temperature, volume fraction	MLP	Esfe et al. [36]
SWCNT-Al_2O_3	EG	Temperature, volume fraction	ANN	Esfe et al. [37]
MWCNT-SiO_2	EG	Temperature, volume fraction	MLP	Esfe et al. [38]
CNT-Fe_3O_4	Water	Temperature, volume fraction	MLP	Shahsavar et al. [39]
SWCNT-MgO	EG	Temperature, volume fraction	ANN	Esfe et al. [40]
Al_2O_3-SiO_2	Water	Temperature, volume fraction	MLP	Kannaiyan et al. [41]
DWCNT-SiO_2	EG	Temperature, volume fraction	MLP	Esfe et al. [42]
rGO/Co_3O_4	DW	Temperature, volume fraction	ANFIS	Said et al. [2]

thermal conductivity of the nanoparticles, and mixing ratio of water and ethylene glycol as input features. Adun et al. [45] provided a predictive model based on 715 experimental data points using MLP and SVR models to predict the thermal conductivity of hybrid nanofluids. They used seven variables as inputs comprising the volume concentration, temperature, acentric factor of the base fluid, nanoparticle bulk density, mixture ratio of particles, thermal conductivity, and size of nanoparticles. Jamei and Ahmadianfar [46] presented a novel evolutionary-based machine learning model, namely, multigene genetic programming (MGGP) for the prediction of viscosity of oil-based hybrid nanofluids based on the volume concentration temperature, mean size and density of nanoparticles, and viscosity of the base fluid as inputs. They validated the MGGP using GEP and MLR methods and provided an accurate relationship to estimate viscosity. More recently, Jamei et al. [47] developed an efficient ML approach called extended Kalman filter-based neural network (EKF-ANN) for accurate prediction of thermal conductivity in the oil-based hybrid nanofluids. In this study, GP and response surface methodology (RSM) were examined for evaluation of EKF-ANN, and the temperature, volume fraction, thermal conductivity of the base fluid, average diameter, and bulk density of nanoparticles were utilized as input variables to provide AI models.

9.2 Modeling structure using AI methods

Basically, the structure of an efficient predictive model involves steps that must be considered to achieve the desired result; in the following, its steps and structure will be introduced. The flowchart of steps for providing a predictive model is shown in Fig. 9.1.

Data collecting

The first step is to prepare a model for collecting datasets that is either the result of the laboratory effort of the current study or from the reliable collecting literature. Datasets should be carefully checked before providing the model, and unrealistic and outlier values should be removed from the dataset.

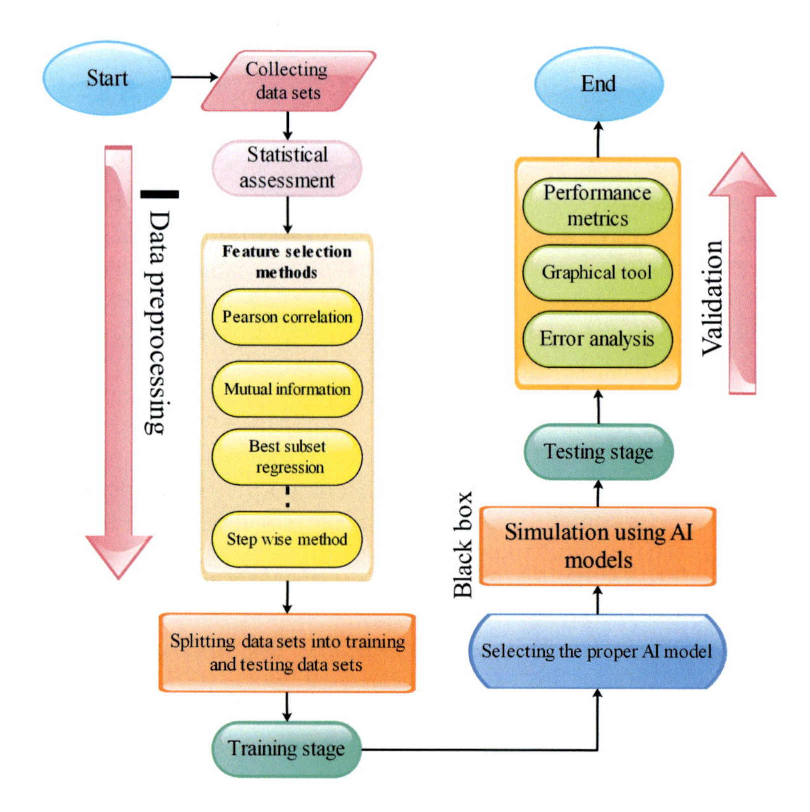

Fig. 9.1 The flowchart of thermophysical properties of hybrid nanofluids.

9.2.1 Data preprocessing

Statistical assessment of datasets

Data preprocessing is one of the most crucial steps in developing a AI-based model to predict hybrid nanofluids' thermophysical properties. After gathering the datasets from experimental investigations or literature, it is necessary to examine them in terms of statistical characteristics to determine the validity of the model. Also, the status of data in terms of skewness and proximity to normal distribution has a significant impact on the choice of AI models.

Feature selection

Feature selection is one of the modeling steps that aims to determine the optimal input composition for modeling and identifying the effective modeling parameters. This important step in modeling is sometimes overlooked. One of the simplest methods to determine the optimal composition is to use the Pearson correlation coefficient, which examines the correlation of all available input parameters with the target parameter. The higher the Pearson coefficient for each input, the more important that parameter is in modeling. The most common use of this method is when the relationship between input variables and the target is linear. In cases where there is a high nonlinear relationship between the input variables and the target, other methods such as mutual information (MI) and best subset regression should be used, which will be discussed below.

Mutual information

Mutual information is one of the most efficient data-preprocessing methods for identifying the best combination and most influential inputs widely used in AI studies. Generally, the MI of two variables W and Z can be evaluated using the following equation [48]:

$$MI(W,Z) = Y(W) - Y(W|Z) \tag{9.1}$$

where the variable entropy can be computed using probability $Q(x)$ as [48].

$$Y(W) = -\sum_{W \in x} \log[Q(x)]\, Q(x) \tag{9.2}$$

$$S(W|Z) = \sum_{x \in W} Q_x(x) \sum_{y \in Z} \log[Q(y|x)]Q(y|x) \tag{9.3}$$

where $p_x(x)$ represents the marginal distribution for x as $Q_x(x) = \sum_{y \in Z} Q(x,y)$ and the conditional distribution for variable y for given x is defined as $Q(y|x) = Q(x,y)/Q_x(x)$. The mutual information perspective is an interesting idea for obtaining the degree of dependence between variables that can efficiently achieve nonlinear dependence between datasets.

Best subset regression (BSR)

BSR, as a statistical-based feature selection method, is one of the most effective ways to reduce input variables to build predictive models in AI systems. This method can efficiently identify the variables that have the most substantial relationship based on the statistical criteria. Here, the minimum variable is defined by the user [47]. Then, the process of producing input combination starts with the most effective parameters (by predefined minimal variables), and by sequentially adding other variables, statistical criteria such as mean squared error (MSE), adjusted-R^2, the Mallow's coefficient [49], Akaike's information criterion [50], and Amemiya's prediction criterion (MPC) [51] are calculated. Of all the possible combinations, the ones with the lowest MSE, Mallow's coefficient [49], Akaike's information criterion [50], and MPC [51] values and the highest adjusted-R^2 values are the best combinations. For further study of feature selection methods such as the "stepwise method," one can refer to Refs. [52, 53].

9.2.2 Introduction to artificial intelligence methods

Artificial neural network

Perhaps the most widely used AI method to estimate nanofluids' thermophysical properties is the artificial neural network (ANN) method, which is widely used in solving nonlinear engineering problems. There are different types of ANN, which are described below.

Feedforward network of neurons (FFNN)

According to the neural network of the brain, a network of neurons is defined like this: a compatible system containing several simple processing origins. The origins of processing, which we know as neurons, gather for completing a processing path. Each neural network has got three layers. The first one is named the layer of "input" that transfers data to a network of neurons. The second one is named the layer of "hidden." It is named "hidden" because there are no connections between this layer and the output world. The third layer is called the "output layer," demonstrating the network output responding to a particular input. The general mathematical model of an artificial neural network is shown in the following [54]:

$$s = \sum_{m=1}^{M} w_m x_m = W^T X \tag{9.4}$$

X is considered as the "input vector," W implies the vector of the "hidden layer" weight, and M represents how many neurons there are in the hidden layer. Afterward, the quantity s goes into a transfer function whose duty is to return the output value:

$$y = f(s) \tag{9.5}$$

The nonlinear transfer function can be mainly defined in the form of sigmoid as it can be seen in the following [54]:

$$f(s) = \frac{1}{1 + \mathrm{e}^{-s}} \tag{9.6}$$

Training a selected network aims to adapt the weights' values and biases to change the existing error between the calculated and observed output values into a minimized level. For training perceptron, artificial neural networks, a multilayer, and a backpropagation algorithm (BP) are generally used [54].

Multilayer perceptron (MLP) neural network

The biological system of neurons in the human brain's cells and mutual relations related to them inspired MLP, known as a prominent ANN method [54]. The neuron with two parameters forms the main component of the ANN: the first one is the "weight" (w), and the second one is the "bias" (β). Information is received, and responses are produced by an artificial neuron, as the biological neuron does. ANN models consist of three sequential layers including input, hidden, and output, which are associated in a way that they are named feedforward. The MLP is considered the most and eminent ANN. It has three layers. The MLP's output layer applies a kind of linear activation function (φ_2) for relating the hidden layer response and the output neuron [1]:

$$y = \varphi_2 \left[\beta_0 + \sum_{j=1}^{n} w_{jk} \left[\varphi_1 \left(\sum_{j=1}^{n} x_i w_{ij} + \beta_j \right) \right] \right] \tag{9.7}$$

where x_i represents an element of the layer of input, w_{ij} represents the existent weights between the first (i) and the second (j)

neurons of hidden or middle, β_j represents the bias of the node of the second layer hidden, φ_1 represents the function of sigmoid activation, the connection weight of neuron j, which is in the second layer (hidden or middle layer), to the neuron near it, which is the k neuron, in the output layer is represented by w_{jk}, and neuron k's bias value in the output is represented by β_0. The backpropagation (BP) training algorithm can specify both weights and biases [55, 56].

Radial basis function neural network (RBF-NN)

RBF-NN is an ANN model, in the form of feedforward, consisting of three layers: only one layer of middle, a layer of input, and a layer of output. The RBF-NN is mainly considered as a universal approximator [57]. The basis function linear combination can specify the output layer neuron. Calculating an RBF-NN output with N neurons, which are hidden, is represented in the following:

$$y_i = \sum_{j=1}^{N} w_{ij}\varphi_j(x) \tag{9.8}$$

where w_{ij} signifies a weight, a linkage one, between the neuron of output and the neuron of hidden and φ is the representative of radial basis performance. The Gaussian or RBF function can be named as one of the activation functions that is normally applied. It is shown in the following [15, 54]:

$$\varphi_i(x) = \exp\left(-\frac{\| x - \mu_i\|^2}{2\sigma_i{}^2}\right), \quad i = 1,2,N \tag{9.9}$$

In the RBF function, σ_i and μ_i are representative of the spread and center [54].

Cascaded forward neural network (CFNN)

One type of advanced artificial neural network (ANN) that provides high-integrity paradigms to deal with various systems is called cascaded forward neural network (CFNN) [58]. Like other types of ANN, neurons are the fundamental origins of information processing in CFNN [59]. They also have three layers: the first one is the input layer, the second one is the hidden layer, and the last one is the output layer. The first layer, input, introduces the system's data, whereas the last layer, the output, is responsible for delivering the outcomes. How many neurons the input layer and output layer have is related to how many input and output parameters the systems have. The number of hidden layers and the neurons related to them alter depending on how complicated

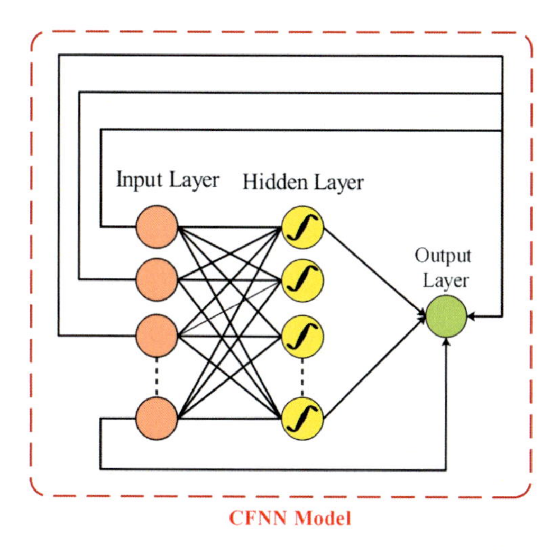

Fig. 9.2 The architecture of cascaded forward neural network (CFNN).

the system is. The hidden layer is responsible for changing the information into higher characteristics using activation functions like logsig and tansig. The cascaded structure is the major aspect that discriminates CFNN from the other sorts of ANN [59]. Due to this, producing extra nodes and relations with all inputs and hidden layers' neurons does the cascade scheme [59], which means that neurons from previous layers are connected to the nodes of the succeeding layers [59].

To achieve an accurate forecasting model, CFNN goes through a training phase whose aim is to find the proper weights for relying upon neurons to each other. For this reason, because of the efficiency and reliability they have, techniques of backpropagation, including the algorithm of Levenberg-Marquardt (LM), are largely suggested. In the present study, LM algorithms have been used in the CFNN learning phase. Research literature includes more details about this kind of algorithm [22, 60]. The structure of a CFNN model is shown in Fig. 9.2.

Generalized regression neural network (GRNN)

The other precise kind of ANN is named generalized regression neural network (GRNN). Specht developed this sort of ANN [61]. Researchers consider GRNN as a type of function network that is radial-based [62, 63]. Its main feature includes its quick and low-costing learning approach, which is guaranteed by the parallel architecture that it has [64]. GRNN is made of four layers. The first one is the input, the second one is the pattern, the next is the summation, and the last one is the output. The same as CFNN, the input layer transfers data to GRNN, and the layer of output releases the outcomes. Pattern and summation layers are located between the input layer and output layer. The pattern layer's goal is to process and transform the data into a nonlinear space to distinguish and identify those relationships that are representative of the system. During using these nonlinear transformations, a spread coefficient should be considered to make sure that there is smoothness. What happens in the summation layer is that the operation of summation is used on outputs of pattern layer nodes. It should be mentioned that they are also multiplied with proper weights.

GRNN's learning strategy is nonrepetitive, and it is introduced based on the theory of nonlinear regression [65, 66]. The training phase was formed in the present study to minimize the squared error between the real data and GRNN prediction.

Extreme learning machine (ELM)

Huang et al. have developed an extreme learning machine (ELM) that is a machine learning technique [67]. ELM is in the single hidden layer feedforward network (SLFN) form [68]. The mentioned technique has several particular benefits, including scalability and learning speed, which causes ELM, in this case, to outperform in comparison to other ML. In fact, the learning concept of ELM guarantees those two benefits. It does so by ascribing accidentally the bias values and the weights that connect the layer of input into the hidden nodes, and it is when it uses an analytical strategy to distinguish the weights that connect the hidden nodes to the layer of output [69]. The SLFN math formula can be seen in the following [70]:

$$f_n(x) = \sum_{j=1}^{n} h_j(x)\alpha_j = h(x)\alpha \tag{9.10}$$

In the formula mentioned above, n indicates how many hidden neurons there are, $\alpha = [\alpha_1, \alpha_2, \ldots, \alpha_n]^T$ implies the matrix of weight that connect the hidden nodes to the layer of output, and $h_j(x)$ represents the jth hidden node output. The last one can be estimated by applying a function that is nonlinear (g) indicating two parameters (c_1, c_2) that must meet the proofs of ELM estimation [70, 71]. The computation below proves that the sigmoid function is a nonlinear one (g):

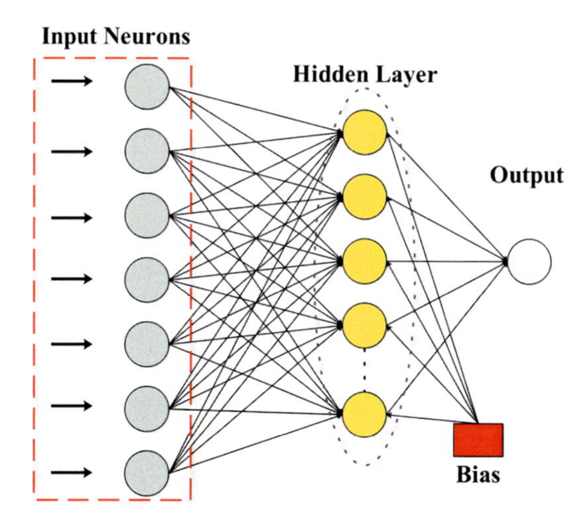

Fig. 9.3 The configuration of the ELM approach.

$$g(c_1, c_2, x) = \frac{1}{1 + \exp[-(c_1 x + c_2)]} \tag{9.11}$$

Through solving the minimization of the approximated squared error, the weight matrix, which is shown as α, can be obtained [71]:

$$\min_{\alpha \in R^{n \times m}} \|H\alpha - t\|^2 \tag{9.12}$$

t and H are indicators of the matrix of target values and the randomized hidden layer output matrix. Here is the ultimate solution of minimizing problem that is formulated as it is shown above:

$$\alpha^* = H^\dagger t \tag{9.13}$$

$H^\dagger$ denotes matrix H's Moore-Penrose generalized reverse [71]. It should be noted that by using techniques like iterative methods and single value decomposition (SVD), this final solution is gained [71]. Fig. 9.3 represents the structure of an ELM model.

Adaptive neuro-fuzzy inferences system (ANFIS)

Adaptive neuro-fuzzy inference system is known as a famous technique, characterized as a smart one. Jang has initially introduced this system [72]. It is known as a sort of MLP network that the benefits of the system of fuzzy and network of neurons have been mixed to construct a model that is of optimized membership functions [73]. It is a kind of fascinating fuzzy logic model technique that is rule based. In this technique, the rules are made

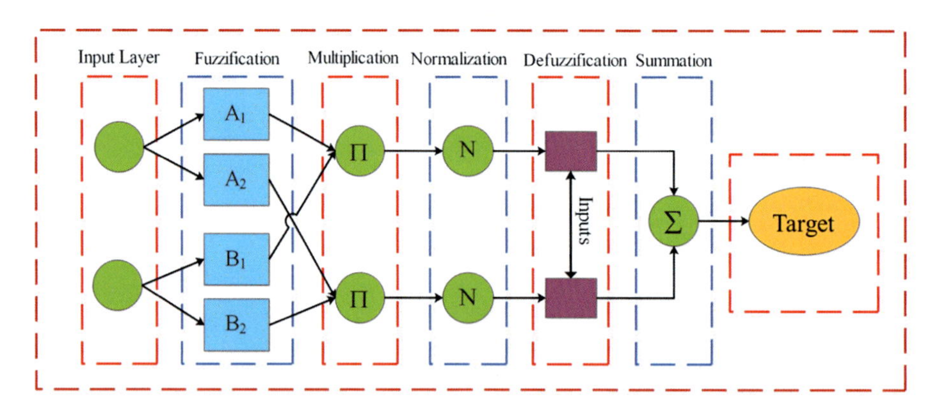

Fig. 9.4 The ANFIS structure.

within the process of training. An ANFIS that has been made perfectly can solve all complex and nonlinear problems with high precision [74, 75].

ANFIS is developed according to the rules that are used to relate several input and output variables [76]. Therefore, the fuzzy inference system (FIS) could be applied as an anticipating model for those situations in which the input and/or output data are very vague [77]; the reason is that, in situations like this, the classical anticipating techniques are not able to inspect the existent ambiguity in data. ANFIS model typically has five layers. Each layer has a specific function. In Fig. 9.4, the structure of ANFIS and also steps of building the model in different layers are demonstrated.

To decrease the assessment errors, ANFIS applies algorithms of backpropagation optimization or a mixture of the least squares and backpropagation known as the hybrid method. In the technique mentioned earlier, the data that has been modeled is the instructor of a fuzzy system; i.e., ANFIS contains both structures in a system of fuzzy and learning capability in neural networks [76, 78, 79].

Group method of data handling (GMDH)

Group method of data handling (GMDH) or polynomial neural network is considered one of the most performant types of neural network family. One of the advantages of GMDH is represented by its user-friendly polynomial formula, which can be applied to the studied system. The GMDH algorithm implies the use of multiple nodes, which are contained by the intermediate layers. Each GMDH node will generate a value that is calculated using a

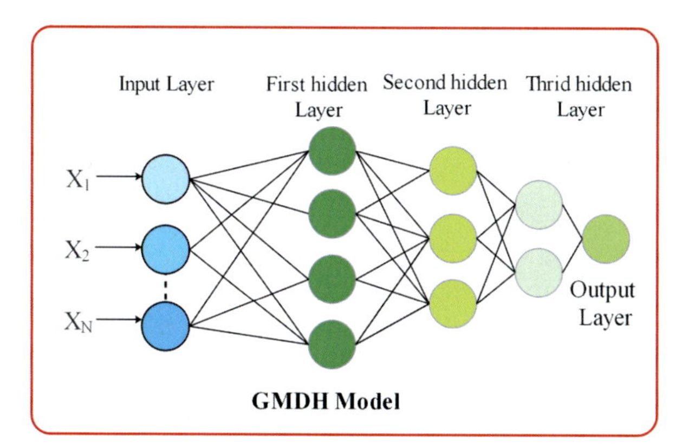

Fig. 9.5 Example of the general framework of the GMDH approach.

quadratic polynomial model, which contains the previous neurons [80, 81]. The GMDH algorithm is based on the following rule:

$$y = a_0 + \sum_{i=1}^{N} a_1 x_1 + \sum_{i=1}^{N}\sum_{j=1}^{N} a_{ij} x_i x_j + \sum_{i=1}^{N}\sum_{j=1}^{N}\sum_{k=1}^{N} a_{ijk} x_i x_j x_k \quad (9.14)$$

where $X = (x_1, x_2, x_3, \ldots, x_N)$ and y are the input vector and output parameters of the model, respectively, $a_{i\ldots k}$ are the polynomial coefficients, and N is the number of the parameters. In the case of GMDH having a structure with two neurons that are related with a quadratic polynomial model, the below relation can be applied to determine the new nodal variables:

$$\text{Quadratic}: \hat{y} = G(x_i, y_j) = a_1 + a_1 x_i + a_2 x_j + a_3 x_i x_j + a_4 {x_i}^2 + a_5 {x_j}^2 \quad (9.15)$$

where G is the nodal variable. An example of the GMDH model is depicted in Fig. 9.5.

Ensemble machine learning methods

M5 tree (M5Tree) models

The M5Tree model has been developed by Quinlan (1992) [82]. It is a kind of tree model of decision for those problems that are regression-based. This model is responsible for dividing a dataset into subdatasets. To examine it conceptually, M5Tree can be considered a regression model called piece-wise linear functions [43, 82]. The M5Tree utilizes some elements such as root, branches, and leaves to produce numbers of equations for various subdatasets. The M5Tree develops a decision tree by applying a splitting criterion. In this research, the M5Tree algorithm has provided the base tree according to the standard deviation reduction (SDR) as a split criterion. As you can see, it is shown below [83]:

$$SDR = \mathrm{SDev}(T^*) - \sum_{i=1}^{N} \left(\frac{T_i^*}{T^*}\right) \mathrm{SDev}(T_i^*) \tag{9.16}$$

where T^* indicates input dataset, T_i^* is the representative of input data subset to the main node (parental) and SDev, and N implies standard deviation and dataset number.

In order to reach the SDR value to its maximized level, for each node, M5Tree chooses parameters. According to the SDR's predefined value, the process of instructing the model finishes. Thereupon, the result is a formula based on regression for each branch (subdataset). Sometimes in such a case, overfitting happens. To prevent it from occurring, the pruning process is applied to merge the lower branches [83]. The last stage, through the M5Tree modeling, is called the smoothing process. If there are discontinuities, this process compensates for it using the nearby branches. The flowchart of M5Tree modeling can be seen in Fig. 9.6.

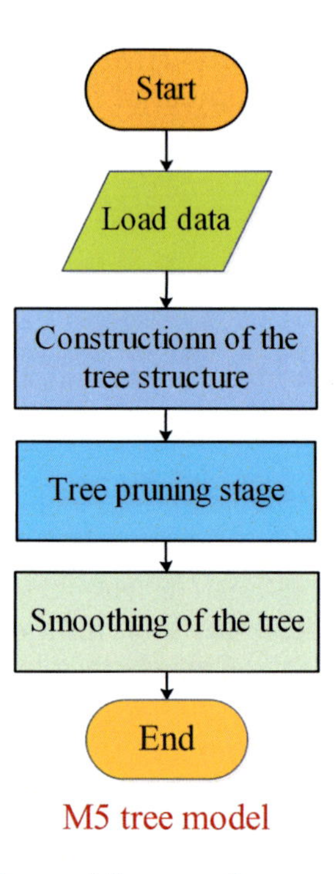

Fig. 9.6 Flowchart of M5Tree modeling procedure.

Random forest (RF)

Random forest (RF) stands for a bootstrap accumulated group model. It has been largely used for those problems in the literature that are based on regression [1, 29, 84, 85]. The RF model is responsible for generating a forest. It does so by applying random double trees and a sample that is bootstrap. To raise the tree, the sample, which is bootstrap, is more applied in an algorithm of classification and regression tree (CART) specified at each node through a lot of variables, random ones [86].

In the subsequent step, the remnant of the sample of training data, which has been excluded from the bootstrap sample, is applied for implying the calculation of out of bag (OOB) error. RF model's accuracy and, meanwhile, optimization of the model parameters such as m_{tree} and m_{try} are normally assessed by OOB values [1, 87]. The RF model is able to work with numerous databases. It is worthy to note that by setting particular parameters like n_{tree} and n_{try}, the anticipation overfitting could be prevented. Applying the RF model has some challenges. Its main challenge can be presented by specifying the optimum number of trees relating to a stabilized OOB error value [87]. To optimize the anticipation, it is needed to apply an inner performance that aims to reduce the node's impurities [87]. The MATLAB MEX function of Andy Liaw et al.'s C code (applied in R package RF) has assisted the RF method to be utilized. It considers applying the following values of default: $n_{tree} = 500$ and $m_{try} = \text{Floor}(\sqrt{M})$ [88]. In the equation, M shows the total of predictors. For each of the decision trees, the mean square error (MSE) of OOB is calculated separately. The way it is done is shown in the below [88, 89]:

$$MSE_{OOB} = \frac{\sum_{i=1}^{n_{tree}} \left(TP_i - \hat{TP}_i^{OOB} \right)^2}{n_{tree}} \tag{9.17}$$

In the equation, $\hat{TP}_i^{OOB}$ denotes the mean predicted seepage and TP_i shows the points of experimental data. The MSE_{OOB} is the one that could be also applied for assessing each variable's importance. The road map of the RF model is depicted in Fig. 9.7.

Support vector machine regression (SVR)

The SVM is known for solving different classification and regression problems first proposed by Vanpik [90]. The SVM for regression or support vector regression (SVR) is responsible for developing an efficient balance between the model's precision and complexity. It also shows a significant ability to anticipate.

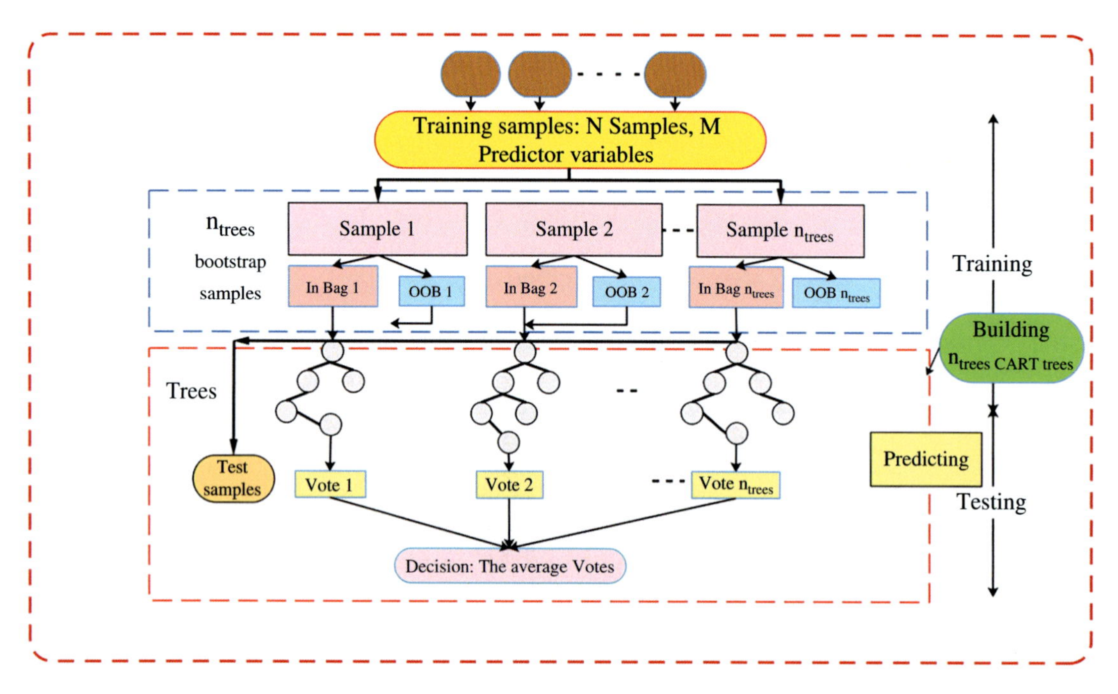

Fig. 9.7 The architecture of the RF model.

Given an input dataset $\{(x_1, z_1), (x_2, z_2), \ldots, (x_m, z_m)\}$, where $x_m \in X \subseteq R$, $z_m \in Y \subseteq R$, and m are considered as the maximum number of the dataset, the function of SVR could be calculated as follows [91]:

$$z(x) = w^T \psi(x) + b \tag{9.18}$$

where $\psi(x)$ implies a nonlinear function, w represents SVR's weight vector, and b indicates a constant. It is possible to minimize the fitness in Eq. (9.18) to characterize w, considering the limitations that are presented in Eq. (9.19) [1]:

$$\text{Fitness} = \frac{1}{2} w^2 + C \sum_{m=1}^{M} \left(\zeta_m + \zeta_m^{(*)} \right) \tag{9.19}$$

Subject to

$$\begin{aligned} & z_m - w^T \psi(x) - b \leq \varepsilon + \zeta_m, m = 1, 2, \ldots., M \\ & \zeta_m, \zeta_m^{(*)} \geq 0, m = 1, 2, \ldots, M \end{aligned} \tag{9.20}$$

Based on the equation above, ζ_m and $\zeta_m^{(*)}$ are the variables of looseness, ε indicates the size of the tube with a fixed value, and

C represents a penalty factor. To solve Eq. (9.18), the Lagrange method having multiplier coefficients (λ_m and λ^*_m) could be applied. This method has been shown below [92]:

$$z(x) = \sum_{m=1}^{M} (\lambda_m - \lambda^*_m) Kr(X, X_i) + b \tag{9.21}$$

In the equation above, $Kr(X,X_i)$ implies a kind of kernel function.

Evolutionary machine learning approaches

Gene expression programming

Ferreira has developed GEP, which is known as a learning method for providing mathematical modeling [93]. This method can be considered a kind of evolutionary calculation concept of natural evolution consisting its basis. The GEP can be considered as genetic programming (GP), but there is a difference between them. That is, it applies fixed-length string, known as chromosome, to do the program modeling, which is ultimately presented in the form of expression trees (ETs). GEP typically has two components, including chromosomes and expression trees. A group of genes makes chromosomes. Each gene is formed of a head and a tail. It is possible to form a head through utilizing math functions (−, +, /, ×) and terminal symbols (x, y, z, −2), and a tail can be formed from final symbols [49, 94]. Chromosomes represent a group of random solutions and hence begin GEP. After that, the chromosomes will be dragged in the ET form and examined considering an objective function and then are chosen according to the fitness to reproduce. It is done by utilizing genetic operators such as crossover and mutation. The same process will be repeated for novel solutions until the fulfillment of the stopping criteria.

Genetic programming

The genetic programming paradigm is known as a technique of evolutionary calculation intelligence. The basis of the random biological algorithm inspires the formation of this technique that its features include reliance and freedom in describing complicated processes. It does so through applying rational performances and math expressions. GP normally expresses decision features. It does so in a tree-structural way with the ability to generate a solution in an expressed form constructed of components such as tree and terminal nodes. In a formulation like this, the tree node signifies performance and the terminal node denotes an operand. In GP, numerical and classification decision

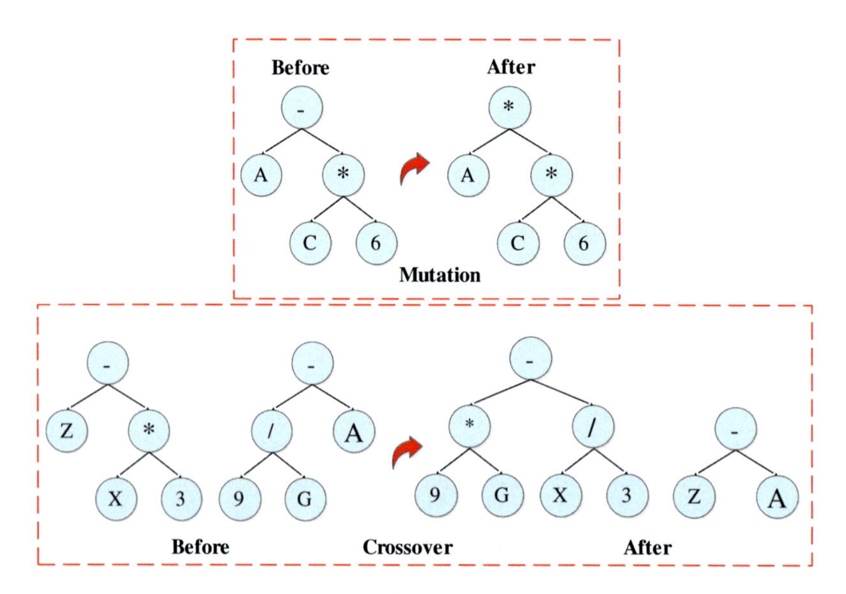

Fig. 9.8 An instance of leap *(upper panel)* and crossover *(lower panel)* process of GP. Copyright Elsevier (Lic # 5065541442272) M. Jamei, I.A. Olumegbon, M. Karbasi, I. Ahmadianfar, A. Asadi, M. Mosharaf-Dehkordi, On the thermal conductivity assessment of oil-based hybrid nanofluids using extended Kalman filter integrated with feed-forward neural network, Int. J. Heat Mass Transf. 172 (2021) 121159.

characteristics have been named "final set," while mathematical symbols ($-$, $+$, $\times$, $\div$), rational operators (e.g., if-then-else), and math functions have been called function sets. The present math operations in Fig. 9.8 (upper panel) are "*" and "$-$" and that shows the expression of "$A-6C$" for tree structure that is located on the left and "$A \times 6C$" that is located on the right. Here is how GP operates:

Step 1: Trees' randomized system starts, and each tree's related tangible functions are assessed.

Step 2: The tree's tangible function gets compared, and the roulette wheel or tournament methods are applied to choose the optimum tangible functions.

Step 3: The next group of trees is regenerated by operators named the crossover and leap operators.

Step 4: The process of generating trees, which is recursive, keeps on to the point that it is maximal enough, and the process finishes. If it does not occur, the iteration will be begun again in the second step.

The crossover operator is responsible for connecting two trees as an individual unit and, afterward, applying the unit to produce the next two units. With producing a novel unit, component change occurs in the individual unit. What that does so is the leap operator. Fig. 9.8 describes the usage of the leap and crossover operator on, first, one and then on two units.

Gaussian process regression (GPR)

The Gaussian process is known as a collection of randomized variables. In the Gaussian process, limited numbers of variables unify with Gaussian distributions [95, 96]. Covariance and mean functions are specified features of the Gaussian process. Actually, this process can be defined as a kind of natural extension of Gaussian distribution. In such a process, the mean is a vector, and the covariance is a matrix. It is hypothesized that in models of Gaussian process regression (GPR), former observations should have information related to each other. In contrast to the Gaussian distribution, the Gaussian process is a sort of over performances. Thereupon, there is no need for a validation process for models based on the Gaussian process to be generalized. Based on the previous information that was mentioned about performance dependency and data, models of the Gaussian regression process can figure out the anticipating distribution relating to the input of the test [96, 97].

Note the dataset of S with observations notified as n. In the ahead equation$=\{(x_i, y_i) | i=1, \ldots, n\}$, x_i represents the vector of input with the dimension of D and y_i implies the vector of the target. The mentioned collection is made of samples such as input and output. To simplify, the collection's inputs are gathered in the matrix of $\Pi=[x_1, x_2, \ldots, x_n]$, and its outputs are gathered in the matrix of $Y=[y_1, y_2, \ldots, y_n]$.

Regression is responsible for generating a novel input x^* for obtaining the anticipated distribution for relating the observational data values y^* according to the dataset of S. Mean $m(x)$ function describes the Gaussian process [97]:

$$m(x)=E(f(x)) \tag{9.22}$$

Covariance $k(x, x')$ functions are demonstrated in the equation below:

$$k(x, x')=E(f(x)-m(x))(f(x')-m(x')) \tag{9.23}$$

In the following, Gaussian process could be seen:

$$f(x) \sim GP(m(x), k(x, x')) \tag{9.24}$$

The value of the mean function is normally adjusted to zero because of making it simplified. The connections of target and input vector in the Gaussian process are presented in the following:

$$y_i = f(x_i) + \psi \tag{9.25}$$

In the equation below, ψ is the Gaussian distribution noise value with σ^2 variance and mean of zero:

$$\psi \sim N(0, \sigma^2) \tag{9.26}$$

Moreover, $f = [f(x_1), \ldots, f(x_n)]^T$ has been hypothesized to have a behavior according to the Gaussian process. In the equation ahead, $p(f|\Pi) = N(0, K)$, K is the matrix of covariance having $k_{i,j} = k(x_i, x_j)$ components.

$$K(x,x) = \begin{bmatrix} K(x_1,x_1) & K(x_1,x_2) & \ldots & K(x_1,x_n) \\ K(x_2,x_1) & K(x_2,x_2) & \ldots & K(x_2,x_n) \\ \vdots & \vdots & \ddots & \vdots \\ K(x_n,x_1) & K(x_n,x_2) & \ldots & K(x_n,x_n) \end{bmatrix} \tag{9.27}$$

Utilizing GPR, one is able to compute f^* at Π^* test points.

The distribution of y, which is providing to the f values, can be seen below:

$$P(y|f, \Pi) = N(f, \sigma_n^2 I) \tag{9.28}$$

In the following equation, I is an indicator of the matrix of the unit $n \times n$. By applying the characteristics of Gaussian distribution, y marginal distribution could be estimated [96]:

$$P(y|\Pi) = \int p(y|f, \Pi) p(f|\Pi) df = N(0, K + \sigma_n^2 I) \tag{9.29}$$

The equation below shows the common distribution of two values including function and the observed target [29]:

$$\begin{bmatrix} y \\ f_* \end{bmatrix} \sim N\left(\begin{bmatrix} K(\Pi,\Pi) + \sigma^2 I & K(\Pi,\Pi_*) \\ K(\Pi_*,\Pi) & K(\Pi_*,\Pi_*) \end{bmatrix}\right) \tag{9.30}$$

In the equation, $K(\Pi_*, \Pi_*)$ signifies test points' matrix of self-covariance (Π_*) and $K(\Pi, \Pi_*)$ denotes $n \times 1$ test points' matrix of covariance Π_* and entire Π input points.

At the time the process of training is finished applying Bayesian theory, the nearest values of output linking to Π_* could be calculated according to the set of training. Observed data and Bayesian theory are applied to update the possibility distribution. The anticipating distribution can be gained (Eq. 9.31) by applying the standard rules to make Gaussian conditioned. The equation is presented in the following form [29, 96]:

$$P(f_*|\Pi, y, \Pi_*) \sim N\left(\overline{f}_*, \text{cov}(f_*)\right) \tag{9.31}$$

Measuring ambiguity (variance) and expected anticipation (mean) of f_* calculate output (function values' anticipation distribution):

$$\overline{f_*} = K(\Pi_*, \Pi)\left[K(\Pi, \Pi) + \sigma^2 I\right]^{-1} y \tag{9.32}$$

$$\text{cov}(f_*) = K(\Pi_*, \Pi_*) - K(\Pi_*, \Pi)\left[K(\Pi, \Pi) + \sigma^2 I\right]^{-1} K(\Pi, \Pi_*) \tag{9.33}$$

The covariance or kernel function composes the major element of regression of a Gaussian process. The resemblance among the observed data can be shown by the function of covariance [97]. The most common covariance function is rational quadratic kernel, matern 3/2, matern 5/2, squared exponential kernel, exponential kernel, and squared exponential kernel [29, 98].

In recent years, several other types of machine learning methods have been considered for simulation of the engineering problems, including MARS [17, 99], extreme gradient boosting (XGBoost) [100], boosted regression tree (BRT) [101], deep neural network (DNN) [21, 102], MGGP [46, 103], etc., which can be used for prediction of thermophysical properties of nanofluids.

9.3 Sensitivity analysis

Sensitivity analysis is one of the most important parts of modeling in which the effective parameters in modeling are identified. Identifying these variables helps to predict the behavior of the phenomena under modeling. In AI-based studies, various methods of sensitivity analysis are used. The simplest method of sensitivity analysis, where the relationship between the input data and the target is relatively linear, is to use the Pearson correlation coefficient between them. The parameter with the highest Pearson coefficient is introduced as the most influential parameter [47]. But when there is a nonlinear relationship between input and output data, methods such as mutual information as a relatively reliable method can be effective. But the safest method of sensitivity analysis is the sequential deletion of input data or zeroing them in the superior model of AI model. The most influential input variable that causes the most significant reduction in model accuracy can be introduced as an effective parameter by examining statistical criteria. This method, unlike other methods, requires the implementation of an AI model and has better compliance with the model. Various other methods can be found to analyze sensitivity in the literature, and one can refer to Refs. [44, 104, 105] for more information.

9.4 Summary

AI-based investigations on thermophysical properties of nanofluids demonstrated that most of the applications of machine learning and DDMs are related to thermal conductivity and viscosity of mono fluids, and limited research has been conducted to model hybrid nanofluids.

Given the increasing capabilities of AI methods and their integration with robust optimization algorithms, it can be hoped to solve nonlinear problems of hybrid nanofluids with a large number of input variables to achieve promising results.

Investigating the thermophysical properties of hybrid nanofluids and their application in industry and energy applications using classical regression-based methods due to high constraints and input variables does not lead to relevant results. At this stage, the importance of AI methods becomes more apparent. Besides, the use of machine learning methods that, in addition to accurate estimation, yield to the production of an efficient and appropriate relationship for estimating a critical parameter such as thermal conductivity, viscosity, or specific heat capacity is of particular importance. The efficient AI methods that include this evaluable feature can be mentioned as GEP and MGGP, and GP rarely implements AI-based recent investigations on nanofluids.

References

[1] M. Gholizadeh, M. Jamei, I. Ahmadianfar, R. Pourrajab, Prediction of nanofluids viscosity using random forest (RF) approach, Chemom. Intell. Lab. Syst. 201 (2020) 104010, https://doi.org/10.1016/j.chemolab.2020.104010.
[2] Z. Said, L.S. Sundar, H. Rezk, A.M. Nassef, H.M. Ali, M. Sheikholeslami, Optimizing density, dynamic viscosity, thermal conductivity and specific heat of a hybrid nanofluid obtained experimentally via ANFIS-based model and modern optimization, J. Mol. Liq. (2020), 114287.
[3] Z. Said, A.A. Hachicha, S. Aberoumand, B.A.A. Yousef, E.T. Sayed, E. Bellos, Recent advances on nanofluids for low to medium temperature solar collectors: energy, exergy, economic analysis and environmental impact, Prog. Energy Combust. Sci. 84 (2021) 100898, https://doi.org/10.1016/j.pecs.2020.100898.
[4] Z. Said, M.A. Abdelkareem, H. Rezk, A.M. Nassef, H.Z. Atwany, Stability, thermophysical and electrical properties of synthesized carbon nanofiber and reduced-graphene oxide-based nanofluids and their hybrid along with fuzzy modeling approach, Powder Technol. 364 (2020) 795–809.
[5] T. Ma, Z. Guo, M. Lin, Q. Wang, Recent trends on nanofluid heat transfer machine learning research applied to renewable energy, Renew. Sust. Energ. Rev. (2020), 110494.
[6] M. Ramezanizadeh, M.H. Ahmadi, M.A. Nazari, M. Sadeghzadeh, L. Chen, A review on the utilized machine learning approaches for modeling the dynamic viscosity of nanofluids, Renew. Sust. Energ. Rev. 114 (2019) 109345.

[7] M. Hojjat, S.G. Etemad, R. Bagheri, J. Thibault, Thermal conductivity of non-Newtonian nanofluids: experimental data and modeling using neural network, Int. J. Heat Mass Transf. 54 (2011) 1017–1023, https://doi.org/10.1016/j.ijheatmasstransfer.2010.11.039.
[8] A. Maleki, A. Haghighi, M.I. Shahrestani, Z. Abdelmalek, Applying different types of artificial neural network for modeling thermal conductivity of nanofluids containing silica particles, J. Therm. Anal. Calorim. 144 (2021) 1613–1622.
[9] A. Komeilibirjandi, A.H. Raffiee, A. Maleki, M.A. Nazari, M.S. Shadloo, Thermal conductivity prediction of nanofluids containing CuO nanoparticles by using correlation and artificial neural network, J. Therm. Anal. Calorim. 139 (2020) 2679–2689.
[10] M. Hemmat Esfe, M. Afrand, S. Wongwises, A. Naderi, A. Asadi, S. Rostami, M. Akbari, Applications of feedforward multilayer perceptron artificial neural networks and empirical correlation for prediction of thermal conductivity of Mg(OH)2-EG using experimental data, Int. Commun. Heat Mass Transf. 67 (2015) 46–50, https://doi.org/10.1016/j.icheatmasstransfer.2015.06.015.
[11] S. Zhang, Z. Ge, X. Fan, H. Huang, X. Long, Prediction method of thermal conductivity of nanofluids based on radial basis function, J. Therm. Anal. Calorim. (2019), https://doi.org/10.1007/s10973-019-09067-x.
[12] M.H. Ahmadi, A. Baghban, M. Sadeghzadeh, M. Hadipoor, M. Ghazvini, Evolving connectionist approaches to compute thermal conductivity of TiO2/water nanofluid, Physica A 540 (2020) 122489.
[13] E. Ahmadloo, S. Azizi, Prediction of thermal conductivity of various nanofluids using artificial neural network, Int. Commun. Heat Mass Transf. 74 (2016) 69–75, https://doi.org/10.1016/j.icheatmasstransfer.2016.03.008.
[14] M. Hemmat Esfe, M. Afrand, W.M. Yan, M. Akbari, Applicability of artificial neural network and nonlinear regression to predict thermal conductivity modeling of Al2O3-water nanofluids using experimental data, Int. Commun. Heat Mass Transf. 66 (2015) 246–249, https://doi.org/10.1016/j.icheatmasstransfer.2015.06.002.
[15] A. Naseri, M. Jamei, I. Ahmadianfar, et al., Nanofluids thermal conductivity prediction applying a novel hybrid data-driven model validated using Monte Carlo based sensitivity analysis, Eng. Comput. (2020), https://doi.org/10.1007/s00366-020-01163-z.
[16] M.H. Esfe, S. Saedodin, N. Sina, M. Afrand, S. Rostami, Designing an artificial neural network to predict thermal conductivity and dynamic viscosity of ferromagnetic nanofluid, Int. Commun. Heat Mass Transf. 68 (2015) 50–57.
[17] A. Maleki, M. Elahi, M.E.H. Assad, M. Alhuyi Nazari, M. Safdari Shadloo, N. Nabipour, Thermal conductivity modeling of nanofluids with ZnO particles by using approaches based on artificial neural network and MARS, J. Therm. Anal. Calorim. (2020) 1–12, https://doi.org/10.1007/s10973-020-09373-9.
[18] A. Barati-Harooni, A. Najafi-Marghmaleki, A. Mohebbi, A.H. Mohammadi, On the estimation of viscosities of Newtonian nanofluids, J. Mol. Liq. 241 (2017) 1079–1090, https://doi.org/10.1016/j.molliq.2017.06.088.
[19] A. Barati-Harooni, A. Najafi-Marghmaleki, An accurate RBF-NN model for estimation of viscosity of nanofluids, J. Mol. Liq. 224 (2016) 580–588, https://doi.org/10.1016/j.molliq.2016.10.049.
[20] E. Heidari, M.A. Sobati, S. Movahedirad, Accurate prediction of nanofluid viscosity using a multilayer perceptron artificial neural network (MLP-ANN), Chemom. Intell. Lab. Syst. 155 (2016) 73–85, https://doi.org/10.1016/j.chemolab.2016.03.031.
[21] S. Changdar, S. Saha, S. De, A smart model for prediction of viscosity of nanofluids using deep learning, Smart Sci. 8 (2020) 242–256.

[22] A. Hemmati-Sarapardeh, A. Varamesh, M.M. Husein, K. Karan, On the evaluation of the viscosity of nanofluid systems: modeling and data assessment, Renew. Sust. Energ. Rev. 81 (2018) 313–329, https://doi.org/10.1016/j.rser.2017.07.049.
[23] M. Mehrabi, M. Sharifpur, J.P. Meyer, Viscosity of nanofluids based on an artificial intelligence model, Int. Commun. Heat Mass Transf. 43 (2013) 16–21, https://doi.org/10.1016/j.icheatmasstransfer.2013.02.008.
[24] I. Olanrewaju, M. Amiruddin, A. Rahman, Y. Yaakob, Application of support vector regression and artificial neural network for prediction of specific heat capacity of aqueous nano fluids of copper oxide, Sol. Energy 197 (2020) 485–490, https://doi.org/10.1016/j.solener.2019.12.067.
[25] I. Olanrewaju, M. Amiruddin, A. Rahman, A. Bagudu, Y. Yaakob, A. Saleh, Development of a predictive model for estimating the specific heat capacity of metallic oxides/ethylene glycol-based nano fluids using support vector regression, Heliyon 5 (2019), https://doi.org/10.1016/j.heliyon.2019.e01882.
[26] M.A. Hassan, D. Banerjee, A soft computing approach for estimating the specific heat capacity of molten salt-based nano fluids, J. Mol. Liq. 281 (2019) 365–375, https://doi.org/10.1016/j.molliq.2019.02.106.
[27] I. Olanrewaju, M. Amiruddin, A. Rahman, T.A. Saleh, Modeling and prediction of the specific heat capacity of Al2O3/water nanofluids using hybrid genetic algorithm/support vector regression model, Nano-Struct. Nano-Obj. 17 (2019) 103–111, https://doi.org/10.1016/j.nanoso.2018.12.001.
[28] A. Ern, Discontinuous Galerkin Methods for the Transport Equation, 2010.
[29] M. Jamei, I. Ahmadianfar, I.A. Olumegbon, M. Karbasi, A. Asadi, On the assessment of specific heat capacity of nanofluids for solar energy applications: application of Gaussian process regression (GPR) approach, J. Energy Storage (2020), 102067.
[30] A.B. Çolak, O. Yıldız, M. Bayrak, B.S. Tezekici, Experimental study for predicting the specific heat of water based Cu-Al2O3 hybrid nanofluid using artificial neural network and proposing new correlation, Int. J. Energy Res. 44 (2020) 7198–7215.
[31] H.H. Balla, S. Abdullah, W.M.F. Wanmahmood, M. Abdul Razzaq, R. Zulkifli, K. Sopian, Modelling and measuring the thermal conductivity of multi-metallic Zn/Cu nanofluid, Res. Chem. Intermed. 39 (2013) 2801–2815, https://doi.org/10.1007/s11164-012-0799-z.
[32] M. Hemmat Esfe, S. Wongwises, A. Naderi, A. Asadi, M.R. Safaei, H. Rostamian, M. Dahari, A. Karimipour, Thermal conductivity of Cu/TiO2-water/EG hybrid nanofluid: experimental data and modeling using artificial neural network and correlation, Int. Commun. Heat Mass Transf. 66 (2015) 100–104, https://doi.org/10.1016/j.icheatmasstransfer.2015.05.014.
[33] S.H. Rostamian, M. Biglari, S. Saedodin, M. Hemmat Esfe, An inspection of thermal conductivity of CuO-SWCNTs hybrid nanofluid versus temperature and concentration using experimental data, ANN modeling and new correlation, J. Mol. Liq. 231 (2017) 364–369, https://doi.org/10.1016/j.molliq.2017.02.015.
[34] M. Vafaei, M. Afrand, N. Sina, R. Kalbasi, F. Sourani, H. Teimouri, Evaluation of thermal conductivity of MgO-MWCNTs/EG hybrid nanofluids based on experimental data by selecting optimal artificial neural networks, Phys. E Low-Dimens. Syst. Nanostruct. 85 (2017) 90–96, https://doi.org/10.1016/j.physe.2016.08.020.
[35] M. Hemmat Esfe, A.A. Abbasian Arani, M. Firouzi, Empirical study and model development of thermal conductivity improvement and assessment of cost and sensitivity of EG-water based SWCNT-ZnO (30%:70%) hybrid nanofluid, J. Mol. Liq. 244 (2017) 252–261, https://doi.org/10.1016/j.molliq.2017.08.087.

[36] M. Hemmat Esfe, S. Esfandeh, S. Saedodin, H. Rostamian, Experimental evaluation, sensitivity analyzation and ANN modeling of thermal conductivity of ZnO-MWCNT/EG-water hybrid nanofluid for engineering applications, Appl. Therm. Eng. 125 (2017) 673–685, https://doi.org/10.1016/j.applthermaleng.2017.06.077.
[37] M.H. Esfe, M. Rejvani, R. Karimpour, A.A. Abbasian Arani, Estimation of thermal conductivity of ethylene glycol-based nanofluid with hybrid suspensions of SWCNT–Al2O3 nanoparticles by correlation and ANN methods using experimental data, J. Therm. Anal. Calorim. 128 (2017) 1359–1371, https://doi.org/10.1007/s10973-016-6002-9.
[38] M. Hemmat Esfe, S. Esfandeh, M. Rejvani, Modeling of thermal conductivity of MWCNT-SiO2(30:70%)/EG hybrid nanofluid, sensitivity analyzing and cost performance for industrial applications: an experimental based study, J. Therm. Anal. Calorim. 131 (2018) 1437–1447, https://doi.org/10.1007/s10973-017-6680-y.
[39] A. Shahsavar, M. Bahiraei, Experimental investigation and modeling of thermal conductivity and viscosity for non-Newtonian hybrid nanofluid containing coated CNT/Fe3O4 nanoparticles, Powder Technol. 318 (2017) 441–450, https://doi.org/10.1016/j.powtec.2017.06.023.
[40] M.H. Esfe, A. Alirezaie, M. Rejvani, An applicable study on the thermal conductivity of SWCNT-MgO hybrid nanofluid and price-performance analysis for energy management, Appl. Therm. Eng. 111 (2017) 1202–1210.
[41] S. Kannaiyan, C. Boobalan, F.C. Nagarajan, S. Sivaraman, Modeling of thermal conductivity and density of alumina/silica in water hybrid nanocolloid by the application of artificial neural networks, Chin. J. Chem. Eng. 27 (2019) 726–736, https://doi.org/10.1016/j.cjche.2018.07.018.
[42] M. Hemmat Esfe, A.A. Abbasian Arani, R. Shafiei Badi, M. Rejvani, ANN modeling, cost performance and sensitivity analyzing of thermal conductivity of DWCNT–SiO2/EG hybrid nanofluid for higher heat transfer: an experimental study, J. Therm. Anal. Calorim. 131 (2018) 2381–2393, https://doi.org/10.1007/s10973-017-6744-z.
[43] M. Jamei, R. Pourrajab, I. Ahmadianfar, A. Noghrehabadi, Accurate prediction of thermal conductivity of ethylene glycol-based hybrid nanofluids using artificial intelligence techniques, Int. Commun. Heat Mass Transf. 116 (2020) 104624.
[44] R. Pourrajab, I. Ahmadianfar, M. Jamei, M. Behbahani, A meticulous intelligent approach to predict thermal conductivity ratio of hybrid nanofluids for heat transfer applications, J. Therm. Anal. Calorim. (2020) 1–18.
[45] H. Adun, I. Wole-Osho, E.C. Okonkwo, O. Bamisile, M. Dagbasi, S. Abbasoglu, A neural network-based predictive model for the thermal conductivity of hybrid nanofluids, Int. Commun. Heat Mass Transf. 119 (2020) 104930.
[46] M. Jamei, I. Ahmadianfar, A rigorous model for prediction of viscosity of oil-based hybrid nanofluids, Physica A (2020), 124827.
[47] M. Jamei, I.A. Olumegbon, M. Karbasi, I. Ahmadianfar, A. Asadi, M. Mosharaf-Dehkordi, On the thermal conductivity assessment of oil-based hybrid nanofluids using extended Kalman filter integrated with feed-forward neural network, Int. J. Heat Mass Transf. 172 (2021) 121159.
[48] Y.-J. Huoh, Sensitivity Analysis of Stochastic Simulators with Information Theory, 2013.
[49] A.H. Gandomi, A.H. Alavi, C. Ryan, Handbook of Genetic Programming Applications, Springer, 2015.
[50] H. Akaike, A new look at the statistical model identification, IEEE Trans. Autom. Control 19 (1974) 716–723.
[51] G. Claeskens, N.L. Hjort, Model Selection and Model Averaging, Cambridge Books, 2008.

[52] Z. Zhao, H. Liu, Spectral feature selection for supervised and unsupervised learning, in: Proc. 24th Int. Conf. Mach. Learn, 2007, pp. 1151–1157.
[53] Z.A. Zhao, H. Liu, Spectral Feature Selection for Data Mining, Taylor & Francis, 2012.
[54] S. Haykin, N. Network, A comprehensive foundation, Neural Netw. 2 (2004) 41.
[55] K. Hornik, M. Stinchcombe, H. White, Multilayer feedforward networks are universal approximators, Neural Netw. 2 (1989) 359–366, https://doi.org/10.1016/0893-6080(89)90020-8.
[56] K. Hornik, Approximation capabilities of multilayer feedforward networks, Neural Netw. 4 (1991) 251–257, https://doi.org/10.1016/0893-6080(91)90009-t.
[57] J. Park, I.W. Sandberg, Universal approximation using radial-basis-function networks, Neural Comput. 3 (1991) 246–257, https://doi.org/10.1162/neco.1991.3.2.246.
[58] M. Nait Amar, Modeling solubility of sulfur in pure hydrogen sulfide and sour gas mixtures using rigorous machine learning methods, Int. J. Hydrog. Energy 45 (2020) 33274–33287, https://doi.org/10.1016/j.ijhydene.2020.09.145.
[59] M.S.S. Abujazar, S. Fatihah, I.A. Ibrahim, A.E. Kabeel, S. Sharil, Productivity modelling of a developed inclined stepped solar still system based on actual performance and using a cascaded forward neural network model, J. Clean. Prod. 170 (2018) 147–159.
[60] A. Hemmati-Sarapardeh, M. Nait Amar, M.R. Soltanian, Z. Dai, X. Zhang, Modeling CO2 solubility in water at high pressure and temperature conditions, Energy Fuel 34 (2020) 4761–4776.
[61] D.F. Specht, A general regression neural Network, IEEE Trans. Neural Netw. 2 (1991) 568–576, https://doi.org/10.1109/72.97934.
[62] H. Mehrjoo, M. Riazi, M. Nait Amar, A. Hemmati-Sarapardeh, Modeling interfacial tension of methane-brine systems at high pressure and high salinity conditions, J. Taiwan Inst. Chem. Eng. (2020), https://doi.org/10.1016/j.jtice.2020.09.014.
[63] Y. Feng, D. Gong, X. Mei, N. Cui, Estimation of maize evapotranspiration using extreme learning machine and generalized regression neural network on the China Loess Plateau, Hydrol. Res. 48 (2017) 1156–1168.
[64] R. Singh, V. Vishal, T.N. Singh, P.G. Ranjith, A comparative study of generalized regression neural network approach and adaptive neuro-fuzzy inference systems for prediction of unconfined compressive strength of rocks, Neural Comput. Applic. 23 (2013) 499–506.
[65] O. Kisi, The potential of different ANN techniques in evapotranspiration modelling, Hydrol. Process. 22 (2008) 2449–2460.
[66] Ö. KISI, Generalized regression neural networks for evapotranspiration modelling, Hydrol. Sci. J. 51 (2006) 1092–1105.
[67] G.-B. Huang, Q.-Y. Zhu, C.-K. Siew, Extreme learning machine: theory and applications, Neurocomputing 70 (2006) 489–501.
[68] M. Mahdaviara, M. Nait Amar, M.H. Ghazanfari, A. Hemmati-Sarapardeh, Modeling relative permeability of gas condensate reservoirs: advanced computational frameworks, J. Pet. Sci. Eng. 189 (2020) 106929.
[69] G.-B. Huang, C.-K. Siew, Extreme learning machine with randomly assigned RBF kernels, Int. J. Inf. Technol. 11 (2005) 16–24.
[70] Z.M. Yaseen, R.C. Deo, A. Hilal, A.M. Abd, L.C. Bueno, S. Salcedo-Sanz, M.L. Nehdi, Predicting compressive strength of lightweight foamed concrete using extreme learning machine model, Adv. Eng. Softw. 115 (2018) 112–125, https://doi.org/10.1016/j.advengsoft.2017.09.004.

[71] G. Huang, G.-B. Huang, S. Song, K. You, Trends in extreme learning machines: a review, Neural Netw. 61 (2015) 32–48.
[72] J.S.R. Jang, ANFIS: adaptive network based fuzzy inference system, IEEE Trans. Syst. Man Cybern. 23 (1993) 665–683.
[73] M.H. Esfe, Thermal conductivity modeling of aqueous CuO nanofluids by adaptive neuro-fuzzy inference system (ANFIS) using experimental data, Period. Polytech. Eng. 62 (2018) 202–208, https://doi.org/10.3311/PPch.9670.
[74] M. Rezakazemi, A. Dashti, M. Asghari, S. Shirazian, H2-selective mixed matrix membranes modeling using ANFIS, PSO-ANFIS, GA-ANFIS, Int. J. Hydrog. Energy 42 (2017) 15211–15225, https://doi.org/10.1016/j.ijhydene.2017.04.044.
[75] A. Mosavi, S. Shamshirband, E. Salwana, K. Chau, J.H.M. Tah, Prediction of multi-inputs bubble column reactor using a novel hybrid model of computational fluid dynamics and machine learning, Eng. Appl. Comput. Fluid Mech. 13 (2019) 482–492, https://doi.org/10.1080/19942060.2019.1613448.
[76] A.A.A.A. Alrashed, M.S. Gharibdousti, M. Goodarzi, L.R. de Oliveira, M.R. Safaei, E.P. Bandarra Filho, Effects on thermophysical properties of carbon based nanofluids: experimental data, modelling using regression, ANFIS and ANN, Int. J. Heat Mass Transf. 125 (2018) 920–932, https://doi.org/10.1016/J.IJHEATMASSTRANSFER.2018.04.142.
[77] A. Zendehboudi, X. Li, B. Wang, Utilisation des modèles ANN et ANFIS pour prédire un compresseur à spirale à vitesse variable avec injection de vapeur, Int. J. Refrig. 74 (2017) 473–485, https://doi.org/10.1016/j.ijrefrig.2016.11.011.
[78] S. Jovic, D. Kalaba, P. Zivkovic, A. Virijevic, Potential of adaptive neuro-fuzzy methodology for investigation of heat transfer enhancement of a minichannel heat sink, Physica A 523 (2019) 516–524, https://doi.org/10.1016/J.PHYSA.2019.02.019.
[79] M. Hemmat Esfe, H. Rostamian, S. Esfandeh, M. Afrand, Modeling and prediction of rheological behavior of Al2O3-MWCNT/5W50 hybrid nano-lubricant by artificial neural network using experimental data, Physica A 510 (2018) 625–634, https://doi.org/10.1016/J.PHYSA.2018.06.041.
[80] N. Amanifard, N. Nariman-Zadeh, M.H. Farahani, A. Khalkhali, Modelling of multiple short-length-scale stall cells in an axial compressor using evolved GMDH neural networks, Energy Convers. Manag. 49 (2008) 2588–2594.
[81] A.G. Ivakhnenko, Polynomial theory of complex systems, IEEE Trans. Syst. Man. Cybern. (1971) 364–378.
[82] J. Quinlan, Learning with continuous classes, in: A. Adams, L. Sterling (Eds.), AI'92: Proceedings of the 5th Australian Joint Conference on Artificial Intelligence, 1992, pp. 343–348.
[83] M. Pal, S. Deswal, M5 model tree based modelling of reference evapotranspiration, Hydrol. Process. 23 (2009) 1437–1443.
[84] I.A. Ibrahim, T. Khatib, A novel hybrid model for hourly global solar radiation prediction using random forests technique and firefly algorithm, Energy Convers. Manag. 138 (2017) 413–425.
[85] S.A. Naghibi, H.R. Pourghasemi, B. Dixon, GIS-based groundwater potential mapping using boosted regression tree, classification and regression tree, and random forest machine learning models in Iran, Environ. Monit. Assess. 188 (2016) 44.
[86] S. Lee, J. Im, J. Kim, M. Kim, M. Shin, H. Kim, L.J. Quackenbush, Arctic Sea ice thickness estimation from CryoSat-2 satellite data using machine learning-based lead detection, Remote Sens. 8 (2016) 698.
[87] L. Breiman, Bagging predictors, Mach. Learn. 24 (1996) 123–140.

[88] A. Liaw, M. Wiener, Classification and regression by randomForest, R News 2 (2002) 18–22.
[89] R. Genuer, J.-M. Poggi, C. Tuleau-Malot, Variable selection using random forests, Pattern Recogn. Lett. 31 (2010) 2225–2236.
[90] V.N. Vapnik, The Nature of Statistical Learning Theory, 1995, https://doi.org/10.1007/978-1-4757-2440-0.
[91] V.H. Quej, J. Almorox, J.A. Arnaldo, L. Saito, ANFIS, SVM and ANN soft-computing techniques to estimate daily global solar radiation in a warm sub-humid environment, J. Atmos. Sol. Terr. Phys. 155 (2017) 62–70, https://doi.org/10.1016/j.jastp.2017.02.002.
[92] M. Zounemat-Kermani, A. Mahdavi-Meymand, M. Alizamir, S. Adarsh, Z.M. Yaseen, On the Complexities of Sediment Load Modeling Using Integrative Machine Learning: Application of the Great River of Loíza in Puerto Rico, Elsevier B.V., 2020, https://doi.org/10.1016/j.jhydrol.2020.124759.
[93] C. Ferreira, Gene Expression Programming: Mathematical Modeling by an Artificial Intelligence, Springer, 2006.
[94] I. Ahmadianfar, M. Jamei, X. Chu, Prediction of local scour around circular piles under waves using a novel artificial intelligence approach, Mar. Georesour. Geotechnol. (2019) 1–12.
[95] R. Grbić, D. Kurtagić, D. Slišković, Stream water temperature prediction based on Gaussian process regression, Expert Syst. Appl. 40 (2013) 7407–7414, https://doi.org/10.1016/j.eswa.2013.06.077.
[96] C.E. Rasmussen, Gaussian processes in machine learning, in: Summer Sch. Mach. Learn, Springer, 2003, pp. 63–71.
[97] C.K.I. Williams, C.E. Rasmussen, Gaussian Processes for Machine Learning, MIT Press, Cambridge, MA, 2006.
[98] I. Ahmadianfar, M. Jamei, M. Karbasi, A. Sharafati, B. Gharabaghi, A novel boosting ensemble committee-based model for local scour depth around non-uniformly spaced pile groups, Eng. Comput. (2021) 1–23.
[99] M. Jamei, I. Ahmadianfar, I.A. Olumegbon, A. Asadi, M. Karbasi, Z. Said, M. Sharifpur, J.P. Meyer, On the specific heat capacity estimation of metal oxide-based nanofluid for energy perspective—a comprehensive assessment of data analysis techniques, Int. Commun. Heat Mass Transf. 123 (2021) 105217, https://doi.org/10.1016/j.icheatmasstransfer.2021.105217.
[100] H. Lu, X. Ma, Hybrid decision tree-based machine learning models for short-term water quality prediction, Chemosphere (2020), https://doi.org/10.1016/j.chemosphere.2020.126169.
[101] H.I. Erdal, O. Karakurt, Advancing monthly streamflow prediction accuracy of CART models using ensemble learning paradigms, J. Hydrol. 477 (2013) 119–128, https://doi.org/10.1016/j.jhydrol.2012.11.015.
[102] F. Yousefi, H. Karimi, M.M. Papari, Modeling viscosity of nanofluids using diffusional neural networks, J. Mol. Liq. 175 (2012) 85–90, https://doi.org/10.1016/j.molliq.2012.08.015.
[103] P. Kanti, K.V. Sharma, K.M. Yashawantha, S. Dmk, Experimental determination for viscosity of fly ash nanofluid and fly ash-Cu hybrid nanofluid: prediction and optimization using artificial intelligent techniques, Energy Sources Part A (2021) 1–20.
[104] M. Jamei, I. Ahmadianfar, X. Chu, Z.M. Yaseen, Estimation of triangular side orifice discharge coefficient under a free flow condition using data-driven models, Flow Meas. Instrum. (2020), 101878.
[105] M. Najafzadeh, M. Rezaie Balf, E. Rashedi, Prediction of maximum scour depth around piers with debris accumulation using EPR, MT, and GEP models, J. Hydroinf. 18 (2016) 867–884, https://doi.org/10.2166/hydro.2016.212.

10

Challenges and difficulties in developing hybrid nanofluids and way forward

Zafar Said[a,b,c,*] and Maham Aslam Sohail[a]

[a]Department of Sustainable and Renewable Energy Engineering, University of Sharjah, Sharjah, United Arab Emirates. [b]Research Institute for Sciences and Engineering, University of Sharjah, Sharjah, United Arab Emirates. [c]U.S.-Pakistan Center for Advanced Studies in Energy (USPCAS-E), National University of Sciences and Technology (NUST), Islamabad, Pakistan

**Corresponding author: zsaid@sharjah.ac.ae, zaffar.ks@gmail.com*

Chapter outline

10.1 Introduction 233
10.2 Foam formation 234
10.3 Stability 237
10.4 Safety and environmental concerns 239
10.5 High cost 242
10.6 Degradation of original properties 244
10.7 Increased friction factor, pumping power, and pressure drop 245
10.8 Selecting suitable hybrid nanofluids 248
10.9 Predicting models for thermophysical properties 248
10.10 Challenges and outlook 252
10.11 Conclusion 253
References 254

10.1 Introduction

Hybrid nanofluids are a novel category of nanofluids, achieved by combining two or more nanoparticles consisting of metal-based or metal oxide particles in various base fluids, displaying exceptional improvement in thermophysical characteristics such as thermal conductivity. Scholars and researchers have worked on the potential properties of hybrid nanofluids, such as thermal conductivity, viscosity, density, and heat transfer coefficient in

Hybrid Nanofluids: Preparation, Characterization and Applications. https://doi.org/10.1016/B978-0-323-85836-6.00010-7

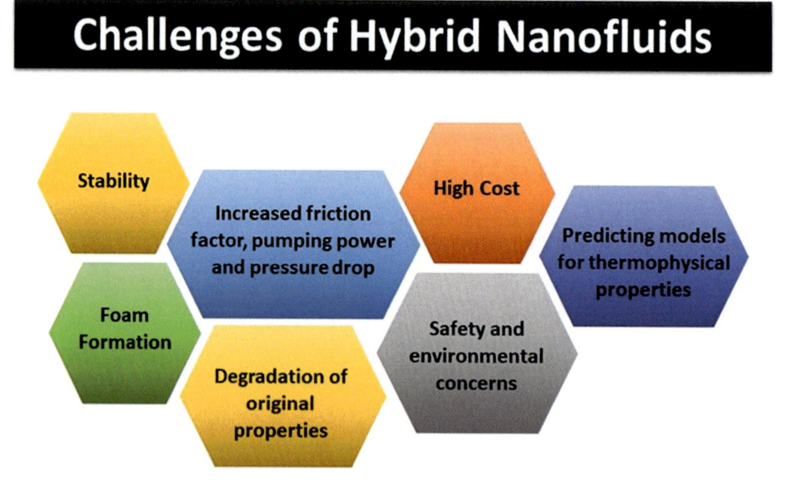

Fig. 10.1 Challenges of hybrid nanofluids.

recent years [1–3]. Still, some critical problems may occur in the commercialization of nanofluids. Several key challenges such as foam formation, stability, high cost, increased friction factor, pressure drop, and pumping power; degradation of original properties; predicting models for thermophysical properties, safety, and environmental concerns; and selection of suitable hybrid materials are presented in Fig. 10.1 and discussed in this chapter [4–6]. Stability is a significant issue for making the research findings more reliable and, especially for practical applications, should be given more research [7–9]. The stability of nanocomposites was achieved with recognized approaches for nanofluids, and for hybrid-based nanofluids, suspending two or more nanoparticles in the base fluid creates positive and negative surface charges, which changes commencing with one molecule then onto the next one [10]. High cost is another significant challenge for various scholars for mono and hybrid nanofluids [11]. The key purpose of hybrid nanofluids is to reduce the production cost with the least effect on the thermal characteristics [12]. Increased viscosity results in increased pumping power and pressure drop that reduces heat transfer rate, which is another serious issue for hybrid nanofluids [13, 14]. These challenges can explain the gradual development of hybrid nanofluids [10, 15].

10.2 Foam formation

Liquid foams containing gas bubbles are tightly filled within a liquid carrier matrix [16]. Combined bubbles are impeded by adding stabilizing agents, such as low molecular mass surfactants, polymers, nanoparticles, or their mixture [17]. Liquid forms are considered significant as templates to produce solid foams, occurring as a by-product in several industrial processes, affecting negative behavior. Several techniques have been discovered to control structural properties such as bubble size distribution and the gas fraction. Such foaming methods have one thing in general that is the production of bubbles within a liquid. They can be either monodisperse or highly polydisperse based on the technique used. Researchers should choose suitable foaming formation technique for industrial applications.

The development of gas/liquid interface in the foam formation techniques creates interfacial tension (γ) defined as an energy input per bubble, expressed as

$$U = 4\gamma r_B^2 \tag{10.1}$$

For interfacial tensions and bubble sizes, the magnitude order is greater than thermal energies (kT), which states that bubble formation is not an impulsive procedure and requires larger energy into a liquid to produce a foam. Foaming techniques are classified into three categories:

- **Physical foaming:** It includes mechanical action (gas sparging, bubbling, and foam generation in porous media) and phase transition (boiling, cavitation, and extrusion).
- **Chemical foaming:** It includes chemical reactions and electrolysis.
- **Biological foaming:** It depends on gas-generating species such as yeast.

The presence of stabilizing agents increases interfacial stresses, affecting bubble generation significantly [18, 19]. Several investigations of foam formation in porous media consider single bubbles. For foam generation under shear, researchers have examined that influence of collective bubbles may function as an essential part in promoting the breakdown of larger bubbles into smaller ones [20]. Porous foam as a heat transfer material has gained vast consideration due to its high surface-to-volume ratio and excellent thermal conductivity [21]. Arbelaez et al. [22] experimentally investigated FC-72 pool boiling heat transfer in aluminum foam and observed that the aluminum foam improved heat transfer with decreased porosity in low-heat flux region whereas the pore density remains the same. It was also noticed that pool boiling heat transfer intensifies with greater pore density at the same porosity. Athreya et al. [23] investigated the effect of aluminum foam alignment and geometry on FC-72's boiling heat transfer properties and observed that higher-pore density of aluminum foam in vertical position deteriorates boiling heat transfer. Nevertheless, as aluminum foam thickness reduces, the lower-pore density of aluminum foam becomes poor and then increases the boiling heat transfer. Raza et al. [24] investigated the impact of foamability on pool boiling critical heat flux with TiO_2/H_2O-based nanofluids. It was observed that by increasing the nanoparticle loading, there is an enhancement in critical heat flux by switching the foamability to the standard wettability regime with reduced surfactant loading or feeble foaming surfactants.

Foamability is governed by the surfactant concentration [25]. Therefore, it is essential to study the impact of surfactant concentration on critical heat flux during nanofluids' boiling. Foamability reduces with the decrease in the concentration of surfactant. Fig. 10.2 represents the bar graph of critical heat flux with sodium dodecyl sulfate (SDS) at different concentrations and the bubble behavior during boiling. It can be seen from the figure that the foaming was significant at the concentration of critical micelle concentration (CMC), CMC/1.4, and CMC/2, concluding that

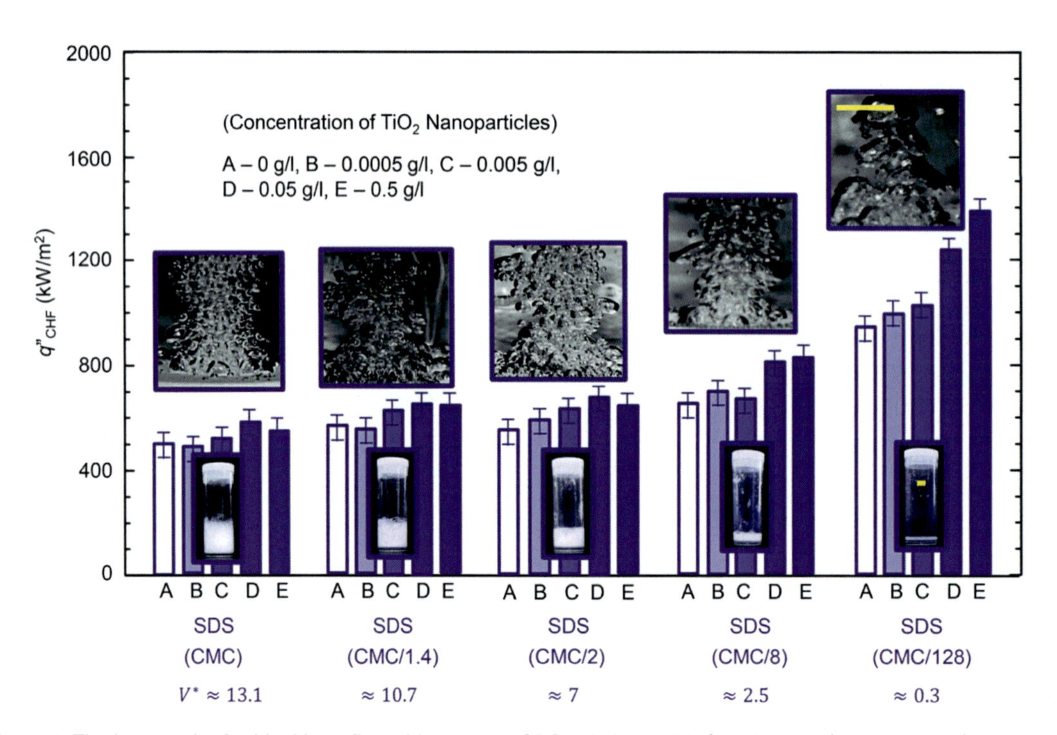

Fig. 10.2 The bar graph of critical heat flux with aqueous SDS solutions with *(shaded bars)* and without *(empty bars)* nanoparticles. During boiling with an aqueous SDS solution without nanoparticles, the bubble dynamics are also displayed in the inset. Adapted with permission from M.Q. Raza, N. Kumar, R. Raj, Effect of foamability on pool boiling critical heat flux with nanofluids, Soft Matter. 15(26) (2019) 5308–5318. Copyright 2019, The Royal Society of Chemistry Lic: (1106955-1).

nanoparticle concentration did not alter the CMC value at different surfactant concentrations. The bubbles dynamics was observed to be considerably changed, and foaming/vapor-crowding was dropped at a low concentration of CMC/128.

10.3 Stability

Research on hybrid nanofluids is becoming a significant concern for many research groups. Researchers introduced the concept of hybrid nanofluids by combining two different kinds of nanoparticles or composites to overcome the challenges encountered by mono nanofluids [26, 27]. As seen from Fig. 10.3, different kinds of hybrid nanofluids consist of nanocomposites, different types of nanoparticles, the addition of nanoparticles to present nanofluids, and suspensions of two mixed nanofluids [28]. Hybrid nanofluids are novel heat transfer fluids for potential industrial applications due to their remarkable thermophysical properties [29–31].

Stability is one of the significant challenges that are responsible for the overall heat transfer improvement of the hybrid nanofluids due to strong Van der Waals forces of nanoparticles, which causes agglomeration [32, 33]. Appropriate dispersion of nanoparticles in the base fluid results in enhanced thermal conductivity of nanofluids [26]. Formation of clusters increases the thermal resistance and pumping power by tightening the flow

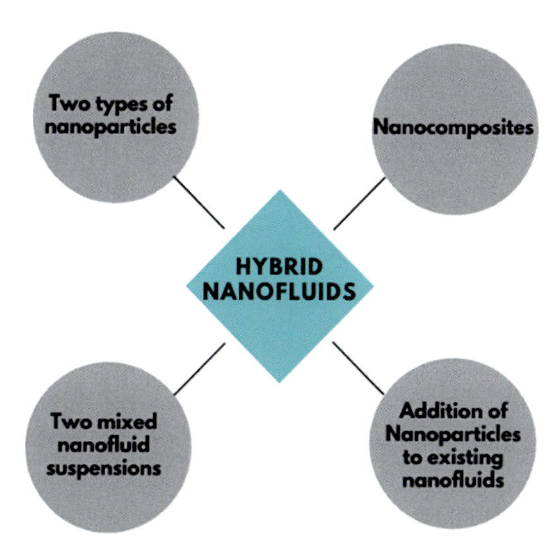

Fig. 10.3 Types of different hybrid nanofluids.

passageways [34]. DLVO theory, known as Derjaguin, Landau, Verway, and Overbeek theory, defines the stability in solution determined by Van der Waals' attractive and electrical double layer forces existing between particles because of Brownian motion [35]. If the attractive force is greater than the repulsive force, the particles will have a collision, and the stability will be poor [36]. On the other hand, if the particles have repulsive forces greater than attractive, the suspension will be in a stable state. Better and long-term stability of nanofluids is the fundamental requirement in industrial uses. Therefore, studies on stability are considered a critical issue that affects the nanofluids' properties, and it is important to determine the parameters affecting nanofluids' stability [37].

Researchers investigated the stability impact of pH value, different particles and base fluids, ultrasonication, and magnetic stirring on the pumping power [38, 39]. In most experimental studies, the stability period of hybrid nanofluids prepared by a two-step method could not surpass 15 days. Akhgar and Toghraie [40] experimentally investigated TiO_2-MWCNTs/H_2O-EG hybrid nanofluids by analyzing the stability of TiO_2/H_2O and MWCNTs/H_2O nanofluids. It was observed that nanofluid containing MWCNTs/H_2O caused sedimentation after 48 h, and TiO_2/H_2O-based nanofluids displayed better dispersion with $pH = 9$ after 48 h. Surfactant CTAB of different concentrations was introduced to enhance the stability of MWCNT-based nanofluids, and then they observed excellent stability even after 72 h with a low concentration of surfactant. Therefore, a detailed analysis of the stability can be promising for selecting nanomaterial. Asadi et al. [41] investigated $Mg(OH)_2$ + MWCNT/engine Oil to maintain stability using the zeta potential analysis technique. This approach was observed to be impractical and unrealistic for these dark colored nanofluids. The literature has shown the following values for zeta potential in different categories, presented in Table 10.1 [42]. Jha and Ramaprabhu [43] investigated

Table 10.1 Distinct categories of stability.

Remarkable stability	Zeta potential value > 60 mV
Good stability	45 mV $<$ zeta potential value < 60 mV
Satisfactory stability	30 mV $<$ zeta potential value < 45 mV
Weak stability	Zeta potential value < 30 mV

Cu-MWCNT/DI-H_2O based nanofluids' stability and observed good stability of about 20 days without any surfactant. Ali et al. [44] observed the stability of CNTs-Fe_3O_4/water hybrid nanofluids and found excellent stability of 60 days.

Several researchers investigated methods to maintain stability, such as pH control, the best possible ultrasonication treatment, and the use of antisedimentation agents and surfactants [45]. For mono nanofluids, the literature has described that the nanofluids' thermal conductivity reduces with time, but it is different for hybrid nanofluids due to the different combinations and bonding of nanoparticles or nanocomposites [46]. The addition of surfactant or dispersant in the two-phase system is a straightforward and cost-effective method to improve stability. Surfactants can affect a system's surface properties in a minor quantity, consisting of a hydrophobic tail section, usually a long-chain hydrocarbon, and a hydrophilic polar head group, enhancing the wettability. Apart from the advantages, surfactants might cause various difficulties that the surfactants may infect the heat transfer modes, producing foams while heating and cooling techniques in heat exchangers. Thermal resistance may widen between nanoparticles and base fluid due to attached surfactant molecules on the nanoparticle's surface, reducing the effective thermal conductivity [47]. Table 10.2 provides information about the investigations on the stability of hybrid nanofluids.

10.4 Safety and environmental concerns

The superior activity of hybrid nanocomposites may cause a serious concern from the perspective of safety and does not always positively impact human health and the environment. Researchers should pay attention to the influence of health and the ecosystem [54]. Nevertheless, all technologies have negative outcomes as well, and by overcoming these challenges, we will be able to utilize the remarkable properties of hybrid nanocomposites for industrial applications [55, 56].

Hybrid nanocomposites can be made from various nanomaterials such as pure metals, metal and nonmetal oxides, and carbon-based nanomaterials [57]. Nanocomposites possess dimensions up to quite a few dozens of nanometers: they look like big complexes of protein molecules in terms of dimension. They are different from proteins because of their chemical composition, shape and size, aggregation, density, and physiochemical

Table 10.2 Investigations on the stability of hybrid nanofluids.

Authors	Hybrid nanofluids	Method	Surfactants used	Performance	Drawback	Refs.
Tiwari et al.	CeO_2 + MWCNT (80:20)/H_2O	Two-step, zeta potential	SDBS, SDS, CTAB, DDC, GA, and PVP	CTAB surfactant shows better stability impact up to 30th day, and after 30th day, the SDBS surfactant shows the maximum stability	Reduction in thermal conductivity for higher amounts of surfactant	[38]
Li et al.	SiC + MWCNTs/EG	Mechanical milling, zeta potential	Polyvinyl pyrrolidone (PVP-K30) and hexane	Excellent stability was observed	High temperature decreased the zeta potential value for 1 wt%	[48]
Cakmak et al.	rGO + Fe_3O_4 + TiO_2/EG	Sol-gel, zeta potential	–	Remarkable stability was observed as compared with rGO	Increase in mass concentration from 0.01% to 0.25% resulted in lower zeta potential	[49]
Xian et al.	GnPs + TiO_2/EG + DW	Two-step, sedimentation, zeta potential, and absorbency test	SDC, CTAB, PVP, Triton X-100, SDS, and SDBS	CTAB represented excellent stability with little sedimentation up to 40 days	Hybrid nanoparticles showed less enhancement in thermal conductivity as compared with mono nanofluid at 60°C	[50]
Said et al.	CNF, rGO, F-CNF, and rGO + F-CNF in deionized water	Two-step, zeta potential	–	F-CNF shows the best stability without any visible sedimentation up to 180 days and CNF would be ideal for high-temperature applications	Lower volume fractions have an insignificant effect on density	[9]
Kumar et al.	Al_2O_3 + SiO_2/H_2O	Two-step, UV-Vis spectrophotometer, zeta potential, and photograph capturing techniques	–	Remarkable stability was observed with 0.4 and 0.6 wt%	Nanofluids with 0.2 wt% show poor stability	[51]
Ma et al.	Al_2O_3 + TiO_2/H_2O Al_2O_3 + CuO/H_2O	Two-step, UV-Vis spectrophotometer, and TEM	SDS, PVP, and CTAB	PVP is observed to be the best surfactant with excellent stability	As the PVP concentration increases, the viscosity increases	[52]
Gulzar et al.	Al_2O_3 + TiO_2/therminol-55 oil	Two-step, zeta potential	–	Remarkable stability was observed up to 7 days of preparation for lower volume concentration	–	[53]

properties. The physicochemical properties of nanosized materials can differ from bulk materials [58]. Hybrid nanofluids are defined as a suspension of nanocomposites in a base fluid. The most widely used and available base fluid is water. It has several advantages such as availability and being nonhazardous, safer, nonflammable, and easy to handle. The significant environmental impacts of nanofluids will be mainly from three parameters:

(1) Type of nanoparticles used (chemical, physical, and environmental properties)
(2) Volume concentration of nanocomposites in the base fluid
(3) Nanofluids' preparation method

Hybrid nanocomposites have been applied in different volume fractions in various base fluids [49]. Esfahani et al. [59] observed maximum improvement in ZnO-Ag/H_2O hybrid nanofluids' thermal conductivity at the highest 2 vol% particle loading. Dalkılıç et al. [60] observed improvement in thermal conductivity of CNT-SiO_2/H_2O hybrid nanofluids at the highest volume concentration of 2 vol%. Large-scale production and utilization of carbon nanotubes (CNTs) have inevitably led to the exposure risks for human beings and animals. It can enter through inhalation, ingestion from food and water, and absorption through scars [61]. Similarly, titanium dioxide (TiO_2) has been widely employed as a promising nanomaterial due to its excellent stability and thermophysical properties. The large production, consumption, and uncontrolled disposal of TiO_2 nanoparticles will inevitably lead to their release into the environment, where they can destroy ecosystems [62]. In this regard, Cakmak et al. [49] noticed an enhancement in thermal conductivity of rGO-Fe_3O_4-TiO_2/EG hybrid nanofluid for 0.25 wt%. These studies noticed that the thermal conductivity increases as the nanoparticle concentration increases, but it should not exceed a certain level of nanoparticle concentration to not have adverse environmental impacts and increased viscosity, which leads to larger pumping power and higher pressure drop, and corrosion in heat exchangers [63, 64].

A life cycle approach must be considered from production to utilization, to manage the potential risk posed by engineered hybrid nanocomposites. Life cycle assessment (LCA) is defined as a technique of evaluating the source management and environmental effects attributable to the whole lifespan of a product, from raw material extraction, to use and end-of-life treatment and final disposal [65]. It needs to be prepared for different nanoparticles' groups to predict the real risk of nano waste. Furthermore, it is difficult to attain a full-spectrum LCA for hybrid nanocomposites due to inadequate information about the detailed system.

By introducing environmentally friendly material, we can simply minimize or reduce nanomaterials' environmental impacts and toxicities. For instance, by decreasing the particle size of nanomaterials, their hazard levels will increase due to the improved activity and surface area. Some natural materials such as silica, iron oxides, alumina, and many others possess lower environmental impacts, so they do not require a synthetic procedure, resulting in lower manufacturing requirements. The interaction between nanoparticles and base fluid should be studied, as it can have further hazardous or perilous environmental impacts. The nanoparticles' type, volume concentration, and preparation technique should be considered while selecting the nanocomposites to prepare hybrid nanofluids [9]. There is a limited investigation in this field that scholars must pay more attention to this subject.

10.5 High cost

The manufacturing amount of hybrid nanofluids varies on the production technique, such as single-step and two-step methods. The devices used in these methods are expensive and complicated. Many investigations have been reported on the economic analysis of hybrid nanofluids. Esfe et al. [66] observed that hybrid nanofluids are more cost-effective than nanofluids. They analyzed nanofluids' cost with its thermal conductivity ratio (*TCR*) by using price-performance factor (*PPF*) expressed as

$$PPF = \frac{TCR}{\text{Price}} \times 100 \tag{10.2}$$

Price-performance analysis (PPA) represents facts of the cost as compared with the nanofluids' thermal performance. The main objective of this assessment is to define the cost ratio to the relative thermal conductivity cost of nanofluids and evaluate their efficiency. The minimum price and the maximum relative thermal conductivity of nanofluids are considered cost-effective nanofluids. Alirezaie et al. [67] observed that the metal oxide-based nanofluids show a greater PPF value and are more economically

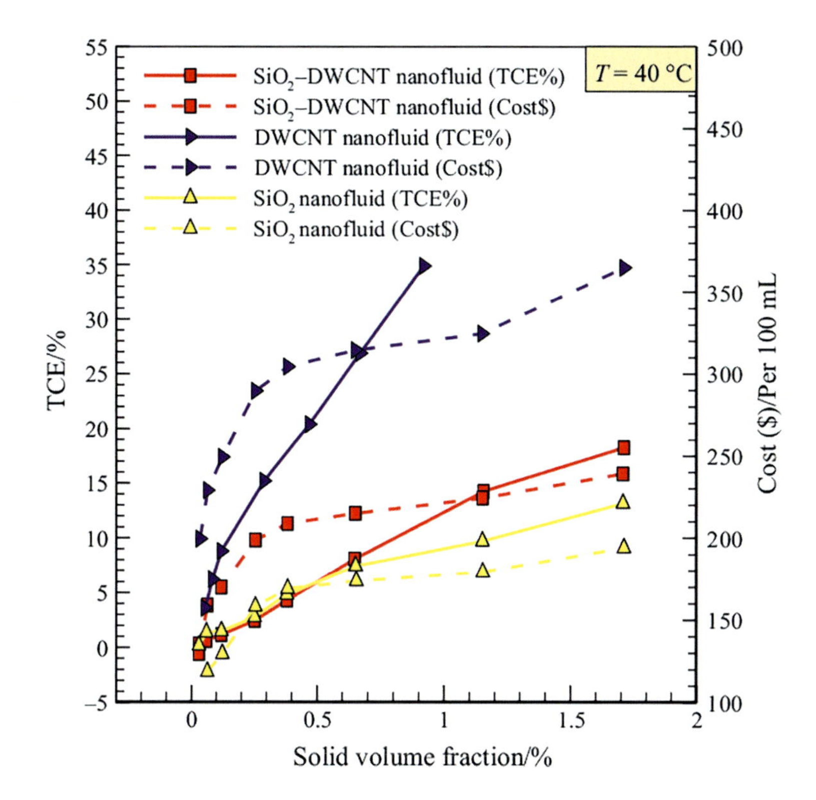

Fig. 10.4 Mono vs hybrid nanofluids in terms of price and TCE. Copyright from Springer LIC: 5030170352485 M. Esfe, et al., ANN modeling, cost performance and sensitivity analyzing of thermal conductivity of DWCNT–SiO2/EG hybrid nanofluid for higher heat transfer: an experimental study, J. Therm. Anal. Calorim. (2017).

effective than metal-based nanofluids. Esfe et al. [68] compared mono and hybrid nanofluids in terms of cost and thermal conductivity improvement. As it is observed from Fig. 10.4, the carbon nanotube-based nanofluids displayed excellent thermal conductivity, but it limits their usage due to high cost. Based on the figure, it was observed that hybrid nanofluids can be a reliable candidate for commercial applications due to excellent efficiency performance compared with their cost with mono nanofluids. Singh et al. [69] investigated the economic analysis of shell and tube condenser using hybrid nanofluids as coolant and observed that Al_2O_3 + MWCNT-based hybrid nanofluids yield maximum cost savings of up to 11.1%. Table 10.3 shows the cost of different nanomaterials with particle configuration taken from NanoAmor (United States) to help researchers by discovering the combinations that prove excellent PPF and by enhancing the production and dispersion approach. It is evident that further investigations are required to minimize the production cost for the concerned applications.

Table 10.3 Costs of the different nanomaterials with particle configuration (NanoAmor).

Nanomaterial	Average particle size	Cost for 100 and 500 g
Aluminum oxide (Al_2O_3)	30 nm	\$70/100 g \$140/500 g
Silicon oxide (SiO_2)	20 nm	\$110/100 g \$180/500 g
Titanium oxide (TiO_2)	5 nm	\$82/100 g \$154/500 g
Iron (II, III) oxide (Fe_3O_4)	30 nm	\$90/100 g \$180/500 g
Iron (III) oxide (Fe_2O_3)	30 nm	\$80/100 g \$160/500 g
Zinc oxide (ZnO)	30 nm	\$85/100 g \$170/500 g
Iron (Fe)	40 nm	\$290/100 g \$1090/500 g
Copper (II) oxide (CuO)	30–50 nm	\$80/100 g \$260/500 g
Multiwalled carbon nanotubes (MWCNTs)	10–30 nm	\$380/1 kg
—COOH functionalized multiwalled carbon nanotubes (MWCNT-COOH)	10–30 nm	\$448/1 kg

10.6 Degradation of original properties

Destabilization or degradation of nanoparticles results in decreased heat transfer performance, which considers the nanofluid insubstantial [70]. The nanoparticles' agglomeration causes obstruction and the degradation of hybrid nanofluids' thermal conductivity over long term [71]. Studies have shown that the nanocomposites' accumulation is time dependent and grows with time, and thermal conductivity reduces sharply with time. The thermal conductivity of mono and hybrid nanofluids increases with volume concentration; however, the large increment in particles leads to their deposition [9]. Researchers have not yet investigated the favorable stability period of hybrid nanofluids. Several researchers and engineers have made efforts to overcome long-term stability issues that influence hybrid nanofluids' properties for practical applications [49, 64].

- Philip and Anushree [72] investigated the long-term stability of water-based metal oxide nanofluids using different stability techniques. The time-dependent absorbance changes were observed to be higher in α-Al_2O_3 and TiO_2 nanofluids, indicating significant aggregation of nanoparticles.

- Akhgar and Toghraie [40] studied TiO_2 + MWCNTs/H_2O + EG hybrid nanofluids for obtaining long-term stability.
- Jha and Ramaprabhu [43] observed stability of 20 days' time span of Cu + MWCNT/H_2O hybrid nanofluid without using any surfactant.

With an increase in the long-term stability of hybrid nanofluids, there will be an increment in the heat transfer rate in industrial applications, reduced processing time, increased life-span of equipment, and saving in energy [57].

10.7 Increased friction factor, pumping power, and pressure drop

Hybrid nanofluids show excellent performance in heat transfer devices because of superior thermal properties [38, 55]. However, in some cases, it gives rise to the viscosity that surges pumping power and friction factor [39]. Pumping power depends on fluid viscosity and density. An inexorable increase in friction factor is a foremost concern for hybrid nanofluids [31]. Sundar et al. [73] investigated Fe_3O_4 + MWCNT hybrid nanofluids at different concentrations and observed an increment of 1.18 times in friction factor for a 0.3% volume concentration at Reynolds number equal to 22,000. Yarmand et al. [27] investigated GNP-Ag hybrid nanofluids and observed an increase of 1.08 times in friction factor with the consequence of increased pumping power at 0.1% concentration and Reynolds number of 17,500, as shown in Fig. 10.5A. Huminic and Huminic [74] investigated the pumping power effect of hybrid nanofluids on elliptical tube performance. It was observed that the pumping power is reduced for lower concentrations (0.1%–0.2%) of hybrid nanocomposites, while for higher concentrations, the pumping power is increased. Eshgarf and Afrand [75] investigated the rheological behavior of COOH-MWCNTs + SiO_2/EG + H_2O hybrid nano-coolant and observed enhancement of about 20,000% in viscosity that augmented the pumping power. Takabi and Shokouhmand [76] investigated the effect of friction factor on Al_2O_3 + Cu/H_2O hybrid nanofluids and observed a 13.76% rise in friction factor when compared with the base fluid, as displayed in Fig. 10.5B. Suresh et al. [77] observed Al_2O_3 + Cu/H_2O hybrid nanofluids and found a 16.97% increment in friction factor when compared with the base fluid, as shown in Fig. 10.5C. Nabil et al. [78] observed an

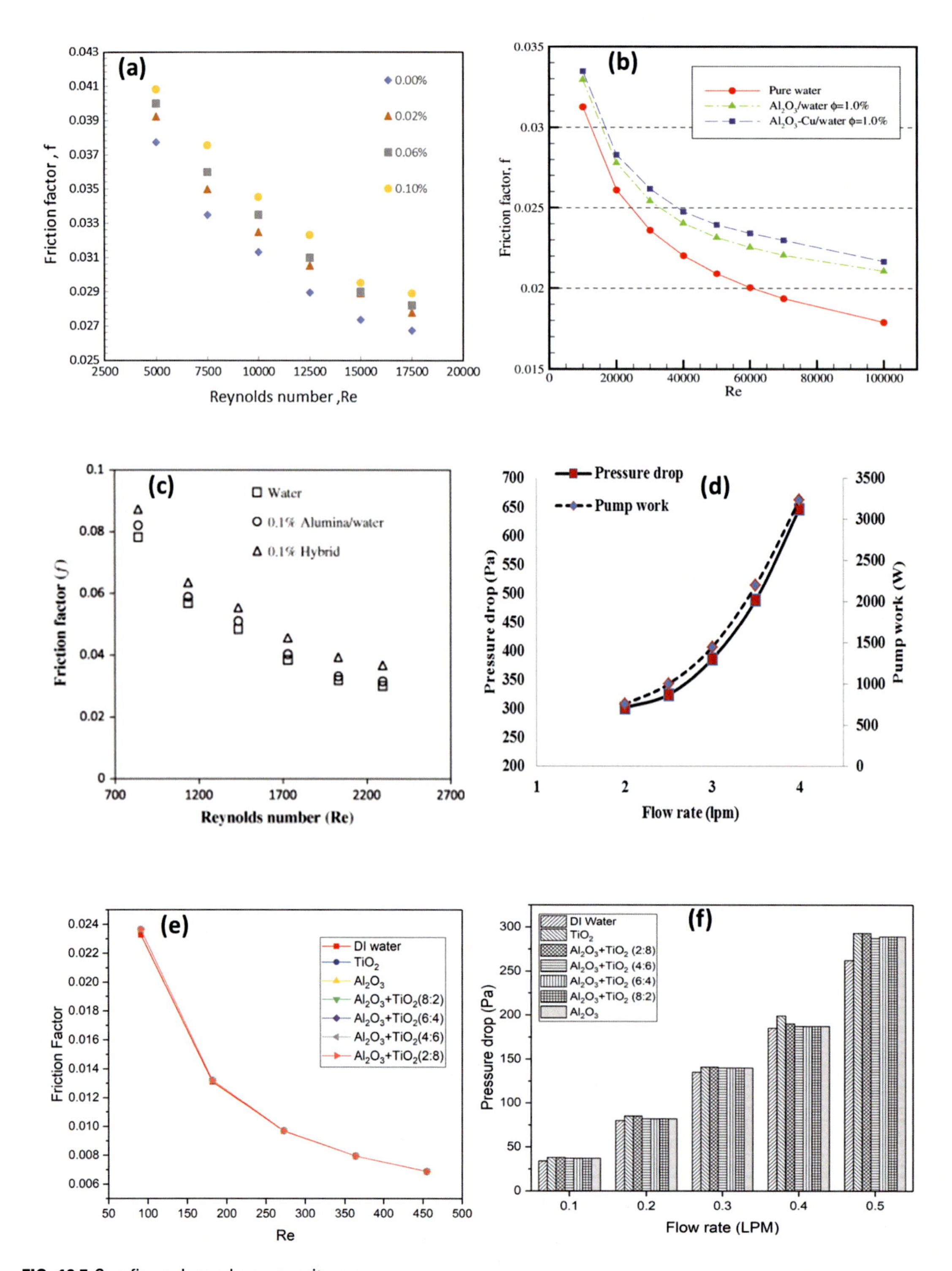

FIG. 10.5 See figure legend on opposite page.

increment in 7%–8% in friction factor of $TiO_2 + SiO_2/EG + H_2O$ hybrid nanofluids at a Reynolds number from 2000 to 10,000. Pressure drop and pumping power are related to each other. With an increment in the nanoparticle volume concentration, the heat transfer rate is increased, as well as the addition of further nanoparticles, leading to higher viscosity, which results in increased pressure drop. This deteriorates the overall efficiency of the system due to agglomeration [79]. Sarkar et al. [80] experimentally studied the heat transfer and pressure drop using hybrid nanofluid containing $Al_2O_3 + MWCNT/H_2O$ in a plate heat exchanger and observed increment in pressure drop with increasing flow rate, as shown in Fig. 10.5D. Sarkar et al. [81] studied friction factor and pressure drop characteristics of $Al_2O_3 + TiO_2/H_2O$ hybrid nanofluids and observed that both pressure drop and friction factor increase with an increased volume fraction of nanoparticles and flow rate, as displayed in Fig. 10.5E and F.

Further experiments have to concentrate on the overall advantages of applying these nanofluids in increasing heat transfer and the downsides presented by the rise in pumping power [31, 82]. An overall energy efficiency parameter is required to be pondered. Pipes made from superhydrophobic materials in solar systems can lessen the impacts of increased friction factor [83]. To lower the pressure drop in thermal systems, simultaneously boosting the heat transfer rate using hybrid nanofluids is one of the major targets for investigators to be accomplished.

Fig. 10.5 (A) Variation of the experimental data for friction factor with different particle concentrations of GNP + Ag hybrid nanofluids. (B) Friction factor against Reynolds number for Al_2O_3/H_2O nanofluids and $Al_2O_3 + Cu/H_2O$ hybrid nanofluids at 1% concentration. (C) Variation of friction factor with Reynolds number for mono and hybrid nanofluids containing Al_2O_3 and $Al_2O_3 + Cu$ at 0.1% concentration. (D) Variation of pressure drop with flow rate. (E) Variation of friction factor with Reynolds number. (F) Variation of pressure drop with mass flow rate at 40°C. (A) Adapted with permission from H. Yarmand, et al., Graphene nanoplatelets–silver hybrid nanofluids for enhanced heat transfer, Energy Convers. Manag. 100 (2015) 419–428, copyright 2015, Elsevier (Lic#5014640303990). (B) (Lic#1104233-1), copyright 1990, World Scientific Publishing B. Takabi, H. Shokouhmand, Effects of Al2O3–Cu/water hybrid nanofluid on heat transfer and flow characteristics in turbulent regime, Int. J. Mod. Phys. C 26(04) (2015) 1550047. (C) Adapted with permission from S. Suresh, et al., Effect of Al2O3–cu/water hybrid nanofluid in heat transfer, Exp. Thermal Fluid Sci. 38 (2012) 54–60, copyright 2012, Elsevier (Lic#5014631308730). (D) Adapted with permission from A. Bhattad, J. Sarkar, P. Ghosh, Discrete phase numerical model and experimental study of hybrid nanofluid heat transfer and pressure drop in plate heat exchanger, Int. Commun. Heat Mass Transfer 91 (2018) 262–273, copyright 2018, Elsevier (Lic#5014640614931). (E, F) Adapted with permission from V. Kumar, J. Sarkar, Numerical and experimental investigations on heat transfer and pressure drop characteristics of Al2O3-TiO2 hybrid nanofluid in minichannel heat sink with different mixture ratio, Powder Technol. 345 (2019) 717–727, copyright 2019, Elsevier (Lic#5014640171621).

10.8 Selecting suitable hybrid nanofluids

The selection of nanocomposites for hybrid nanofluids still continues with an open discussion in the literature. The main objective of utilizing hybrid nanofluids is to increase thermal conductivity and reduce viscosity [30]. For the synthesis, appropriate selection of the manufacturing process and hybrid nanoparticles must take place by investigating their possible synergy and thermophysical characteristics like Al_2O_3 + MWCNT, Cu + TiO_2, CNT-Cu, CNT, Cu, etc. [10, 29]. Each nanoparticle has its promising properties; by comparing these nanofluids' effectiveness, their behavior should be investigated to make a suitable selection. Moldoveanu et al. [84] observed that by applying similar conditions on two hybrid nanofluids containing water, alumina, silica, and titania, alumina + silica hybrid nanofluids showed superior heat transfer performance than alumina + titania because of synergetic effects of alumina + silica. Optimized investigations of nanofluids for their size and shape, type, temperature, stability, and synthesis techniques must be considered [12, 13]. In regard to the parameters like hazardous chemical reactions, solubility, and specific heat, PPF is a factor that specifies the excellent combinations of nanoparticles and base fluid. Labib et al. [85] investigated the CNTs + Al_2O_3/H_2O hybrid nanofluids and observed an enhancement in the convective heat transfer coefficient due to the shear-thinning effect of CNTs.

Suitable selection of base fluids is dependent on the specificity of a particular technical area. Water- and EG-based nanofluids are widely used in cooling applications, whereas oil-based nanofluids are engaged in manufacturing procedures for lubrication or high-temperature use [34]. For the base fluid, this desiderate depends on the ultimate functioning of the nanofluid and the temperature interval required for a specific application [86]. Suitable selection of hybrid nanomaterials provides a reliable base for the work and allows us to provide significant details in the area of heat transfer.

10.9 Predicting models for thermophysical properties

The dilemma of general correlation for the thermophysical properties of mono nanofluids has not however settled. Similar problems have been reported for hybrid nanofluids [11]. Thermal conductivity is a valuable property of heat transfer fluid [87, 88]. Following are the classical models by different researchers that were unsuccessful to estimate the thermal conductivity of hybrid nanofluids in acceptable inaccuracy range:

- Hamilton and Crosser [89] considered the effect of nanoparticle volume fraction and the thermal conductivity of both the nanoparticles and base fluid by the following expression:

$$k_{eff} = k_{bf}\left[\frac{k_{nc} + (n-1)k_{bf} - (n-1)\varphi(k_{bf} - k_{nc})}{k_{nc} + (n-1)k_{bf} + \varphi(k_{bf} - k_{nc})}\right] \tag{10.3}$$

- Maxwell [90] considered the effect of thermal conductivity on the particle volume concentration by the following expression:

$$k_{eff} = k_{bf}\left[\frac{k_{nc} + 2k_{bf} + 2\varphi(k_{nc} - k_{bf})}{k_{nc} + 2k_{bf} - \varphi(k_{nc} - k_{bf})}\right] \tag{10.4}$$

- Nan et al. [91] modified the Maxwell model by considering the influence of interfacial layer resistance with thermal conductivity and particle loading as the expression below:

$$k_{eff} = k_{bf}\left[\frac{2k_{bf} + (1+2\alpha)k_{nc} + 2\varphi[(1-\alpha)k_{nc} - k_{bf}]}{2k_{bf} + (1+2\alpha)k_{nc} - \varphi[(1-\alpha)k_{nc} - k_{bf}]}\right] \tag{10.5}$$

- Yu and Choi [92] expressed the following model by taking into consideration the interfacial nanolayer thickness "γ":

$$k_{eff} = k_{bf}\left[\frac{k_{pe} + 2k_{bf} + 2(k_{pe} - k_{bf})(1+\beta)^3\varphi}{k_{pe} + 2k_{bf} - (k_{pe} - k_{bf})(1+\beta)^3\varphi}\right] \tag{10.6}$$

$$k_{pe} = k_p\left[\frac{\left[2(1-\gamma) + (1+\beta)^3(1+2\gamma)\right]\gamma}{-(1-\gamma) + (1+\beta)^3(1+2\gamma)}\right] \tag{10.7}$$

Researchers have introduced the models that are presumed within the boundary that draws in their studies on which an analysis had already been pondered in the subdivision of thermophysical properties. For hybrid nanofluids, the authors have developed complex and technical models. There is an urgent need to provide simple models for hybrid nanofluids. Chougule and Sahu [93] represented the current thermal conductivity model developed by Kumar et al. [94], which is mainly for an individual kind of nanoparticle suspension, expressed as

$$k_{eff} = k_m\left[1 + \frac{k_p \varepsilon r_m}{k_m(1-\varepsilon)r_p}\right] \tag{10.8}$$

$$k_{eff} = k_m \left[1 + \frac{k_{p1}\varepsilon_1 r_m}{k_m \left(1 - (\varepsilon_1 + \varepsilon_2) r_{p1} + \frac{k_{p2}\varepsilon_2 r_m}{k_m \left(1 - (\varepsilon_1 + \varepsilon_2) r_{p2} \right]} \right.} \right. \tag{10.9}$$

The upgraded model is favorable for hybrid nanofluids and considers the effect of surface area, size, and particle loading and thermal conductivity of single material. Esfe et al. [95] proposed a complicated analysis model to calculate the thermal conductivity of hybrid nanofluid containing Cu + TiO_2/H_2O + EG by considering the effect of particle concentration and temperature using an artificial neural network (ANN), and it is expressed as

$$\frac{k_{nf}}{k_{bf}} = 1.07 + 0.000589T + \frac{-0.000184}{T\varphi} + 4.44T\varphi \cos(6.11 + 0.00673T + 4.41T\varphi - 0.0414 \sin(T)) - 32.5\varphi \tag{10.10}$$

Esfe et al. [96] and Rostamian et al. [97] discovered complicated expressions for thermal conductivity and fitted well with the experimental results by considering temperature and particle concentration impacts, expressed by the following expressions:

$$\frac{k_{nf}}{k_{bf}} = 0.8707 + 0.179\varphi^{0.179} \exp(0.09624\varphi^2) + \varphi T \times 8.883 \times 10^4 + \varphi^{0.252} T \times 4.435 \times 10^{-3} \tag{10.11}$$

$$\frac{k_{nf}}{k_{bf}} = 1 + (0.04056 \times \varphi T) - \left(0.003252 \times (\varphi T)^2\right) + \left(0.0001181 \times (\varphi T)^3\right) - \left(0.000001431 \times (\varphi T)^4\right) \tag{10.12}$$

Yildiz et al. [98] presented the following expression for Al_2O_3 + -SiO_2/H_2O hybrid nanofluids and concluded that the theoretical models in the literature might underestimate the effect of using nanofluids, especially at higher volume concentrations:

$$k_{hnf} = k_{bf}(1.0563 + 0.0410\phi) \tag{10.13}$$

Further thermal conductivity models are needed to understand and utilize hybrid nanofluid's thermal conductivity. With the better thermal conductivity model, researchers and engineers can predict hybrid nanofluids' effectiveness as heat transfer fluid before fully being used in practical applications.

Unlike thermal conductivity, viscosity is also an essential property for heat transfer enhancement. Einstein [99] was the first one who proposed a viscosity model with the restrictions of spherical shaped nanoparticles with less than 0.02 vol%, expressed as below:

$$\mu_{nf} = (1 + 2.5\varphi_{nc})\mu_{bf} \tag{10.14}$$

Brinkman [100] proposed the extended version of Einstein with the restriction of particle volume concentration from 0.02 vol% to less than 4 vol%, expressed as below:

$$\mu_{nf} = \frac{\mu_{bf}}{(1 - \varphi_{nc})^{2.5}} \tag{10.15}$$

Batchelor [101] presented the existing Einstein model by presenting Brownian motion expressed as

$$\mu_{nf} = \left(1 + 2.5\varphi_{nc} + 6.2\varphi_{nc}^2\right)\mu_{bf} \tag{10.16}$$

Corcione [102] predicted the viscosity of hybrid nanofluid and noticed that the following model could not be applied for the estimation of viscosity at larger volume concentration ($\varphi > 0.01$ vol%).

$$\frac{\mu_{hnf}}{\mu_{bf}} = \left[\frac{1}{1 - 34.87 \times (d_f)^{0.3}(d_{p1})^{-0.3}(\varphi_1)^{1.03} + (d_{p2})^{-0.3}(\varphi_2)^{1.03}}\right] \tag{10.17}$$

$$d_f = 0.1\left(\frac{6M}{N\pi\rho_{bf}}\right)^{\frac{1}{3}}$$

where d_f, M, d_p, and N are the molecular diameter of base fluid, molecular weight of fluid, particle diameter, and Avogadro's number, respectively.

Esfe et al. [103] presented a complicated correlation to predict the dynamic viscosity for different volume concentrations and noticed a good correspondence between experimental data and the proposed model, expressed as

$$\frac{\mu_{nf}}{\mu_f} = \left(1 + 32.795\varphi_p - 7214\varphi_p^2 + 714{,}600\varphi_p^3 - 0.1941 \times 10^8\varphi_p^4\right) \tag{10.18}$$

Available viscosity models presented in the literature are considered obsolete and undesirable. Researchers are doing their best toward the formulation of universal correlation with great accuracy by considering the following factors:

- Temperature
- Particle shape

- Base fluid
- pH value
- Clustering effect
- Nanoparticle concentration
- Brownian motion of nanoparticles

10.10 Challenges and outlook

The research initiated with micro-sized fluids has advanced to hybrid nanofluids. Hybrid nanofluids exhibit promising characteristics through various phases such as nanoparticles' dispersion, addition of surfactant or dispersant, pH adjustment, and particle coatings. The key challenges such as stability, increased friction factor or pumping power, and high cost are still under investigation. Most of the researchers and engineers cannot employ hybrid nanofluids in applications due to serious challenges. The upcoming years will be significant to the researchers for the future of heat transfer fluids. For better sedimentation, researchers and scholars may adopt the idea to decrease further the size, which is $<10^{-9}$ or $<10^{-12}$ of particles. Fig. 10.6 shows the schematic illustration of

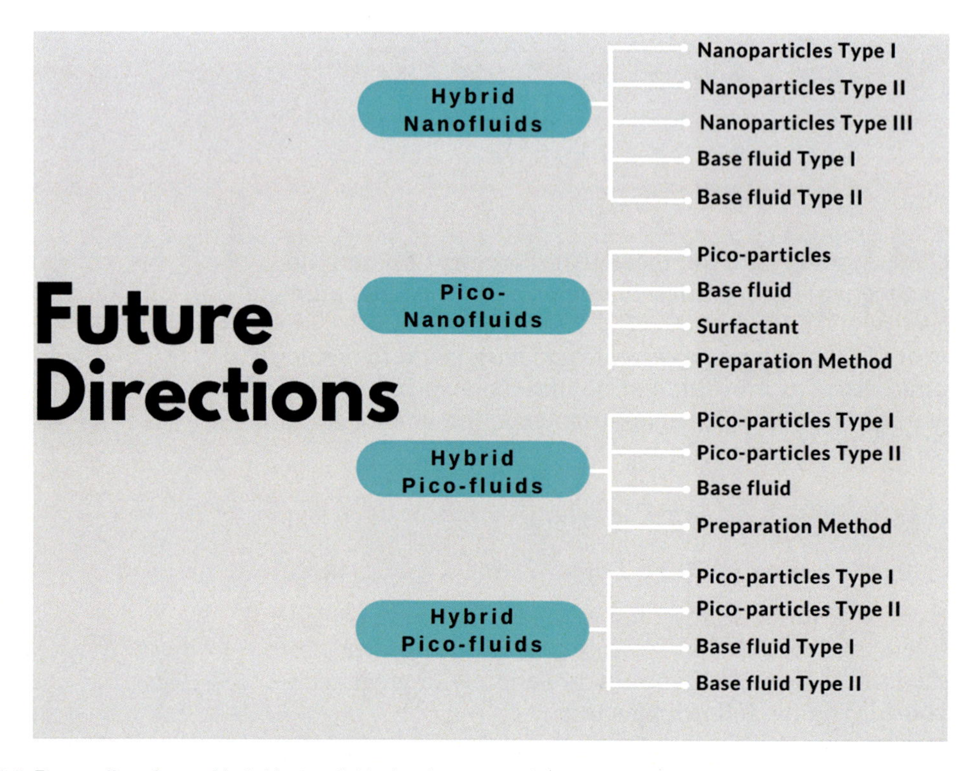

Fig. 10.6 Future directions of hybrid nanofluids for the research community.

the future directions of heat transfer fluids. Although several studies have been reported on nanofluid behavior, researchers still do not know the complicated mechanisms of rheological variations and heat transfer of hybrid nanofluids. Extensive experimental research needs to be analyzed to comprehend the science behind the inconsistent improvement in thermophysical properties and to introduce models for performance-impacting factors for the sustainable commercialization of hybrid nanofluids in numerous engineering applications.

10.11 Conclusion

Unneglectable significance of hybrid nanofluids is their remarkable thermal conductivity, which makes them superior to mono nanofluids. Hybrid nanofluids being superior heat transfer fluids may hinder some serious challenges in their development. Extensive investigations are required to show their potential in commercial applications. The following significant conclusions were portrayed regarding the challenges of hybrid nanofluids.

(1) Foam formation is challenging to attain sustain hybrid nanofluids. Further techniques should be developed to reduce foam formation.

(2) Stability is one of the major parameters affecting the performance of hybrid nanofluids, and lack of good stability negatively affects the nanofluid. Suitable methods and surfactants are developed to achieve excellent stability. Further research should be analyzed to improve hybrid nanofluids' stability as it is an essential step for the commercialization of hybrid nanofluids.

(3) The use of nanoparticles has been drastically improved in recent years in various areas. Constant exposure of nanoparticles to the environment also increases in both positive and negative aspects. Green or eco-friendly technologies must be introduced to prepare hybrid nanocomposites.

(4) To promote nanofluid applications, more research should be focused on economical production methods with enhanced nanofluids' stability.

(5) Hybrid nanofluids have displayed better performance in heat transfer devices because of the nanomaterials' dispersion that improves their thermal properties. It also gives rise to pumping power and friction factor, which restrains the heat transfer rate. Increased pumping power, friction factor, and pressure drop must be considered for better performance of hybrid nanofluids.

(6) Selection of hybrid nanoparticles is important to synthesize nanofluids, and investigations should be studied with respect to their application.

(7) Available thermophysical models for mono nanofluids do not provide a better prediction of hybrid nanofluids There is a lack of research on theoretical models that can foresee the behavior of hybrid nanofluids. Universal correlations are today's requirement that can be utilized for all hybrid nanofluids.

References

[1] Z. Said, et al., Recent advances on nanofluids for low to medium temperature solar collectors: energy, exergy, economic analysis and environmental impact, Prog. Energy Combust. Sci. 84 (2021) 100898.
[2] M. Sheikholeslami, et al., Recent progress on flat plate solar collectors and photovoltaic systems in the presence of nanofluid: a review, J. Clean. Prod. (2021) 126119.
[3] A.K. Tiwar, et al., A review on the application of hybrid nanofluids for parabolic trough collector: recent progress and outlook, J. Clean. Prod. (2021) 126031.
[4] X. Wang, et al., Vegetable oil-based nanofluid minimum quantity lubrication turning: academic review and perspectives, J. Manuf. Process. 59 (2020) 76–97.
[5] A. Kumar, Z. Said, E. Bellos, An up-to-date review on evacuated tube solar collectors, J. Therm. Anal. Calorim. (2020) 1–17.
[6] Z. Said, et al., A comprehensive review on minimum quantity lubrication (MQL) in machining processes using nano-cutting fluids, Int. J. Adv. Manuf. Technol. 105 (5) (2019) 2057–2086.
[7] A.A. Hachicha, et al., A review study on the modeling of high-temperature solar thermal collector systems, Renew. Sust. Energ. Rev. 112 (2019) 280–298.
[8] Z. Said, et al., Acid-functionalized carbon nanofibers for high stability, thermoelectrical and electrochemical properties of nanofluids, J. Colloid Interface Sci. 520 (2018) 50–57.
[9] Z. Said, et al., Stability, thermophysical and electrical properties of synthesized carbon nanofiber and reduced-graphene oxide-based nanofluids and their hybrid along with fuzzy modeling approach, Powder Technol. 364 (2020) 795–809.
[10] L. Sundar, et al., Heat transfer and second law analysis of ethylene glycol based ternary hybrid nanofluid under laminar flow, J. Thermal Sci. Eng. Appl. (2021) 1–45.
[11] L.S. Sundar, et al., Energy, efficiency, economic impact, and heat transfer aspects of solar flat plate collector with Al2O3 nanofluids and wire coil with core rod inserts, Sustainable Energy Technol. Assess. 40 (2020) 100772.
[12] L.S. Sundar, et al., Combination of Co3O4 deposited rGO hybrid nanofluids and longitudinal strip inserts: thermal properties, heat transfer, friction factor, and thermal performance evaluations, Therm. Sci. Eng. Progr. 20 (2020) 100695.
[13] Z. Said, et al., Optimizing density, dynamic viscosity, thermal conductivity and specific heat of a hybrid nanofluid obtained experimentally via ANFIS-based model and modern optimization, J. Mol. Liq. 321 (2021) 114287.

[14] A.K. Tiwari, et al., 3S (sonication, surfactant, stability) impact on the viscosity of hybrid nanofluid with different base fluids: an experimental study, J. Mol. Liq. (2021) 115455.
[15] P. Kanti, et al., Entropy generation and friction factor analysis of fly ash nanofluids flowing in a horizontal tube: experimental and numerical study, Int. J. Therm. Sci. 166 (2021) 106972.
[16] D. Exerowa, P.M. Kruglyakov, Foam and Foam Films: Theory, Experiment, Application, Elsevier, Amsterdam, 1998.
[17] S. Tcholakova, N.D. Denkov, A. Lips, Comparison of solid particles, globular proteins and surfactants as emulsifiers, Phys. Chem. Chem. Phys. 10 (12) (2008) 1608–1627.
[18] D. Langevin, Influence of interfacial rheology on foam and emulsion properties, Adv. Colloid Interf. Sci. 88 (1–2) (2000) 209–222.
[19] L. Saulnier, et al., Comparison between generations of foams and single vertical films—single and mixed surfactant systems, Soft Matter 10 (29) (2014) 5280–5288.
[20] R. Liontas, et al., Neighbor-induced bubble pinch-off: novel mechanisms of in situ foam generation in microfluidic channels, Soft Matter 9 (46) (2013) 10971–10984.
[21] J. Xu, et al., Pool boiling heat transfer of ultra-light copper foam with open cells, Int. J. Multiphase Flow 34 (11) (2008) 1008–1022.
[22] F. Arbelaez, S. Sett, R.L. Mahajan, An experimental study on pool boiling of saturated FC-72 in highly porous aluminum metal foams, in: 34th National Heat Transfer Conference, 2000, pp. 759–767.
[23] B. Athreya, R. Mahajan, S. Sett, Pool boiling of FC-72 over metal foams: effect of foam orientation and geometry, in: 8th AIAA/ASME Joint Thermophysics and Heat Transfer Conference, 2002.
[24] M.Q. Raza, N. Kumar, R. Raj, Effect of foamability on pool boiling critical heat flux with nanofluids, Soft Matter 15 (26) (2019) 5308–5318.
[25] H. Cao, et al., Synergistic action of TiO2 particles and surfactants on the foamability and stabilization of aqueous foams, RSC Adv. 7 (71) (2017) 44972–44978.
[26] M. Gupta, et al., Up to date review on the synthesis and thermophysical properties of hybrid nanofluids, J. Clean. Prod. 190 (2018) 169–192.
[27] H. Yarmand, et al., Graphene nanoplatelets–silver hybrid nanofluids for enhanced heat transfer, Energy Convers. Manag. 100 (2015) 419–428.
[28] A.K. Tiwari, et al., Experimental and numerical investigation on the thermal performance of triple tube heat exchanger equipped with different inserts with WO3/water nanofluid under turbulent condition, Int. J. Therm. Sci. 164 (2021) 106861.
[29] N.S. Pandya, et al., Influence of the geometrical parameters and particle concentration levels of hybrid nanofluid on the thermal performance of axial grooved heat pipe, Therm. Sci. Eng. Progr. 21 (2021) 100762.
[30] Z. Said, et al., Heat transfer, entropy generation, economic and environmental analyses of linear fresnel reflector using novel rGO-Co3O4 hybrid nanofluids, Renew. Energy 165 (2021) 420–437.
[31] L.S. Sundar, et al., Experimental investigation of thermo-physical properties, heat transfer, pumping power, entropy generation, and exergy efficiency of nanodiamond+ Fe3O4/60: 40% water-ethylene glycol hybrid nanofluid flow in a tube, Therm. Sci. Eng. Progr. 21 (2021) 100799.
[32] Z. Said, et al., Energy and exergy efficiency of a flat plate solar collector using pH treated Al2O3 nanofluid, J. Clean. Prod. 112 (2016) 3915–3926.

[33] Z. Said, R. Saidur, N. Rahim, Energy and exergy analysis of a flat plate solar collector using different sizes of aluminium oxide based nanofluid, J. Clean. Prod. 133 (2016) 518–530.
[34] H. Babar, H.M. Ali, Towards hybrid nanofluids: preparation, thermophysical properties, applications, and challenges, J. Mol. Liq. 281 (2019) 598–633.
[35] T. Missana, A. Adell, On the applicability of DLVO theory to the prediction of clay colloids stability, J. Colloid Interface Sci. 230 (1) (2000) 150–156.
[36] Z. Said, et al., Performance enhancement of a flat plate solar collector using titanium dioxide nanofluid and polyethylene glycol dispersant, J. Clean. Prod. 92 (2015) 343–353.
[37] E. Bellos, Z. Said, C. Tzivanidis, The use of nanofluids in solar concentrating technologies: a comprehensive review, J. Clean. Prod. 196 (2018) 84–99.
[38] A.K. Tiwari, et al., 4S consideration (synthesis, sonication, surfactant, stability) for the thermal conductivity of CeO2 with MWCNT and water based hybrid nanofluid: an experimental assessment, Colloids Surf. A Physicochem. Eng. Asp. 610 (2021) 125918.
[39] L. Syam Sundar, et al., Heat transfer of rGO/CO3O4 hybrid nanomaterial-based nanofluids and twisted tape configurations in a tube, J. Thermal Sci. Eng. Appl. 13 (3) (2021).
[40] A. Akhgar, D. Toghraie, An experimental study on the stability and thermal conductivity of water-ethylene glycol/TiO2-MWCNTs hybrid nanofluid: developing a new correlation, Powder Technol. 338 (2018) 806–818.
[41] A. Asadi, et al., An experimental and theoretical investigation on heat transfer capability of mg (OH)2/MWCNT-engine oil hybrid nano-lubricant adopted as a coolant and lubricant fluid, Appl. Therm. Eng. 129 (2018) 577–586.
[42] B. Wei, C. Zou, X. Li, Experimental investigation on stability and thermal conductivity of diathermic oil based TiO2 nanofluids, Int. J. Heat Mass Transf. 104 (2017) 537–543.
[43] N. Jha, S. Ramaprabhu, Synthesis and thermal conductivity of copper nanoparticle decorated multiwalled carbon nanotubes based nanofluids, J. Phys. Chem. C 112 (25) (2008) 9315–9319.
[44] H.M. Ali, M.U. Sajid, A. Arshad, Heat transfer applications of TiO2 nanofluids, in: Application of Titanium Dioxide, 2017.
[45] P. Chattopadhyay, R.B. Gupta, Production of griseofulvin nanoparticles using supercritical CO2 antisolvent with enhanced mass transfer, Int. J. Pharm. 228 (1) (2001) 19–31.
[46] P.K. Das, A review based on the effect and mechanism of thermal conductivity of normal nanofluids and hybrid nanofluids, J. Mol. Liq. 240 (2017) 420–446.
[47] W. Yu, H. Xie, A review on nanofluids: preparation, stability mechanisms, and applications, J. Nanomater. 2012 (2012) 435873.
[48] X. Li, G. Zeng, X. Lei, The stability, optical properties and solar-thermal conversion performance of SiC-MWCNTs hybrid nanofluids for the direct absorption solar collector (DASC) application, Sol. Energy Mater. Sol. Cells 206 (2020) 110323.
[49] N.K. Cakmak, et al., Preparation, characterization, stability, and thermal conductivity of rGO-Fe3O4-TiO2 hybrid nanofluid: an experimental study, Powder Technol. 372 (2020) 235–245.
[50] H.W. Xian, N.A.C. Sidik, R. Saidur, Impact of different surfactants and ultrasonication time on the stability and thermophysical properties of hybrid nanofluids, Int. Commun. Heat Mass Transfer 110 (2020) 104389.
[51] P.C. Mukesh Kumar, K. Palanisamy, V. Vijayan, Stability analysis of heat transfer hybrid/water nanofluids, Mater. Today 21 (2020) 708–712.

[52] M. Ma, et al., Effect of surfactant on the rheological behavior and thermophysical properties of hybrid nanofluids, Powder Technol. 379 (2021) 373–383.
[53] O. Gulzar, A. Qayoum, R. Gupta, Experimental study on stability and rheological behaviour of hybrid Al2O3-TiO2 Therminol-55 nanofluids for concentrating solar collectors, Powder Technol. 352 (2019) 436–444.
[54] Z. Said, S. Arora, E. Bellos, A review on performance and environmental effects of conventional and nanofluid-based thermal photovoltaics, Renew. Sust. Energ. Rev. 94 (2018) 302–316.
[55] L.S. Sundar, et al., Properties, heat transfer, energy efficiency and environmental emissions analysis of flat plate solar collector using nanodiamond nanofluids, Diam. Relat. Mater. 110 (2020) 108115.
[56] Z. Said, et al., Heat transfer enhancement and life cycle analysis of a Shell-and-Tube Heat Exchanger using stable CuO/water nanofluid, Sustainable Energy Technol. Assess. 31 (2019) 306–317.
[57] A.K. Tiwari, et al., Experimental comparison of specific heat capacity of three different metal oxides with MWCNT/water-based hybrid nanofluids: proposing a new correlation, Appl. Nanosci. (2020) 1–11.
[58] R. Aitken, Nanoparticles: an occupational hygiene review, IOM Res. Rep. 274 (2004) 41–44.
[59] N.N. Esfahani, D. Toghraie, M. Afrand, A new correlation for predicting the thermal conductivity of ZnO–Ag (50%–50%)/water hybrid nanofluid: an experimental study, Powder Technol. 323 (2018) 367–373.
[60] A.S. Dalkılıç, et al., Experimental study on the thermal conductivity of water-based CNT-SiO2 hybrid nanofluids, Int. Commun. Heat Mass Transfer 99 (2018) 18–25.
[61] C. Buzea, I.I. Pacheco, K. Robbie, Nanomaterials and nanoparticles: sources and toxicity, Biointerphases 2 (4) (2007) MR17–MR71.
[62] K. Khosravi-Katuli, et al., Effects of nanoparticles in species of aquaculture interest, Environ. Sci. Pollut. Res. 24 (21) (2017) 17326–17346.
[63] F. Abbas, et al., Nanofluid: potential evaluation in automotive radiator, J. Mol. Liq. 297 (2020) 112014.
[64] M. Gupta, V. Singh, Z. Said, Heat transfer analysis using zinc ferrite/water (hybrid) nanofluids in a circular tube: an experimental investigation and development of new correlations for thermophysical and heat transfer properties, Sustainable Energy Technol. Assess. 39 (2020) 100720.
[65] W. Klöpffer, et al., Nanotechnology and Life Cycle Assessment: A Systems Approach to Nanotechnology and the Environment, Woodrow Wilson International Center for Scholars, 2007.
[66] M. Hemmat Esfe, S. Esfandeh, M. Rejvani, Modeling of thermal conductivity of MWCNT-SiO2 (30:70%)/EG hybrid nanofluid, sensitivity analyzing and cost performance for industrial applications, J. Therm. Anal. Calorim. 131 (2) (2018) 1437–1447.
[67] A. Alirezaie, et al., Price-performance evaluation of thermal conductivity enhancement of nanofluids with different particle sizes, Appl. Therm. Eng. 128 (2018) 373–380.
[68] M. Esfe, et al., ANN modeling, cost performance and sensitivity analyzing of thermal conductivity of DWCNT–SiO2/EG hybrid nanofluid for higher heat transfer: an experimental study, J. Therm. Anal. Calorim. (2017).
[69] S.K. Singh, J. Sarkar, Energy, exergy and economic assessments of shell and tube condenser using hybrid nanofluid as coolant, Int. Commun. Heat Mass Transfer 98 (2018) 41–48.

[70] Z. Said, et al., Fuzzy modeling and optimization for experimental thermophysical properties of water and ethylene glycol mixture for Al2O3 and TiO2 based nanofluids, Powder Technol. 353 (2019) 345–358.
[71] D. Dhinesh Kumar, A.V. Arasu, A comprehensive review of preparation, characterization, properties and stability of hybrid nanofluids, Renew. Sust. Energ. Rev. 81 (2018) 1669–1689.
[72] A. Chintaparthi, J. Philip, Assessment of long term stability of aqueous nanofluids using different experimental techniques, J. Mol. Liq. 222 (2016).
[73] L.S. Sundar, et al., Enhanced thermal conductivity and viscosity of nanodiamond-nickel nanocomposite nanofluids, Sci. Rep. 4 (2014).
[74] G. Huminic, A. Huminic, The influence of hybrid nanofluids on the performances of elliptical tube: recent research and numerical study, Int. J. Heat Mass Transf. 129 (2019) 132–143.
[75] H. Eshgarf, M. Afrand, An experimental study on rheological behavior of non-Newtonian hybrid nano-coolant for application in cooling and heating systems, Exp. Thermal Fluid Sci. 76 (2016) 221–227.
[76] B. Takabi, H. Shokouhmand, Effects of Al2O3–Cu/water hybrid nanofluid on heat transfer and flow characteristics in turbulent regime, Int. J. Mod. Phys. C 26 (04) (2015) 1550047.
[77] S. Suresh, et al., Effect of Al2O3–cu/water hybrid nanofluid in heat transfer, Exp. Thermal Fluid Sci. 38 (2012) 54–60.
[78] M.F. Nabil, et al., An experimental study on the thermal conductivity and dynamic viscosity of TiO2-SiO2 nanofluids in water: ethylene glycol mixture, Int. Commun. Heat Mass Transfer 86 (2017) 181–189.
[79] H.W. Xian, N.A.C. Sidik, G. Najafi, Recent state of nanofluid in automobile cooling systems, J. Therm. Anal. Calorim. 135 (2) (2019) 981–1008.
[80] A. Bhattad, J. Sarkar, P. Ghosh, Discrete phase numerical model and experimental study of hybrid nanofluid heat transfer and pressure drop in plate heat exchanger, Int. Commun. Heat Mass Transfer 91 (2018) 262–273.
[81] V. Kumar, J. Sarkar, Numerical and experimental investigations on heat transfer and pressure drop characteristics of Al2O3-TiO2 hybrid nanofluid in minichannel heat sink with different mixture ratio, Powder Technol. 345 (2019) 717–727.
[82] M. Sheikholeslami, S.A. Farshad, Z. Said, Analyzing entropy and thermal behavior of nanomaterial through solar collector involving new tapes, Int. Commun. Heat Mass Transfer 123 (2021) 105190.
[83] Y. Tanaka, et al., Gateway vectors for plant genetic engineering: overview of plant vectors, application for bimolecular fluorescence complementation (BiFC) and multigene construction, in: Genetic Engineering—Basics, New Applications and Responsibilities, 2012, pp. 35–58.
[84] G.M. Moldoveanu, A.A. Minea, Specific heat experimental tests of simple and hybrid oxide-water nanofluids: proposing new correlation, J. Mol. Liq. 279 (2019) 299–305.
[85] M. Nuim Labib, et al., Numerical investigation on effect of base fluids and hybrid nanofluid in forced convective heat transfer, Int. J. Therm. Sci. 71 (2013) 163–171.
[86] P. Kanti, et al., Numerical study on the thermo-hydraulic performance analysis of fly ash nanofluid, J. Therm. Anal. Calorim. (2021) 1–13.
[87] Z. Said, et al., Thermophysical properties using ND/water nanofluids: an experimental study, ANFIS-based model and optimization, J. Mol. Liq. 330 (2021) 115659.

[88] M. Jamei, et al., On the specific heat capacity estimation of metal oxide-based nanofluid for energy perspective—a comprehensive assessment of data analysis techniques, Int. Commun. Heat Mass Transfer 123 (2021) 105217.
[89] R.L. Hamilton, O.K. Crosser, Thermal conductivity of heterogeneous two-component systems, Ind. Eng. Chem. Fundam. 1 (3) (1962) 187–191.
[90] J.C. Maxwell, A treatise on electricity and magnetism, in: Cambridge Library Collection—Physical Sciences, vol. 1, Cambridge University Press, Cambridge, 2010.
[91] C.-W. Nan, et al., Effective thermal conductivity of particulate composites with interfacial thermal resistance, J. Appl. Phys. 81 (10) (1997) 6692–6699.
[92] W. Yu, S.U.S. Choi, The role of interfacial layers in the enhanced thermal conductivity of nanofluids: a renovated Maxwell model, J. Nanopart. Res. 5 (1) (2003) 167–171.
[93] S.S. Chougule, S.K. Sahu, Model of heat conduction in hybrid nanofluid, in: 2013 IEEE International Conference ON Emerging Trends in Computing, Communication and Nanotechnology (ICECCN), 2013.
[94] D.H. Kumar, et al., Model for heat conduction in nanofluids, Phys. Rev. Lett. 93 (14) (2004) 144301.
[95] M. Hemmat Esfe, et al., Thermal conductivity of Cu/TiO2–water/EG hybrid nanofluid: experimental data and modeling using artificial neural network and correlation, Int. Commun. Heat Mass Transfer 66 (2015) 100–104.
[96] M. Hemmat Esfe, A.A. Abbasian Arani, M. Firouzi, Empirical study and model development of thermal conductivity improvement and assessment of cost and sensitivity of EG-water based SWCNT-ZnO (30%:70%) hybrid nanofluid, J. Mol. Liq. 244 (2017) 252–261.
[97] S.H. Rostamian, et al., An inspection of thermal conductivity of CuO-SWCNTs hybrid nanofluid versus temperature and concentration using experimental data, ANN modeling and new correlation, J. Mol. Liq. 231 (2017) 364–369.
[98] Ç. Yıldız, M. Arıcı, H. Karabay, Comparison of a theoretical and experimental thermal conductivity model on the heat transfer performance of Al2O3-SiO2/water hybrid-nanofluid, Int. J. Heat Mass Transf. 140 (2019) 598–605.
[99] A. Einstein, Eine neue Bestimmung der Moleküldimensionen, Ann. Phys. 324 (2) (1906) 289–306.
[100] H.C. Brinkman, The viscosity of concentrated suspensions and solutions, J. Chem. Phys. 20 (4) (1952) 571.
[101] G.K. Batchelor, The effect of Brownian motion on the bulk stress in a suspension of spherical particles, J. Fluid Mech. 83 (1) (2006) 97–117.
[102] M. Corcione, Empirical correlating equations for predicting the effective thermal conductivity and dynamic viscosity of nanofluids, Energy Convers. Manag. 52 (1) (2011) 789–793.
[103] M. Hemmat Esfe, et al., Experimental determination of thermal conductivity and dynamic viscosity of Ag–MgO/water hybrid nanofluid, Int. Commun. Heat Mass Transfer 66 (2015) 189–195.

Index

Note: Page numbers followed by *f* indicate figures, and *t* indicate tables.

A
Adaptive neuro-fuzzy inferences system (ANFIS), 215–216
Aluminium oxide (Al_2O_3), 7
ANN. *See* Artificial neural network (ANN)
Artificial intelligence (AI)
  artificial neural network, 210–217
  data preprocessing, 209–210
  ensemble machine learning methods, 217–219
  Gaussian process, 223–225
  machine learning approaches, 221–223
  nanofluids
    heat capacity, 206*t*
    thermal conductivity, 205*t*, 207*t*
    viscosity, 206*t*
  sensitivity analysis, 225
  support vector machine regression, 219–221
Artificial neural network (ANN), 210–217

B
Best subset regression (BSR), 210

C
Cascaded forward neural network (CFNN), 212–213
Centrifugation method, 40–41
Correlations, 150–151

D
DASC. *See* Direct absorption solar collector (DASC)
Data preprocessing
  feature selection, 209–210
  mutual information, 209–210
  statistical assessment, 209
Density, 160
Desalination, 22, 192–194
Dielectric property, 86–87
Direct absorption solar collector (DASC), 180

E
Electronics cooling, 19–20, 172–177
Electron microscopy, 50–51
ELM. *See* Extreme learning machine (ELM)
Engine cooling, 21, 185–187
Enhanced heat transfer, 111–112
Ensemble machine learning methods, 217–219
Extreme learning machine (ELM), 214–215

F
Feedforward network of neurons (FFNN), 210–211
Foam formation, 234–237
Fouling factor, 101–105
Friction factor, 96–99, 245–247
Fuzzy logic model technique, 215–216

G
Gaussian process regression (GPR), 223–225
Gene expression programming (GEP), 221
Generalized regression neural network (GRNN), 213–214
Genetic programming (GP), 204–208, 221–223
GMDH. *See* Group method of data handling (GMDH)
GPR. *See* Gaussian process regression (GPR)
GRNN. *See* Generalized regression neural network (GRNN)
Group method of data handling (GMDH), 216–217

H
Heat capacity, 154–160
Heat exchangers, 20–21, 181–185
Heat transfer, 2–4
Hybrid nanofluids, 173*f*
  applications, 19–22
  challenges, 194–195, 252–253
  degradation, 244–245
  density, 17
  development, 7–10
  electrical, magnetic, dielectric, 18–19
  energy efficiency, 13–14
  heat and density, 79–85
  heat capacity, 17
  high cost, 242–243
  magnetic property, 86
  preparation, 10–12
  properties, 13–19
  rheology, 117–122
  safety and environmental concerns, 239–242
  stability, 153
  synthesis processes, 33–34

Hybrid nanofluids *(Continued)*
thermal conductivity, 14–16
thermal diffusivity, 17–18
thermophysical properties, 13–14
viscosity, 16, 74–79, 111–112

M
M5 tree (M5Tree) models, 217–218
Machining, 21, 190–192
Magnetic analysis, 86
Maxwell-Garnett approximation, 136–137
Mie scattering theory, 137–140
Minimum quantity lubrication (MQL), 191–192
Mono nanofluids, 7
Multilayer perceptron (MLP) neural network, 211–212

N
Nanofluids, 4–7
fouling factor, 101–105
stability, 37–57
Nanomaterials, 4–7
Nanoparticles, 55–56
Nuclear pressurized water reactor (PWR), 21

O
One-step method, 34
vs. two-step method, 34–36
Optical properties, 132–140
particle size, 143
volume fraction, 143–144
Optimization algorithms, 226
Oscillating heat pipe (OHP) cooling system, 19–20

P
Photovoltaic/thermal (PV/T) systems, 2
pH value, 56–57
Pressure drop, 99–101, 245–247
Pressurized water reactor (PWR), 21
Price-performance analysis (PPA), 242–243
Pumping power, 101, 245–247
PWR. *See* Pressurized water reactor (PWR)

R
Radial basis function neural network (RBF-NN), 212
Radiative transfer, 140–143
Random forest (RF), 219
Rayleigh scattering approximation, 133–136
RBF-NN. *See* Radial basis function neural network (RBF-NN)
Refrigeration systems, 21, 188–190
RF. *See* Random forest (RF)
Rheology
experimental and numerical studies, 113–117
particle size and shape, 120–121
temperature, 119–120
volume concentration, 121–122

S
Scanning electron microscopy (SEM), 50–51
Sedimentation method, 38–40
Sensitivity analysis, 225
Solar collectors, 20, 177–181
Solar energy, 2
Solar photovoltaics, 2
Sonication, 51–52
Spectral absorbance analysis, 44–50
Stability, 10–12, 237–239
enhancement methods, 51–57
Support vector machine regression (SVR), 219–221
Surface tension, 94–96
Surfactants, 53–55
SVR. *See* Support vector machine regression (SVR)

T
TEM. *See* Transmission electron microscopy (TEM)
Thermal conductivity, 50, 67–74, 152–154
Thermophysical properties, 65–87, 165–166, 204, 248–252
Transmission electron microscopy (TEM), 50–51
Two-step method, 34–36
vs. one-step method, 34

U
Ultrasonication, 51–52

V
Viscosity, 161–164

Z
Zeta potential method, 41–43

Printed in the United States
by Baker & Taylor Publisher Services